Engineering Problems - Uncertainties, Constraints and Optimization Techniques

Edited by Marcos S.G. Tsuzuki,
Rogério Y. Takimoto, André K. Sato,
Tomoki Saka, Ahmad Barari,
Rehab O. Abdel Rahman and Yung-Tse Hung

Published in London, United Kingdom

IntechOpen

Supporting open minds since 2005

Engineering Problems – Uncertainties, Constraints and Optimization Techniques
http://dx.doi.org/10.5772/intechopen.93447
Edited by Marcos S.G. Tsuzuki, Rogério Y. Takimoto, André K. Sato, Tomoki Saka, Ahmad Barari, Rehab O. Abdel Rahman and Yung-Tse Hung

Contributors
Ju Qiu, Chaofeng Liu, Ermelinda Serena Sanseviero, Marco Rao, Tomoki Saka, Marcos S.G. Tsuzuki, Tiago G. Goto, Hossein R.Najafabadi, Guilherme C. Duran, Edson K. Ueda, André K. Sato, Thiago C. Martins, Rogério Y. Yugo Takimoto, Hossein Gohari, Ahmad Barari, Afef Salhi, Fahmi Ghozzi, Ahmed Fakhfakh, Houcine Meftahi, Zakaria Belhachmi, Amel Ben Abda, Belhassen Meftahi, Chimmiri Venkateswarlu, Bennasr Hichem, M'Sahli Faouzi, Azzabi Lotfi, Azzabi Dorra, Abdessamad Kobi, Alvaro Humberto Salas, Samir A. Abd El-Hakim El-Tantawy, William A. Crossley, Satadru Roy, Samarth Jain, Rehab O. Abdel Rahman, Yung-Tse Hung, Annalina Lombardi, Valentina Colaiuda, Barbara Tomassetti, Mario Papa, Frank Silvio Marzano, Carla Giansante, Federica Di Giacinto, Maria Paola Manzi, Nicola Ferri, Carla Ippoliti, Seyedkiarash Sharifiilierdy, Howard H. Paul, Christopher R. Huhnke

Notice
Statements and opinions expressed in the chapters are these of the individual contributors and not necessarily those of the editors or publisher. No responsibility is accepted for the accuracy of information contained in the published chapters. The publisher assumes no responsibility for any damage or injury to persons or property arising out of the use of any materials, instructions, methods or ideas contained in the book.

First published in London, United Kingdom, 2022 by IntechOpen
IntechOpen is the global imprint of INTECHOPEN LIMITED, registered in England and Wales, registration number: 11086078, 5 Princes Gate Court, London, SW7 2QJ, United Kingdom
Printed in Croatia

British Library Cataloguing-in-Publication Data
A catalogue record for this book is available from the British Library

Additional hard and PDF copies can be obtained from orders@intechopen.com

Engineering Problems – Uncertainties, Constraints and Optimization Techniques
Edited by Marcos S.G. Tsuzuki, Rogério Y. Takimoto, André K. Sato, Tomoki Saka, Ahmad Barari, Rehab O. Abdel Rahman and Yung-Tse Hung
p. cm.
Print ISBN 978-1-83969-367-0
Online ISBN 978-1-83969-368-7
eBook (PDF) ISBN 978-1-83969-369-4

We are IntechOpen,
the world's leading publisher of
Open Access books
Built by scientists, for scientists

6,000+
Open access books available

147,000+
International authors and editors

185M+
Downloads

Our authors are among the

156
Countries delivered to

Top 1%
most cited scientists

12.2%
Contributors from top 500 universities

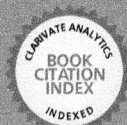

Interested in publishing with us?
Contact book.department@intechopen.com

Numbers displayed above are based on latest data collected.
For more information visit www.intechopen.com

Meet the editors

Marcos de Sales Guerra Tsuzuki is a senior associate professor in the Department of Mechatronics and Mechanical Systems Engineering, Escola Politécnica da Universidade de São Paulo, Brazil. Dr. Tsuzuki is an expert in CAD/CAM, optimization methods, cutting and packing, inverse problems, ultrasound, and computer graphics. He developed methodologies for various industrial applications related to ultrasound, software development, and CAD/CAM systems. He is a senior member of the Institute of Electrical and Electronics Engineers (IEEE) and a member of the International Federation of Automatic Control (IFAC). Dr. Tsuzuki has been the chair of several events held by the International Federation of Automatic Control (IFAC) and IEEE. He was elected chair of the IFAC Technical Committee on Manufacturing Plant Control and vice chair of the IFAC Technical Committee on Biological and Medical Systems.

Ahmad Barari is an associate professor in the Department of Mechanical and Manufacturing Engineering, and director of Advanced Digital Design, Manufacturing, and Metrology (AD-2MLabs), Ontario Tech University, Canada. Dr. Barari is an expert in the digitalization of the product life cycle including design, manufacturing, inspection, and maintenance. He developed methodologies for various industrial applications in product development, process control, prognostics, health monitoring, and prescriptive and predictive maintenance. His contributions include software solutions, new machines and processes, graduate course curricula, and more than 180 publications in highly ranked journals and conference proceedings. Dr. Barari has closely collaborated with various industries on projects funded by national and international sources. He has been a member of numerous international scientific committees and organizations, including the American Society of Mechanical Engineering (ASME) and the International Federation of Automatic Control (IFAC). He has been an editorial board member for various journals and proceedings. He has organized, chaired, or administrated more than twenty academic national and international events, conference topics, invited sessions, and seminars. Dr. Barari was previously the chair of IFAC's 13th international workshop on Intelligent Manufacturing Systems. He is currently the vice chair of the IFAC Technical Committee on Manufacturing Plant Control and the chair of the IFAC Intelligent Manufacturing Systems Working Group.

Tomoki Saka teaches Machine Learning, Computer Languages, and Information Processing Laboratory at the Department of Media Informatics, College of Informatics and Human Communication, Kanazawa Institute of Technology, Japan. He graduated from the Department of Mechanical Engineering and Materials Science, Faculty of Engineering, Yokohama National University, Japan, in 2009. In 2011, he received a master's degree in Information Media and Environment Sciences from the Graduate School of Environment and Information Science at the same university. In the same year, he was engaged in PC mechanical design work at Fujitsu Limited. In 2017, he received Dr. info. Sci. in the same department at the same university. He has been a research associate since 2021.

André Kubagawa Sato is a postdoctoral researcher in Mechatronics Engineering at the University of Sao Paulo, Brazil. His research concerns the study of optimization and inverse problems, mostly in biomedical applications. Recently, his work is focused on identifying forest fires from sensor data.

Rogério Yugo Takimoto received his bachelor's degree in Electrical Engineering (Computer Engineering) from the University of Sao Paulo, Brazil. He received his MSc and Ph.D. in Ocean and Space Systems Engineering from Yokohama National University, in 2006 and 2009, respectively. He is currently a postdoctoral fellow at the Mechatronics Engineering Department, University of Sao Paulo, Brazil. His research interests include signal and image processing, filtering, 3D reconstruction, segmentation, pattern recognition, optimization, and biomedical signal processing.

Rehab O. Abdel Rahman is an Associate Professor of Chemical Nuclear Engineering, Radioactive Waste Management Department, Hot Laboratories & Waste Management Center, Atomic Energy Authority of Egypt. She obtained a Ph.D. in Nuclear Engineering from Alexandria University, Egypt. She has more than forty peer-reviewed scientific papers, twenty-five book chapters, and ten books to his credit. She has taught and supervised postgraduate research in chemistry, physics, petrochemical, and environmental chemical engineering departments. She serves as a verified reviewer for several journals and is managing editor for the *International Journal of Environment and Waste Management* and the *International Journal of Engineering Education*.

Dr. Yung Tse Hung is a professor in the Department of Civil and Environmental Engineering, Cleveland State University, USA. He received his Ph.D. in Environmental Engineering from the University of Texas at Austin in 1970 and an MS and BS in Civil Engineering from Cheng Kung University, Taiwan in 1964. He has 15 books, 450 reports and journal publications, and numerous conference presentations in water and wastewater treatment to his credit. He is a fellow of the American Society of Civil Engineers, a diplomate of the American Academy of Environmental Engineers, a fellow of the Ohio Academy of Science, and a member of the Association of Environmental Engineering and Water Environment Federation. He is the executive director of the Overseas Chinese Environmental Engineers and Scientists Association. His research interests include water supply and water treatment, municipal wastewater treatment, industrial waste treatment, biological waste treatment, water and wastewater treatment plant design, water pollution control, and water quality engineering.

Contents

Preface

Engineering by nature is the art of creation, that is, creating the most efficient solution for an existing problem given a set of constraints. Engineering techniques allow navigating toward a design solution through the set of all the existing *constraints* within an acceptable level of *uncertainties*. *Optimization techniques* are employed widely by engineers in various disciplines to conduct this navigation task.

The combination of existing constraints, uncertainties, and corresponding optimization techniques demonstrate the intrinsic nature of engineering. An example of a simple engineering problem is the design of a bookshelf, wherein the constraints include the limited available space and the objective is to maximize the number of books that can be shelved.

As the key requirement of the fourth industrial revolution, known as Industry 4.0, intelligence needs to be added to engineering solutions. This allows dynamic refinement of solutions, self-calibration, self-configuration, and self-adjustment according to the feedback received from the design environment. Dynamic feedback-based solutions have been used significantly in production and manufacturing systems in many cases, such as machining adjustment based on the intermittent data received from sensors, inspection surface fitting based on the dynamic nature of the systematic and non-systematic error patterns in the capture data points, and adaptive process planning for additive manufacturing of a product.

This book addresses a range of state-of-the-art methodologies and solutions to engineering problems. Considering the three important elements of constraints, uncertainties, and optimization techniques, the book classifies certain types of circumstances that engineering problems may be formulated for. It also provides a collective view to all these topics for a variety of engineering disciplines.

Marcos S.G. Tsuzuki, André K. Sato and Rogério Y. Takimoto
Computational Geometry Laboratory,
Escola Politénica da USP,
São Paulo, Brazil

Ahmad Barari
Ontario Tech University,
Oshawa, Canada

Tomoki Saka
Kanazawa Institute of Technology,
Kanazawa, Japan

Rehab O. Abdel Rahman
Hot Lab. Center,
Egyptian Atomic Energy Authority,
Egypt

Yung-Tse Hung
Department of Civil and Environmental Engineering,
Cleveland State University,
USA

Optimization Problems in Engineering

Chapter 1

Introductory Chapter: Optimization Problems in Engineering

Marcos S.G. Tsuzuki, Ahmad Barari, André K. Sato, Rogrio Y. Takimoto and Tomoki Saka

1. Introduction

Engineers are devoted to solving real-life problems. The solutions to these problems require the minimization (or maximization) of a function (usually called an objective function or cost function). The problem might have some constraints, turning the problem solving to a challenging task. The current set of chapters deals with different applications: supersonic flutter, motion estimation, chemical and environmental processes, complex nonlinear systems, cutting and packing, topology optimization, curve interpolation, etc. Despite the different domains for each application, they have some essential features in common. They have some parameters representing the solution, the parameters are the inputs to the objective function. The objective function must be minimized through the optimization process to eventually maximize the reliability and efficiency of the resulting solution [1].

Two major categories of methods can be used to determine the set of parameters which supplies the minimal (or maximal) value for an objective. The first category relies on deterministic methods [2]. They require a seed solution (also called initial solution) and iteratively determine a solution. Once the initial solution is given, the deterministic method will end up always the same final solution. However, it might be not the global optimum. The second category relies on probabilistic methods (also called metaheuristics) [3]. They do not rely on the initial solution and each execution can reach a different solution.

In the sequence, it explains the deterministic and probabilistic methods. It will be shown that the different methods have different properties. Surely, each problem has the most appropriate method. However, this association might not be an easy task.

2. Deterministic methods

Among the deterministic methods, the Gauss-Newton method is one of the most commonly used methods and it can be formulated as

$$\min_{x} \|f(x)\|_2^2, \tag{1}$$

where some initial solution x_0 must be provided. The next solution is iteratively determined by

$$x_{k+1} = x_k - \frac{\nabla f(x_k)}{\nabla^2 f(x_k)} \cdot \alpha_k, \tag{2}$$

where α_k is the step size, $\nabla f(x_k)$ denotes the gradient and $\nabla^2 f(x_k)$ is the Hessian, both at x_k. This iteration can happen only if the Hessian is invertible at x_k.

The definition of the step size is not trivial, some approaches, like the Armijo rule, using nonmonotone line search were proposed [4]. A very well-known approach uses the Levenberg-Marquardt modification. It can be used even if $\nabla^2 f(x_k)$ is not a descent direction and if $\nabla^2 f(x_k)$ is singular. The Levenberg-Marquardt was applied in different applications [5, 6]. These two methods have continuous parameters. In case the parameters are integers, or just a subset of the parameters are integers, another class of methods should be used: integer or mixed-integer programming [7]. The determination of the next solution is a key step in deterministic methods. The current solution is summed by a vector defined by the sensitivity of the method, guiding the algorithm towards the convergence.

This set of chapters has several deterministic methods, in the following they are briefly presented. The chapter "Verification and validation of supersonic flutter of the rudder model for experiment" uses a deterministic approach combined with a gradient evaluation to determine the next solution. The parameters are exclusively continuous. The problem has some constraints related to the first and second frequencies of the rudder.

The chapter "An optimization procedure of the model's base construction in multimodel representation of complex nonlinear systems" uses a deterministic method. Artificial Neural Networks must be trained before usage. During the training phase, their parameters are determined using some cost function and gradient information. After trained, the artificial neural network will provide the same output for a given input.

The chapter "Approximation algorithm for scheduling a chain of tasks for motion estimation on heterogeneous systems MPSoC" uses a deterministic approach. This research is focused on processing speed. As explained by the authors, motion estimation is a key feature in a video codec. Temporal redundancy can be estimated and used in interframe video compression. The authors explored a parallel approach suitable for embedded platforms.

The chapter "Analytical solutions of some strong oscillators" presents an analytical solution for cubic and quintic Diffing's oscillators. The authors consider arbitrary initial solutions. The authors point out that the Duffing equation has a large area of application: mechanical engineering, electrical engineering, plasma physics, etc.

The chapter "Optimal heat distribution using asymptotic analysis techniques" uses a deterministic method. The next solution is determined using the concept of topological derivative. The topological derivative defines a vector that iteratively defines the path to the final solution (optimal heat distribution).

3. Metaheuristics

Metaheuristics are a wide range of methods, usually inspired from nature. Every metaheuristic has a random component and this is why they are also known as probabilistic methods. The chapter "Optimization multicriteria scheduling criteria through analytical hierarchy process and lexicography goal programming modeling" is the first example that uses a probabilistic method. Multicriteria decision analysis is a research area where multiple possibilities are considered to determine the best choices. It creates an aleatory matrix of judgments which will determine the coherence among the criteria. The authors present an example about electric cable cut. The following criteria were considered: qualification of workers, safety knowledge, equipment performance, scheduling equipment, production technology, and cost of upgrading waste.

Some metaheuristics used in this set of chapters are briefly introduced: simulated annealing, tabu search, and genetic algorithms.

3.1 Simulated annealing

Simulated annealing is based on metal annealing [8]. It was originally proposed to solve combinatorial problems. However, several proposals to incorporate continuous parameters soon appeared [9–11]. Simulated annealing has clearly two different phases: exploratory and refinement [12]. Initially, at higher temperature, the domain is explored. At lower temperatures, the solution is refined and convergence happens. The simulated annealing has only one current solution.

The simulated annealing with crystallization factor described in the chapter "Versatility of simulated annealing with crystallization heuristic: its application to a great assortment of problems" is an example of how continuous parameters can be approached. The crystallization factor supplies a sensitivity to the algorithm. It was successfully applied to irregular packing problems with fixed containers, in this situation, the cost function is discrete and some parameters are continuous [13–15]. This chapter shows different applications, the curve interpolation application has integer and continuous parameters [16–20].

The objective function of some problems might be costly computationally. For example, when it is required to solve a finite element problem. The simulated annealing can be modified to incorporate a partial evaluation of the objective function. The objective function returns an interval which surely contains the function's precise value. This approach was successfully applied to the electrical impedance tomography problem where 32 finite element problems were solved every time to evaluate the objective function [21–28].

Several books have been published about simulated annealing [29, 30]. Simulated annealing has also solved multi-objective problems [31]. A well known simulated annealing for multi-objective is AMOSA [32, 33]. The simulated annealing with crystallization factor has also solved multi-objective problems [34–36].

The simulated annealing with crystallization heuristic has shown to be very versatile, regarding the type of parameters (integer, continuous, combinatorial) and the type of objective function (continuous, discrete, intervalar, and multiple).

3.2 Tabu search

Tabu search was originally proposed to solve combinatorial problems (like simulated annealing). Tabu search can be viewed as a special case of simulated annealing. However, not with all possibilities available to the simulated annealing. Tabu search uses a local search and it has a structure storing visited solutions. One already visited solution is avoided. The word Tabu means that something cannot be touched, in the sense that already visited solutions must not be visited again. Three types of memories are commonly used to store the visited solutions: short-term, intermediate-term, and long-term [37]. The tabu search, like the simulated annealing, has only one current solution.

The chapter "A methaheuristic tabu search optimization algorithm: applications to chemical and environmental process" shows that tabu search can be used in a wide spectrum of engineering problems.

3.3 Genetic algorithms

Genetic algorithms are based on natural selection and they create a population of solutions. Based on this population of solutions, a new population is created using some operators: cross-over and mutation [38]. Genetic algorithm has a strong exploratory feature. Currently, it can be used with integer and continuous parameters. However, it does not have any convergence criteria. After a certain number of iterations, the best so far solution is defined as the *converged solution*. This weakness motivates one chapter in this set of chapters.

The chapter "Mixed-discrete nonlinear programming engineering problems" combines genetic algorithms with sequential quadratic programming. The sequential quadratic programming is a well-known gradient-based search algorithm.

4. Conclusions

As described, there are many types of engineering problems and this is the reason for the great number of methods developed to solve them. In this set of chapters, we intend to enumerate a few of them, showing their pros and cons and discuss the types of their parameters, type of the objective functions, computational cost, implementation facility, etc.

Acknowledgements

M. S. G. Tsuzuki is partially supported by CNPq (process 311.195/2019-9). A. K. Sato is supported by FUSP/Petrobras.

Author details

Marcos S.G. Tsuzuki[1*†], Ahmad Barari[2†], André K. Sato[1†], Rogrio Y. Takimoto[1†] and Tomoki Saka[3†]

1 Computational Geometry Laboratory, Escola Politínica da USP, São Paulo, Brazil

2 Ontario Tech University, Oshawa, Canada

3 Kanazawa Institute of Technology, Kanazawa, Japan

*Address all correspondence to: mtsuzuki@usp.br

† These authors contributed equally.

IntechOpen

References

[1] Peyvandi H. Computational Optimization in Engineering—Paradigms and Applications. London, UK: IntechOpen; 2017. p. 4

[2] Auroux D, Caillau J-B, Duvigneau R, Habbal A, Pantz O, Pronzato L, et al. Survey of sequential convex programming and generalized Gauss-Newton methods. ESAIM: Proceedings and Surveys. 2021;**71**:64-88

[3] Boussa I, Lepagnot J, Siarry P. A survey on optimization metaheuristics. Information Sciences. 2013;**237**(7): 82-117

[4] Grippo L, Lampariello F, Lucidi S. Nonmonotone line search technique for Newton's method. SIAM Journal on Numerical Analysis. 1986;**23**:707-716

[5] Ichikawa M, Gotoh T, Kagei S, Iwasawa T, Tsuzuki MSG. Pulmonary blood flow analysis based on two input model with aorta and pulmonary artery contribution using contrast-enhanced MRI. In: Proceeding of the 11th IEEE ISBI; Beijing, China. 2014. pp. 882-885

[6] Saka T, Ichikawa M, Kagei S, Gotoh T, Iwasawa T, Tsuzuki MSG. Perfusion analysis for lung MR images considering non-monotonic response of Gd-contrast agent. IFAC-PapersOnLine. 2014;**47**(3): 3587-3592

[7] Burer S, Letchford AN. Non-convex mixed-integer nonlinear programming: A survey. Surveys in Operations Research and Management Science. 2012;**17**(7):97-106

[8] Kirkpatrick S, Gelatt CD, Vecchi MP. Optimization by simulated annealing. Science. 1983;**220**(4598):671-680

[9] Corana A, Marchesi M, Martini C, Ridella S. Minimizing multimodal functions of continuous variables with the simulated annealing algorithm. ACM Transactions on Mathematical Software. 1987;**13**:262-280

[10] Ingber L. Very fast simulated re-annealing. Mathematical and Computer Modelling. 1989;**12**(8):967-973

[11] Duran GC, Sato AK, Ueda EK, Takimoto RY, Bahabadi HG, Barari A, et al. Using feedback strategies in simulated annealing with crystallization heuristic and applications. Applied Sciences. 2021;**11**:11814

[12] Tavares RS, Martins TC, Tsuzuki MSG. Simulated annealing with adaptive neighborhood: A case study in off-line robot path planning. Expert Systems with Applications. 2011;**38**(4): 2951-2965

[13] Martins TC, Tsuzuki MSG. Simulated annealing applied to the rotational polygon packing. IFAC Proceedings Volumes. 2006;**39**(3): 475-480

[14] Martins TC, Tsuzuki MSG. Rotational placement of irregular polygons over containers with fixed dimensions using simulated annealing and no-fit polygons. Journal of the Brazilian Society of Mechanical Sciences. 2008;**30**(3):205-212

[15] Sato AK, Martins TC, Tsuzuki MSG. Rotational placement using simulated annealing and collision free region. IFAC Proceedings Volumes. 2010;**43**: 234-239

[16] Ueda EF, Martins TC,Tsuzuki MSG. Planar curve fitting by simulated annealing with feature points determination, IFAC-PapersOnLine. Jan 2018;**51**:290-295

[17] Ueda EK, Tsuzuki MS, Barari A. Piecewise Bézier curve fitting of a point cloud boundary by simulated annealing,

in INDUSCON 2018, Jan 2019. pp. 1335-1340

[18] Ueda EK, Barari A, Sato AK, Tsuzuki MSG. Detection of defected zone using 3D scanning data to repair worn turbine blades. IFAC-PapersOnLine. 2020;**53**:10531-10535

[19] Ueda EK, Barari A, Tsuzuki MSG. Determination of open boundaries in point clouds with symmetry. In: Advances in Intelligent Systems and Computing. Proceeding of the ICGG 2020. Vol. 1296. 2021. pp. 332-342

[20] Ueda EK, Sato AK, Martins TC, Takimoto RY, Rosso RSU Jr, Tsuzuki MSG. Curve approximation by adaptive neighborhood simulated annealing and piecewise Bézier curves. Soft Computing. 2020;**24**: 18821-18839

[21] Martins TC, Camargo EDLB, Lima RG, Amato MBP, Tsuzuki MSG. Electrical impedance tomography reconstruction through simulated annealing with incomplete evaluation of the objective function. In: 33rd IEEE EMBC; Boston, USA. 2011. pp. 7033-7036

[22] Martins TC, Tsuzuki MSG. Simulated annealing with partial evaluation of objective function applied to electrical impedance tomography. IFAC Proceedings Volumes. 2011;**44**: 4989-4994

[23] Tavares RS, Martins TC, Tsuzuki MSG. Electrical impedance tomography reconstruction through simulated annealing using a new outside-in heuristic and GPU parallelization. Journal of Physics: Conference Series. Dec 2012;**407**:012015

[24] Martins TC, Tsuzuki MSG. Electrical impedance tomography reconstruction through simulated annealing with total least square error as objective function. In: 34th IEEE EMBC; San Diego, USA. 2012. pp. 1518-1521

[25] Martins TC, Tsuzuki MSG. Electrical impedance tomography reconstruction through simulated annealing with multi-stage partially evaluated objective functions. In: 35th IEEE EMBC; Osaka, Japan. 2013. pp. 6425-6428

[26] Martins TC, Tsuzuki MSG, Camargo EDLBD, Lima RG, Moura FSD, Amato MBP. Interval simulated annealing applied to electrical impedance tomography image reconstruction with fast objective function evaluation. Computers & Mathematcs with Applications. 2016;**72**(5):1230-1243

[27] Martins TC, Tsuzuki MSG. Investigating anisotropic EIT with simulated annealing. IFAC-PapersOnLine. 2017;**50**:9961-9966

[28] Tavares RS, Sato AK, Martins TC, Lima RG, Tsuzuki MSG. GPU acceleration of absolute EIT image reconstruction using simulated annealing. Biomedical Signal Processing and Control. 2019;**52**:445-455

[29] Tan CM. Simulated Annealing. London, UK: IntechOpen; 2008

[30] Tsuzuki MSG. Simulated Annealing—Advances, Applications and Hybridizations. London, UK: IntechOpen; 2012

[31] Tsuzuki MSG. Simulated Annealing—Single and Multiple Objective Problems. London, UK: IntechOpen; 2012

[32] Bandyopadhyay S, Saha S, Maulik U, Deb K. A simulated annealing-based multiobjective optimization algorithm: AMOSA. IEEE Transactions on Evolutionary Computation. 2008;**12**: 269-283

[33] Martins TC, Fernandes AV, Tsuzuki MSG. Image reconstruction by electrical impedance tomography using multi-objective simulated annealing. In: IEEE

ISBI 2014; Beijing, China. July 2014.
pp. 185-188

[34] Martins TC, Tsuzuki MSG. EIT
image regularization by a new multi-
objective simulated annealing
algorithm. In: Proceeding of the
37th IEEE EMBC; Milan, Italy. 2015.
pp. 4069-4072

[35] Ueda EK, Tsuzuki MSG, Takimoto
RY, Sato AK, Martins TC, Miyagi PE,
et al. Piecewise Bézier curve fitting by
multiobjective simulated annealing.
IFAC-PapersOnLine. 2016;**49**:49-54

[36] Najafabadi HR, Goto TG, Martins
TC, Barari A, Tsuzuki MSG. Multi-
objective topology optimization using
simulated annealing method. In:
Advances in Intelligent Systems and
Computing, Proceeding of the ICGG
2020. Vol. 1296. 2021. pp. 343-353

[37] Jaziri W. Tabu Search. Vienna,
Austria: I-Tech Education and
Publishing; 2008

[38] Popa R. Genetic Algorithms in
Applications. London, UK: IntechOpen;
2012

Verification and Validation of Supersonic Flutter of Rudder Model for Experiment

Ju Qiu and Chaofeng Liu

Abstract

The abrupt and explosive nature of flutter is a dangerous failure mode, which is closely related to the structural modes. In this work, the principal goal of the study is to produce the model, which is used very accurately for flutter predictions. Mode correctness of the model can correct the test deflects by the optimization technique——Sequential Quadratic Programming (SQP). The optimization of two finite element models for two flight conditions, transonic and supersonic speeds, had the different objectives which were defined by the nonlinear and linear eigenvector errors. The first and second frequencies were taken as constraints. And the stiffness of the rotation shaft was also restricted to some limits. The stiffness of the rudder axle was defined as design variables. Experiments were performed for considering springs both in plunge and in torsion of the rudder shaft. When the comparison between experimental information and analyzed calculations is described, generally excellent agreement is obtained between experimental and calculated results, and aeroelastic instability is predicted that agrees with experimental observations. Comments are also given concerning improvements of the flutter speed to be made to the model with changing stiffness of the rudder axle. Most importantly, V&V Method is used to provide the confidence in the results from simulation in this paper. Firstly, it introduces experimental data from Ground Vibration Test to build up or modify the Finite Element Model, during the Verification phase, which makes simulated models closer to the real world and guarantees satisfaction of final computed results to requirements, such as airworthiness. Secondly, the flutter consequence is validated by wind tunnel test. These enhancements could find potential applications in industrial problems.

Keywords: Classical flutter, Mode verification, Rudder, Wind tunnel, Verification and Validation (V&V)

1. Introduction

Generally, a wing or tail of an aircraft will be damped by damping when the speed of it is low; when the speed of flight exceeds a certain value, small disturbance will cause vibration divergence and the structure to collapse in a matter of seconds or even tens of milliseconds. This phenomenon is called flutter. Flutter is the most important subject in aeroelasticity. It is a kind of self-excited vibration which can maintain the constant amplitude oscillation under the interaction of elastic force, inertia force and aerodynamic force when the lifting surface moves at a certain speed in the air flow. At this time, the damping (g) equals 0. Due to the existence of system damping (g < 0), the vibration of the aircraft soon attenuates or even disappears completely with a small flight speed after being disturbed, but when the flight speed increases to a certain value, the amplitude caused by the disturbance just keeps the same. This speed is called the critical flutter speed, the vibration frequency at this time is called the critical flutter frequency and g = 0. In order to prevent flutter, the critical flutter velocity must be larger than the maximum flight velocity under all flight conditions and there must be some margin. **Figure 1** taken from a Chinese book written by Liu, C., F. and Qiu, J. [1] is an example of the wing flutter calculation: when the speed of flight, VCR, is 850 m/s, the wing is in constant amplitude vibration, and if the speed of flight is less than or greater than it, the wing vibration will attenuate or disperse.

Classical flutter calculations in frequency domain are performed using either the K method proposed by Bisplinghof, R. L., and Ashley, H. [2] or the P-K method proposed by Hassig, H. [3] and Lawrence, J. A., and Jackson, P. [4]. The K method is generally very fast and quite simple, but it has a downfall in that sometimes the frequency and damping values "loop" around themselves and generate multi-value frequency and damping as a function of velocity. The K method solution is only valid when g = 0 and the structural motion is neutrally stable and matches the aerodynamic motion with the neutral stability. The P-K method is acknowledged to provide more accurate modal damping values after the K method. Gradually, the P-K method has become one of the most widely used methods in aeroelastic engineering. After that, the p method, proposed by Abel, I. [5], improves the damping and frequency trends by taking into account the effect of nonzero damping by means of generalized aerodynamic forces, which are approximately valid for the damping-frequency area under consideration. At the end of the 1990s, the μ method studied by Lind, R., and Brenner, M. [6], was used to fitting procedures to transform the aerodynamics to the state space. At the beginning of the new century, a g method proposed by P. C. Chen [7] is used in the analytic property of unsteady aerodynamics and a damping perturbation approach. These two methods of P-K method and g method are different in the

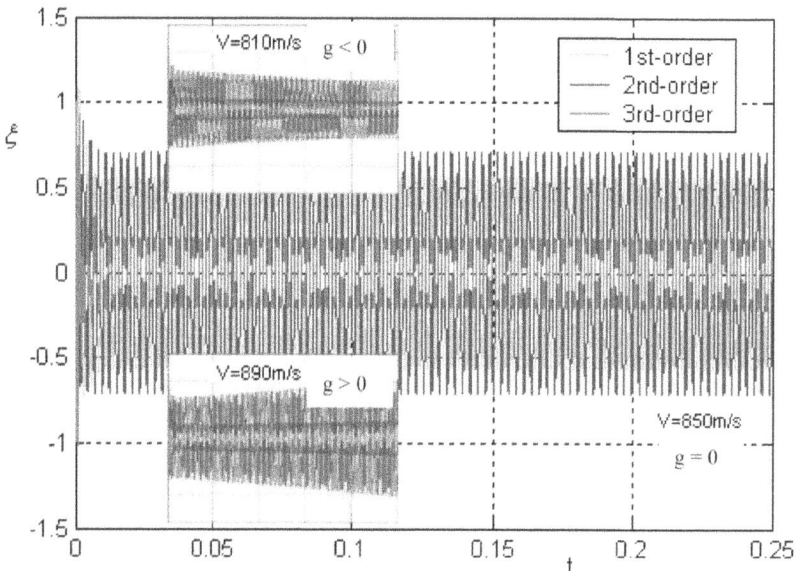

Figure 1.
The wing flutter calculation, $V_{CR} = 850$ m/s.

equation form, but they share the same stability criterion, i.e., an eigen root of aeroelastic equation is solved and the root with positive real part indicates flutter. Additionally, the g method uses a reduced-frequency sweep technique to search for the roots of the flutter solution and a predictor corrector scheme to ensure the robustness of the sweep technique. The g method includes a first-order damping term in the flutter equation which is rigorously derived from the Laplace-domain aerody-namics. And then, an improved g method proposed by Ju Qiu and Qin Sun [8], increases a second-order damping term in the flutter equation. It is also valid in the entire reduced frequency domain and up to the second order of damping. Recently, the H method proposed by Michaël H. L. [9], automatically extends the aerodynamic data obtained for purely oscillatory motions to damping and diverging oscillatory motions by means of a direct harmonic interpolation method, thereby improving the prediction of damping and frequencies. This procedure may assist the aeroelastician in making improved estimates of aerodynamic damping at g < 0 conditions to support flight flutter testing and probably offers potential for flight control system design or analysis. Brian P. Danowsky and etc. [10] showed that three different flutter suppression controllers were designed using a flight-test-validated aeroelastic aircraft model. The flight tests demonstrated that the flutter boundary could be successfully expanded using active control. Eli [11] indicated that AFS (Active flutter suppression) technology had the potential to lead to significant weight savings and performance gains.

In the new and challenging field of energy harvesting through fluid–structure instabilities, such as analysis in time domain, the coupled-mode flutter mechanism has been recently scrutinized. A lot of complicated aeroelastic characteristics are predicted by a structural-aerodynamic fully-coupled formulation, such as rotorcraft written by Bernardini G., Serafini J., Molica Colella M., and Gennaretti M. [12], a transport wing's wingtip introduced by Peng Cui and Jinglong Han [13], a transonic wing analyzed by Xiang Zhao, Yongfeng Zhu and Sijun Zhang [14]. Francisco Palacios, Michael R. Colonno, and etc. [15] showed an integrated platform for multi-physics simulation and design, e.g. flutter predictions by a loosely-coupled method, remained open source and serves as a starting point for new capabilities that will hopefully be contributed by users in both academic and industrial envi-ronments. Most recently, Leclercq T., Peake N., and de Langre E. [16] pointed out that flutter did not prevent drag reduction by reconfiguration. Eirikur Jonsson [17] put flutter and post-flutter constraints into the process of the aircraft design opti-mization. Sergey Shitov and Vasily Vedeneev [18] investigated the flutter bound-aries of rectangular panels simply supported at all edges, and used potential flow theory to calculate the unsteady pressure.

Naturally, for every simulation and every test, it is desired to produce a model that could be accurate, studied by Samuel C. McIntosh Jr., Robert E. Reed Jr. T. and William P. Rodden [19], so as to permit meaningfully calculated and experimental comparisons and to provide data for evaluating new theoretical or numerical tech-niques for treating aeroelastic problems. The studies of all cases suggest that at increased airspeeds the aircraft may develop unstable oscillations leading to a cata-strophic failure, described by Jieun Song, Seung Jin Song and Taehyoun Kim [20], Jie Zeng and Sunil L. Kukreja [21], Thomas Andrianne and Grigorios Dimitriadis [22]. At all times, on one hand, there exist analytical models and tools that allow estimating the onset of the dynamic instability, but they are often subject to modeling uncertainties and limitations that can produce inaccurate, unreliable results. On the other hand, the test data are a direct reflection of the actual aircraft, to some degree, and hence can be used with confidence. Also, it is a common and required practice in aerospace industry to conduct flight flutter tests and validate the simulation effect of the analysis model before aircraft enter into service.

Since the beginning of the century, American Society of Mechanical Engineers guide to Verification and Validation (V&V) defined the goal of V&V process as to develop standards for assessing the correctness and credibility of modeling and simulation in computational science. According to the paper, for Validation, and Predictive Capability in Computational Engineering and Physics, written by William L. Oberkampf, Timothy G. Trucano, Charles Hirsch [23], a proposed question was how confidence in modeling and simulation should be critically assessed. Verification and validation (V&V) of computational simulations are the primary methods for building and quantifying this confidence. Briefly, verification is the assessment of the accuracy of the solution to a computational model. Validation is the assessment of the accuracy of a computational simulation by comparison with experimental data. In verification, the relationship of the simulation to the real world is not an issue. In validation, the relationship between computation and the real world, i.e., experimental data, is the issue. Furthermore, Verification is defined as following: the process of determining that a model implementation accurately represents the developer's conceptual description of the model and the solution to the model, while Validation means that the process of determining the degree to which a model is an accurate representation of the real world from the perspective of the intended uses of the model. Supanee Arthasartsri and He Ren [24], in A380 Aircraft Reliability Program also pointed out that the validation is a process determining whether the mathematical model describes sufficiently well the reality with respect to the decision to be made, which includes requirement validation and product validation, whereas, the purpose of Validation is to ensure that the requirements for a product are sufficiently correct and complete to achieve safety and to satisfy the needs of the customer within program constraints (e.g. cost, schedule). The Product Validation is to check if product meets the implicit needs of the customer. It is also to ensure the final product meets the requirement, for example, airworthiness certification of commercial aircraft. Additionally, V & V method was described in Guide for Verification and Validation in Computational Solid Mechanics, written by Schwer L. E. and etc. [25].

In **Figure 2**, it is evident to see the test has not been involved in the first stage of the V&V process, Verification and only is employed in the second step, Validation.

When V & V technique is widely used to aerospace engineering, civil engineering, auto engineering, etc., it obtains fruitful achievements. For instance, the example of V&V method implementation on GP7200 by Engine Alliance, which the results of reliability before entry into service exceeded the expectation requirements from FAA. With the entry into service of Airbus A380, the application of validation and verification method in safety and reliability program has been proven successful. The other example is Validation and Verification of the Lunar Atmosphere Dust Environment Explorer (LADEE) mission models and software. This program identified and prioritized a set of V&V technologies that significantly reduced the development cost and compressed the development schedule of emerging safety critical flight control systems. It is predicted that it could also be applied to other aerospace system design and to other industry that require the high level of safety and V&V Method would be used to provide the confidence in the results from simulation.

The present research uses V & V technologies to handle the flutter problem of the rudder. There is, however, a significant difference, that is to say, the experimental data were added to modify the simulated models, besides codes and software verification.

In the Verification process, the stiffness of the rudder axis could be adjusted during the flight or the experiment. The accuracy of the flutter-prediction results of these methods heavily depends on how good the model or mode estimations are.

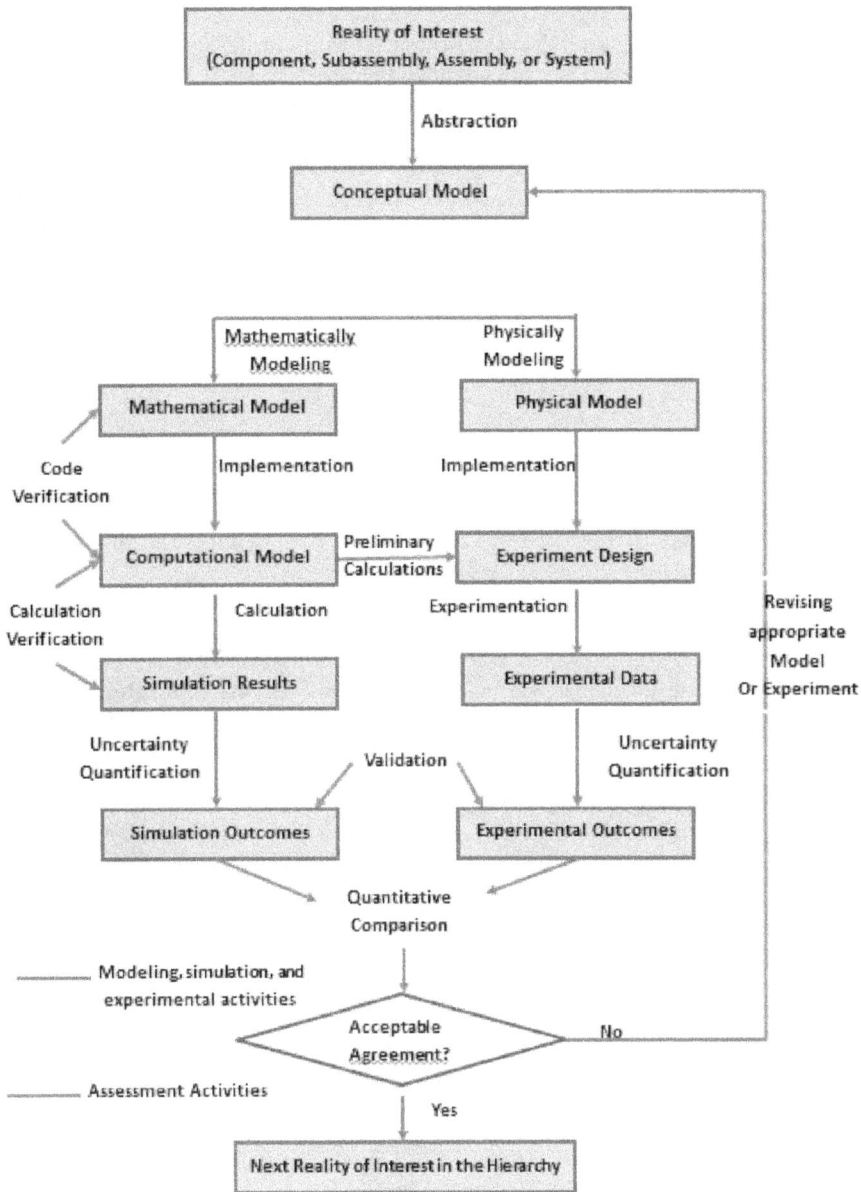

Figure 2.
Verification and Validation's activities and products.

Therefore, the modes of the rudder directly influencing flutter speeds become very critical. In order to obtain a precise model to simulate two cases of supersonic flutter, the appropriate axis structures were provided by optimization process. Unlike traditionally aeroelastic systems of maximum performance and minimum weight, studied by Melike Nikbay, and Muhammet N. Kuru, [26], the objective was a minimal 1-order or 2-order error which was a ratio of test eigenvectors and calculated ones, constraints were the first and second natural frequencies, and design variables were stiffness of the different rudder axles. Finally, modes of the first bending and second twisting of the rudder were found out in Case one, and these of the first torsion and second bending were extracted in Case two. In addition, it focused on finding a right analysis model by optimization method, according to experimental data of mode shapes studied by Zimmerman, N. H., and Weissenburer, J. T. [27]. Test investigations were first presented, followed by verifying the structural model and mode by optimization. Although two flutter cases both had the same flutter mechanism, classical bending-torsion flutter, they had their own critical flutter points, respectively.

The present work also aims at a deeper understanding of the phenomena by characterizing the tight interaction between the unsteady flow patterns in the flow-field and the response of the structure. After numerical aeroelastic simulations were finished, the test data were employed to validate flutter prediction to confirm the observed critical velocity and coupled-mode flutter in the validation stage of V& V technologies.

2. Experimental set-up

Flight flutter testing is an expensive and hazardous task, but it is required to verify if the aircraft is truly free from flutter within the margin of the aircraft's operational envelope.

In order to validate the supersonic flutter calculation method and the verification of the flutter model accuracy, done by Bingyuan Yang, Weili Song [28], in FD-06 wind tunnel of 701 Institute, a test of a rudder was conducted. The data of eigenvector records taken in ground vibration test (GVT) were being preserved and would be available for further analysis. All test conditions and model configurations for available data are summarized in **Table 1**.

Case 1 Test Number: 2	Mach number = 1.53 Angle of attack = 2°10′ 1st order frequency:35.86 2nd order: 64.74	1st order mode: −0.8039 −0.3889 −0.2023 −0.0046 0.1788 0.3497 0.7712 −0.7353 −0.3562 −0.1222 0.0454 0.2275 0.3726 0.8007 −0.5425 −0.2471 −0.0853 0.1294 0.3065 0.4967 0.9575 −0.4314 −0.1592 0.0239 0.1654 0.2892 0.5000 0.9444 −0.2748 −0.0588 0.1150 0.2128 0.3399 0.5490 0.9902 −0.1085 0.0660 0.1856 0.3301 0.4804 0.6307 1.0000 2nd order mode: 0.8040 0.4517 0.3125 0.1040 0.0054 −0.1389 −0.3665 0.9347 0.5426 0.3040 0.1744 0.0713 −0.0599 −0.2503 0.9034 0.5284 0.3636 0.2710 0.1739 0.0514 −0.1298 0.9688 0.5682 0.4318 0.3324 0.2310 0.1679 0.0744 1.0000 0.6307 0.5256 0.3977 0.3296 0.2778 0.2401 0.9517 0.7358 0.6136 0.5682 0.5398 0.4517 0.4347
Case 2 Test Number: 13	Mach number = 2.51 Angle of attack = 4°50′ 1st order frequency:29.68 2nd order: 58.86	1st order mode: −0.7386 −0.4281 −0.1837 −0.0046 0.2301 0.4895 0.8628 −0.6863 −0.3033 −0.1137 0.0739 0.2680 0.4994 0.8431 −0.5170 −0.2281 −0.0268 0.1288 0.3222 0.5373 0.9085 −0.3915 −0.0987 0.0510 0.1954 0.3706 0.6183 1.0000 −0.2529 −0.0229 0.1346 0.2804 0.4556 0.6314 0.9281 −0.0791 0.0869 0.2183 0.3523 0.5170 0.7190 0.9739 2nd order mode: 0.6652 0.5773 0.3856 0.1864 0.0265 −0.1591 -0.3258 0.7455 0.6159 0.4364 0.2515 0.1402 −0.0568 -0.2470 0.8030 0.6576 0.4644 0.3538 0.2447 0.1061 −0.0902 0.8712 0.7189 0.5871 0.4992 0.3689 0.2470 0.1205 1.0000 0.8561 0.7106 0.5811 0.5061 0.4265 0.2197 1.0000 0.8939 0.8258 0.7349 0.6689 0.5386 0.4311

Table 1.
Data of vibration test.

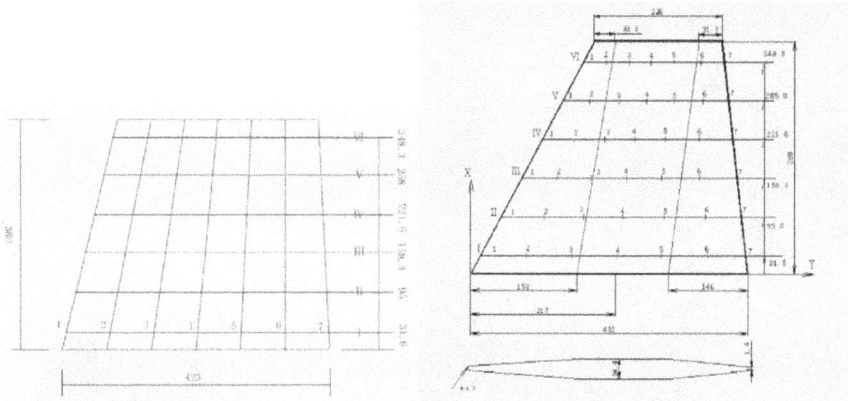

Figure 3.
Location of monitoring points for eigenvectors (unit: Millimeter).

Case	Points*1	Mach	Angle of attack	Static pressure*2 (Pa)	Temperature (K)	Density (Kg/m³)	Speed of sound (m/s)
1	2	1.53	2°10′	33230	197.215	0.5873	281.44
2	13	2.51	4°50′	15457	127.720	0.40132	226.49

*1 Points refer to the number of the test.
*2 the initial angle causes static pressure.

Table 2.
Test parameters.

In the **Table 1** above, the control points of vibration shapes are shown in **Figure 3**. **Table 2** shows the flutter results via this test.

3. Model/mode verification

The following section introduces the details of the rudder model used in the test and the generation of an accurate FEM (Finite Element Method), which matches experimental data.

3.1 Rudder model

The rudder axle in **Figure 4** was attached to the pitch and twist springs. The outer leaves of the springs were rigidly mounted to the tunnel walls. The pitch springs were commercially available flexure pivots that allowed ±7.5 deg. of rotation through the elastic bending of rudder flexures. The twist springs had a flexural rigidity range. Usually, the spring retained its linear characteristics over large displacements. Also, there were no sliding surfaces at support points which would produce damping. The springs were designed to deflect elastically approximately 10 mm. During flight or in the wind tunnel, rudder stiffness uncertainties are related to spring rotation and operating conditions. In particular, some record data from the experiment are not so reliable, maybe, from immature operations or instrumental limitations. So, it is naturally involved in designing and optimizing a realistic structure for a required level of reliability and efficiency for supersonic flutter validation.

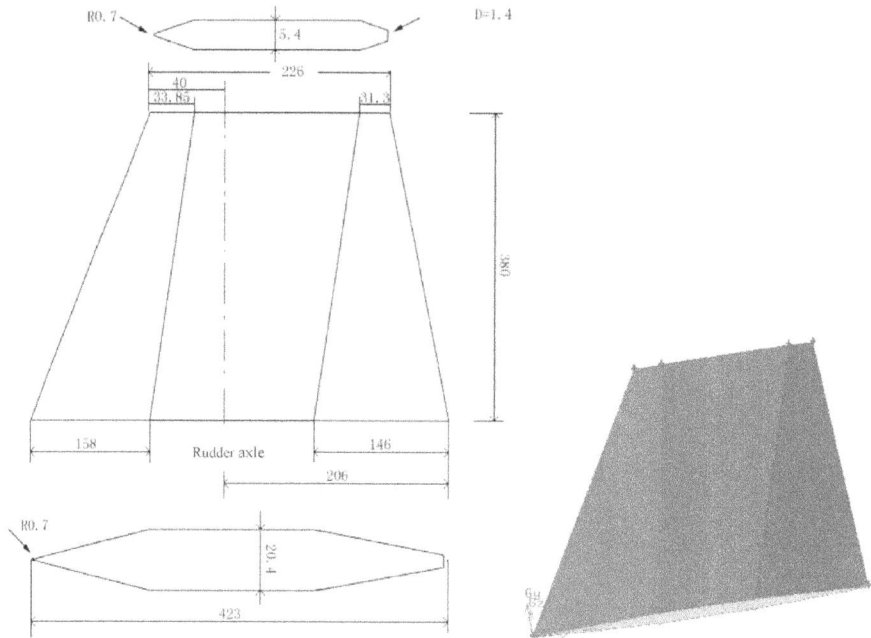

Figure 4.
Rudder model (unit: Millimeter).

3.2 FEM and boundary conditions

The rudder was composed of a Ti-alloy material with the rudder torque tube that was made of a steel material. The material parameters are listed in **Table 3**.

The finite element model was composed of 540 hexahedral structural elements. The rudder torque tube was modeled with 10 linear beam elements in **Figure 5**.

The node at the end of the rudder shaft was fixed in 3 translations and 2 rotations, but was free to rotate about the axis of the shaft.

3.3 Optimization method and parameters

In the present research, Sequential Quadratic Programming (SQP), also known as Quadratic Approximation, is applied. It has arguably become the most successful method for solving nonlinearly constrained optimization problems. Its outstanding strongpoint is the less number of function and gradient evaluation, and the higher computational efficiency, especially for the rudder structural optimization objective function being a linear or nonlinear function of the design variables, and constraints such as frequencies for a function of the design variables. Using the method to optimize its stiffness, applying the mode shape error derivative concept to calculate

Parameters	Young module(MPa)	Poisson ratio	Density (Kg/m³)
Rudder	4.1e10	0.34	1810.
Axle	2.11e11	0.3	7800.

Table 3.
Structural material parameters.

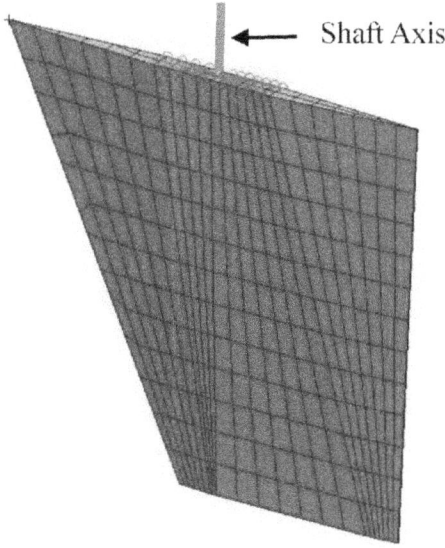

Figure 5.
FEM of rudder.

the sensitivity of the stiffness, and employing this Method in the iterative process to make rapid the optimization convergence, and reliable the computational results.

When the optimal model was established, it was possible to define the optimization conditions and perform mode analysis.

1. The objective function:

    From **Table 1**, the vibration amplitude distribution of the first frequency of the rudder was nonlinear in Case 1, and therefore, the eigenvector error equation was written by.

    $$error1 = \sqrt{\sum_{i=1}^{n}(x_i - y_i)^2/n} \qquad (1)$$

    N sample nodes were located at the same location of the test points from the upper rudder surface in **Figure 6**.

    For Case 2, the eigenvector error equation was defined by linear, due to the linear changing of the vibration displacements.

    $$error2 = \sum_{i=1}^{n}|x_i - y_i|/n \qquad (2)$$

    The two errors in modal analysis of the rudder were the objectives to be minimized in this research, respectively.

2. Constraints:

    The first and second frequencies were taken as constraints. At the same time, the stiffness of the rotation shaft was also restricted to some limits.

3. Design variables:

    The stiffness of the rudder axle was defined as design variables.

Figure 6.
Monitoring nodes of FEM.

3.4 Optimization results

The optimization typically took 9 and 26 iterations, for Case 1 and Case 2 respectively to converge to the precision required for the gradient optimization in **Figures 7** and **8**.

In Case 1, the first mode surface was fitted in the test data and the optimization monitoring points in **Figure 9**.

The **Figure 9** above shows that the experimental points change dramatically and disorderedly, while the optimized ones transit gradually and softly. It also demonstrates the 2nd-order error for the first mode shape in Case 1 is selected very right.

In Case 2, the first mode surface is fitted in the test data and the optimization monitoring points in **Figure 10**.

It is indicated in **Figure 10** that more test data deviate from the fit left surface, but almost all of optimization points lie in the right surface.

In cases 1 and 2, the natural frequencies and mode shapes are shown in **Figures 11–14**.

Compared with the experimental frequencies, the errors are shown in **Table 4**.

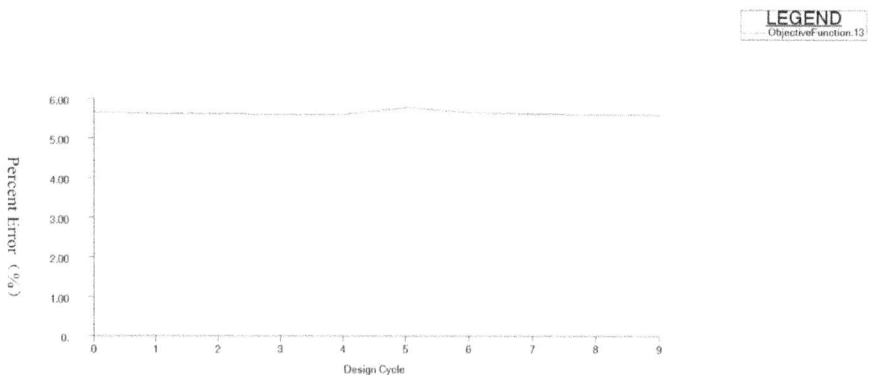

Figure 7.
History of case 1 (vertical axis shows percent error in Eq. (1)).

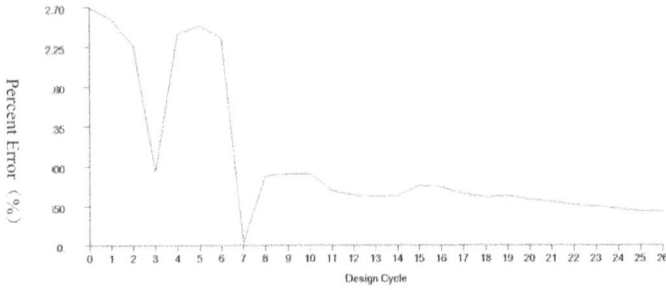

Figure 8.
History of case 2 (vertical axis shows percent error in Eq. (2)).

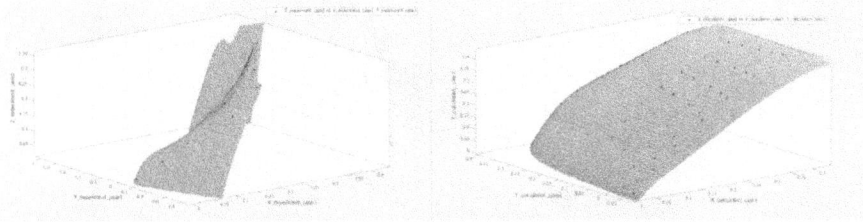

Figure 9.
First mode surface of the test data (left) VS. first mode surface of the optimization monitoring points (right).

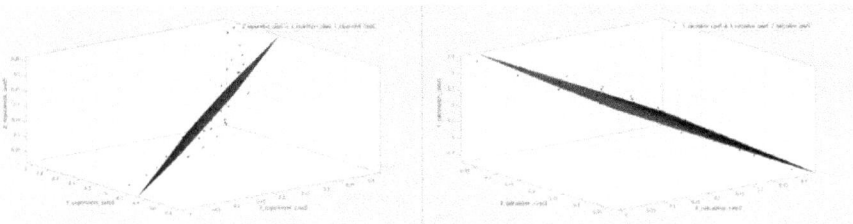

Figure 10.
First mode surface of the test data (left) VS. first mode surface of the optimization monitoring points (right).

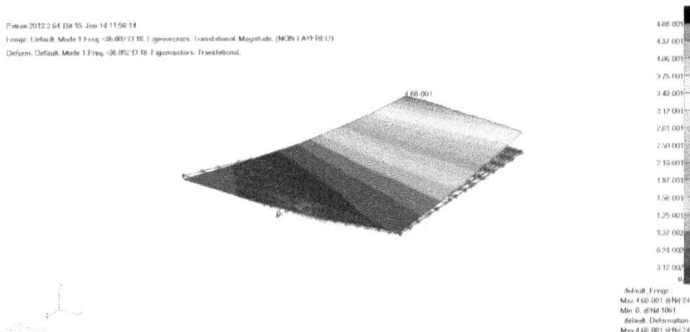

Figure 11.
First bending mode (case 1).

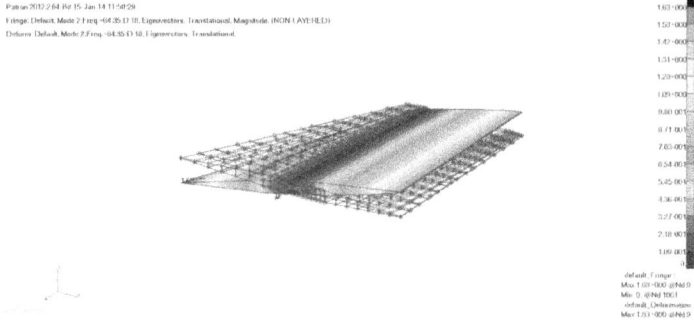

Figure 12.
Second torsion mode (case 1).

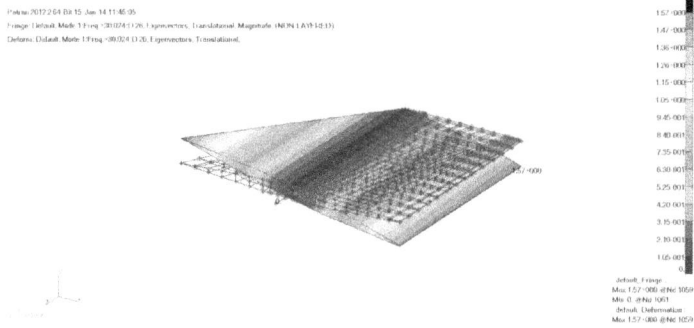

Figure 13.
First torsion mode (case 2).

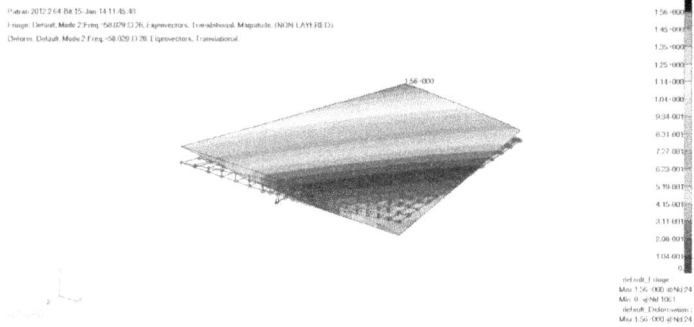

Figure 14.
Second bending mode (case 2).

Cases	First frequency		Error (%)	Second frequency		Error (%)
	Test	Opti.		Test	Opti.	
1	35.86	36.002	0.396	64.74	64.35	0.602
2	29.68	30.024	1.16	58.86	58.029	1.41

Table 4.
Frequency errors between test and optimization.

From **Table 4**, the error of the first frequency is less than 1%, and that of the second one is not more than 1.5%. Again, it represents optimization design is very successful.

4. Flutter prediction

In this section, two flutter-prediction method, namely Zona51of Nastran from MSC Software Corporation [29], and Local piston theory, which is performed by home-made software are employed to obtain the flutter speeds.

4.1 Zona51

ZONA51, written by MSC Software Corporation, is a supersonic lifting surface theory that accounts for the interference among multiple lifting surfaces. It is similar to the Doublet-Lattice method (DLM) in that both are acceleration potential methods that need not consider flow characteristics in any wake. An outline of the development of the acceleration-potential approach for ZONA51 and its outgrowth from the harmonic gradient method (HGM) are described. ZONA51 is a linearized aerodynamic small disturbance theory that assumes all interfering lifting surfaces lie nearly parallel to the airflow, which is uniform and either steady or vibrating harmonically. As in the DLM, the linearized supersonic theory does neglect any thickness effects of the lifting surfaces.

For aeroelastic analysis, the unsteady aerodynamic forces are obtained using Doublet Lattice for supersonic flight. The rudder section was subdivided into a lattice of 20 chordwise $\times$20 spanwise space vortex panels, yielding a total of 400 vortex panels. **Figure 14** describes aerodynamic trapezoidal panels of the rudder in **Figure 15**.

Through the flutter analysis by Nastran's ZONA51, the V-g and V-f curves of Case 1 are shown in **Figures 16** and **17**, when M = 1.35 and Density Ratio = 0.479.

From **Figure 16**, g of the first bending mode changes from the negative value to the positive at the speed of 380 m/s, and **Figure 17** presents frequencies of the second torsion mode and the first bending mode try to go toward the same value at the speed of 380 m/s, that is, 1.35 M. At this point, flutter occurs.

Through the flutter analysis, the V-g and V-f curves of Case 2 are shown in **Figures 18** and **19**, when M = 2.4 and Density Ratio = 0.327.

Figure 15.
DLM grid.

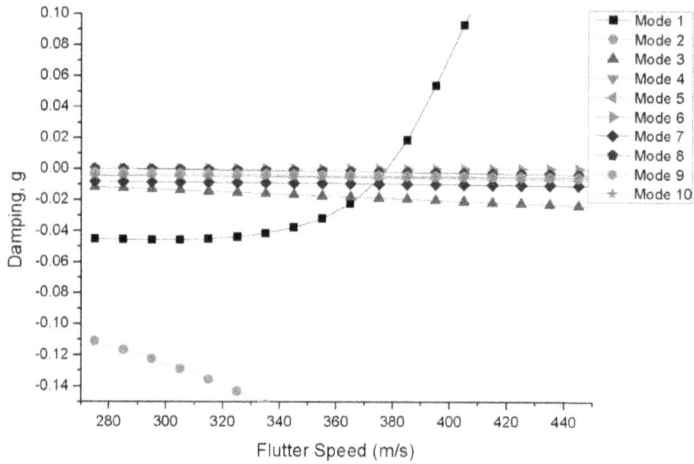

Figure 16.
V-g curve.

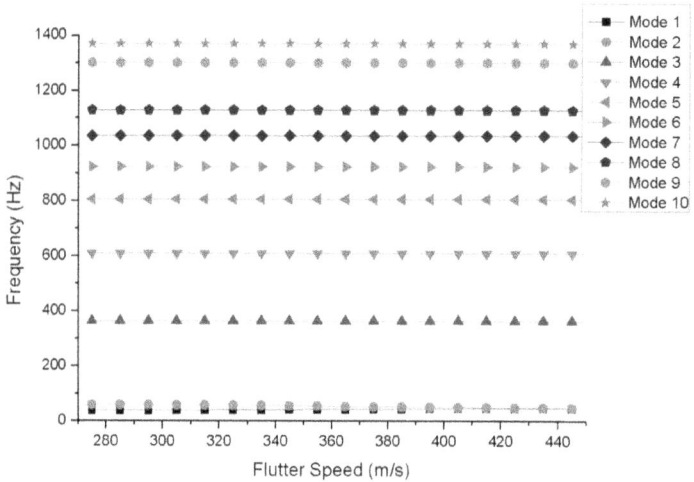

Figure 17.
V-f curve.

From **Figure 18**, g of the 2st-order bending mode changes from the negative value to the positive at the speed of 550 m/s, and **Figure 19** presents frequencies of the second bending mode and the first torsion mode try to go toward the same value at the speed of 550 m/s, e.g. 2.4 M. At this point, flutter occurs.

As can be seen from the preceding **Figures 16–19**, two cases present the same bending-torsion coupling modes that lead to flutter failure in terms of the same flutter mechanisms. However, aeroelastic flutter speeds have somewhat obvious differences, though the first two frequencies are slightly similar.

4.2 Comparison of calculated methods and tests

Due to the different aerodynamic expressions, Zona51, and Local piston theory, the flutter results are indicated in **Table 5**.

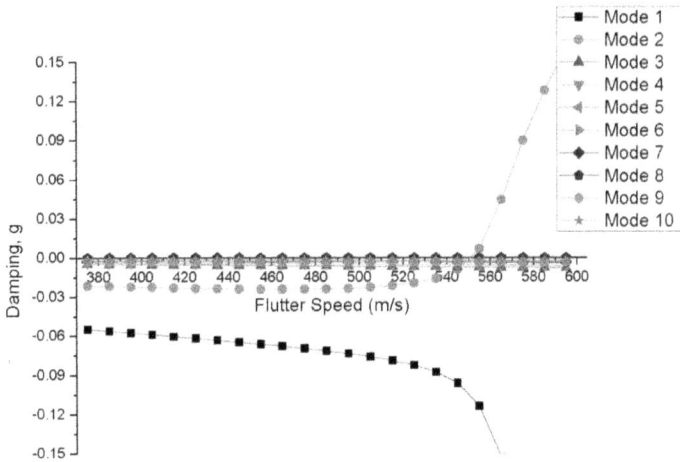

Figure 18.
V-g curve.

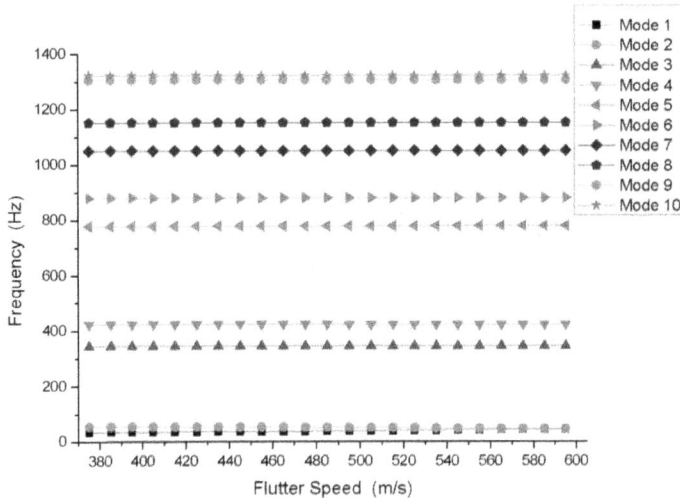

Figure 19.
V-f curve.

Cases	Flutter speed (Mach)			Error (%)	
	Test	Zona51	Local piston theory	Zona51	Local piston theory
1	1.53	1.35	1.33	11.76	13.07
2	2.51	2.4	2.46	4.38	1.99

Table 5.
Comparison of flutter prediction.

From the above table, the error is somewhat large in Case 1, because it is different for getting unsteady aerodynamic forces near transonic flight, and the other reason is not to consider the static pressure, which is caused by an initial angle and test points are much fewer.

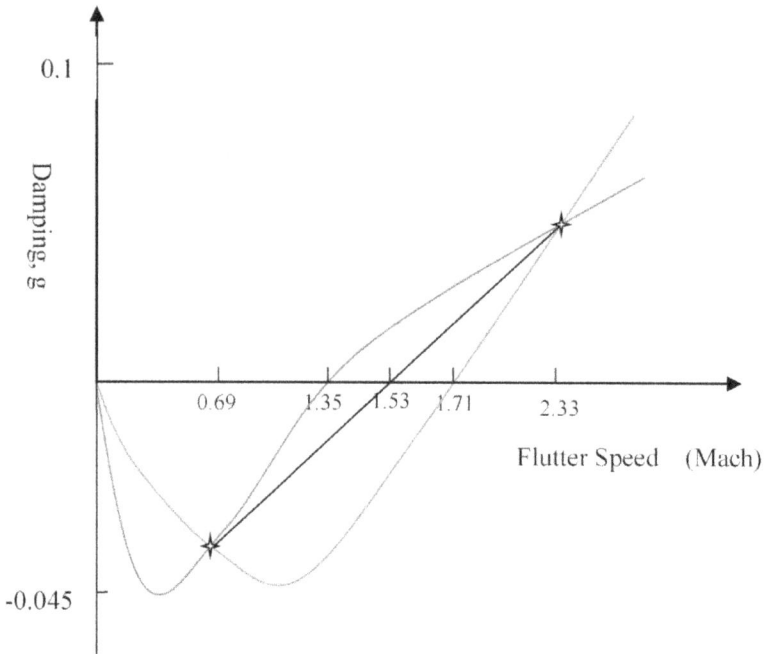

Figure 20.
Flutter V-g curve.

In **Figure 20**, there is only two test points, which is connected to a straight line transcend the zero point, and makes us find the flutter speed. However, the other two colorful curves also go through the horizontal axis, which gets the different flutter speeds, greater than the former or less than it.

But in Case 2, in supersonic flight, the test result is in good agreement with all three methods, due to better optimization model and much more test points.

5. Conclusions

From experimental mode data to mode verification, optimized FEM is much closer to the test rudder. Through flutter predictions, the predicated results are basically similar to the test data. We can draw some conclusions listed as follows:

1. According to the experimental eigenvectors in two supersonic flight cases, optimization technique with the sensitivity-based approximation approaches, and the first-order and second-order errors, helps to find the different finite element models with the first bending and second torsion, or the first torsion and second bending, by modifying the stiffness of the rudder axle.

2. During the optimization procedure, **Figures 11** and **12** represent the optimized mode surface can correct some mistake of the test data, perhaps, coming from manual errors or tool limitations.

3. Via the flutter prediction in frequency domain, it is in agreement with the mechanism of coupled mode instability, and the classical bending-twist coupled flutter failure is presented, once more.

4. By two flutter prediction methods, their results shown in **Table 5** and supplied by the current solutions agree well with the test values. Once again, it reveals the optimized model is reliable and robust.

5. The predicated flutter speeds are less than the test ones. It shows flutter predictions are safe at most of times. But, being a robust analysis of its flutter margin solution might be too conservative to be realistic.

6. The most important is that V&V technologies are utilized to the flutter prediction of flight vehicles, the GVT figures are added to the modeling for simulation in the Verification, and the flutter test validates the simulated results. Experimental data are involved in the first phase, which greatly increases confidence and reliability for validation in the second one.

6. Final remarks

From the preceding discussions, an interesting phenomenon is observed, when we rotate the rudder axle for getting the diverse stiffness, the flutter speed is increased from 1.53 M to 2.51 M, but the structural weight or the rudder itself does not vary. Does it mean that the rudder itself can suppress flutter as the rudder is operated or controlled appropriately, and thereby, the flutter boundary is expanded?

Besides, V&V are tools for assessing the accuracy of the conceptual and computerized models. It can be extended to airworthiness certification of civil aircraft (see **Figure 21**).

From the above flowchart, if we modify the calculated model by test data, which ensures the analyzed precision, does it mean that analysis prediction could directly pass by Airworthiness Certification? Especially, for much of the Operational Reliability work, other highly nonlinear dynamic problems or very high frequency of acoustic ones, the assessment is so difficult, if it is not possible to copy in the experiment, that V&V became more associated with the issue of credibility, i.e., the quality, capability, or power to elicit belief, when we use the test to modify the model.

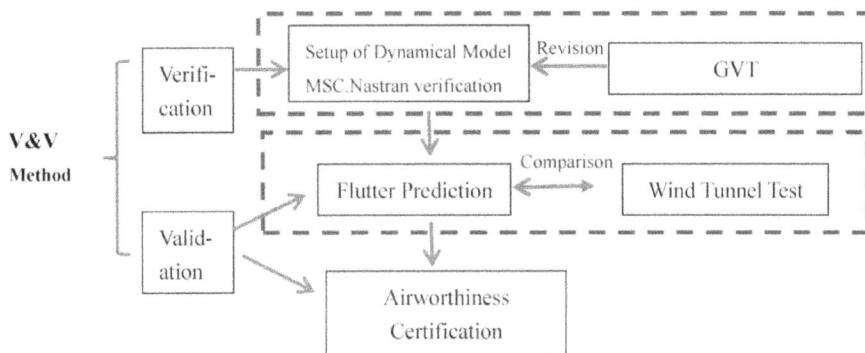

Figure 21.
V & V for airworthiness certification.

Acknowledgements

My deepest gratitude goes first and foremost to Mr. Erwin Johnson and Mr. Mohan Barbela, my previous coworkers in MSC, for their revisions of this paper.

I would like to extend my heartfelt gratitude to MSC Software Corporation which has given me precious chances and experiences to improve my knowledge of FEM analysis.

This research work was financially supported by Shanghai Science and Technology Committee (Grant agreement No. 13QB1401500).

Nomenclature

g = Damping
n = Sample number
x_i = Calculated value of eigenvector corresponding to point i
y_i = Test value of eigenvector corresponding to point i

Author details

Ju Qiu[1,2*] and Chaofeng Liu[3]

1 Composites Center, COMAC Shanghai Aircraft Manufacturing Co., Ltd., Shanghai, China

2 Beijing Key Laboratory of Civil Aircraft Structures and Composite Materials, COMAC Beijing Aircraft Technology Research Institute, Beijing, China

3 Mechanics Engineering School, Shanghai University of Engineering Science, Songjiang, Shanghai City, China

*Address all correspondence to: qiu_x_j@126.com

IntechOpen

References

[1] Liu, C., F. and Qiu, J. (2016). "A fluid-structure coupling analysis method of Aeroelasticity, Beihang University Press", 11. (In Chinese)

[2] Bisplinghof, R. L. and Ashley, H. (1962). "Principles of Aeroelasticity", *Wiley*, New York.

[3] Hassig, H. (1971). "An Approximate True Damping Solution of the Flutter Equation by Determinant Iteration", *Journal of Aircraft*, Vol. 8, No. 11, 885–890.

[4] Lawrence, J. A., and Jackson, P. (1968). "Comparison of Different Methods of Assessing the Free Oscillatory Characteristics of Aeroelastic systems", *Aeronautical Research Council*, London, England.

[5] Abel, I. (1979). "An Analytical Technique for Predicting the Characteristics of a Flexible Wing Equipped with an Active Flutter-Suppression System and Comparison with Wind-Tunnel Data", *NASA*, TP-1367.

[6] Lind, R., and Brenner, M. (1999)." Robust Aeroservoelastic Stability Analysis", *Springer–Verlag*, New York.

[7] Chen, P. (2000). "A Damping Perturbation Method for Flutter Solution: The g-Method", *AIAA Journal*, Vol. 38, No. 9, 1519-1524, doi:10.2514/2.1171.

[8] Ju Qiu and Qin Sun (2009). "New Improved Method for Flutter Solution", *Journal of Aircraft, AIAA*, Vol. 46, No. 6, 2184-2186.

[9] Michaël H. L. Hounjet (2010). "Verification of H Flutter Analysis", *Journal of Aircraft*, Vol. 47, No. 6, 2168.

[10] Brian P. Danowsky, Aditya Kotikalpudi, et al. (2018). "Flight Testing Flutter Suppression on a Small Flexible Flying-Wing Aircraft", AIAA AVIATION Forum, June 25-29, 2018, Atlanta, Georgia, 2018 Multidisciplinary Analysis and Optimization Conference, 1-13.

[11] Eli Livne (2018). "Aircraft Active Flutter Suppression: State of the Art and Technology Maturation Needs", JOURNAL OF AIRCRAFT, Vol. 55, No. 1, 2018, 410-450.

[12] Bernardini G., Serafini J., et al. (2013). "Analysis of a structural-aerodynamic fully-coupled formulation for aeroelastic response of rotorcraft", *Aerospace Science and Technology*, 29, 175-184.

[13] Peng Cui and Jinglong Han (2012). "Prediction of flutter characteristics for a transport wing with wingtip devices", *Aerospace Science and Technology*, 23, 461-468.

[14] Xiang Zhao, Yongfeng Zhu, et al. (2012). "Transonic wing flutter predictions by a loosely-coupled method", *Computers & Fluids*, 58, 45-62.

[15] Francisco Palacios, Michael R. Colonno, et al. (2013). "Stanford University Unstructured (SU2): An open-source integrated computational environment for multi-physics simulation and design", *51st AIAA Aerospace Sciences Meeting including the New Horizons Forum and Aerospace Exposition, Grapevine (Dallas/Ft. Worth Region), Texas, AIAA 2013-0287*, 1-60.

[16] Leclercq T., Peake N. and de Langre E.(2018), Does flutter prevent drag reduction by reconfiguration? Proc. R. Soc. A 474: 20170678. doi:10.1098/rspa.2017.0678

[17] Eirikur Jonsson, Cristina Riso, et al. (2019). "Flutter and Post-Flutter Constraints in Aircraft Design

Optimization", Progress in Aerospace Sciences, 1-77. https://www.researchga te.net/publication/333513498 doi: 10.1016/j.paerosci.2019.04.001.

[18] Sergey Shitov and Vasily Vedeneev (2017). "Flutter of rectangular simply supported plates at low supersonic speeds", Journal of Fluids and Structures 69 (2017), 154–173.

[19] Samuel C. Mclntosh Jr., Robert E. Reed Jr. T., et al. (1981). "Experimental and Theoretical Study of Nonlinear Flutter", VOL. 18, NO. 12, *AIAA 80-0791R, Journal of Aircraft,* 1057-1062.

[20] Jieun Song, Seung Jin Song, et al. (2010). "Experimental Determination of Unsteady Aerodynamic Coefficients and Flutter Behavior of a Rigid Wing", 51st AIAA/ASME/ASCE/AHS/ASC *Structures, Structural Dynamics, and Materials Conference,* AIAA 2010-2875, 1-9.

[21] Jie Zeng and Sunil L. Kukreja (2013). "Flutter Prediction for Flight/ Wind-Tunnel Flutter Test under Atmospheric Turbulence Excitation", *Journal of Aircraft,* Vol. 50, No. 6, 1696-1699.

[22] Thomas Andrianne and Grigorios Dimitriadis (2013). "Experimental and numerical investigations of the torsional flutter oscillations of a 4: 1 rectangular cylinder", *Journal of Fluids and Structures*, 41, 64–88.

[23] William L. Oberkampf, Timothy G. Trucano, et al. (2013). "Verification, Validation, and Predictive Capability in Computational Engineering and Physics", SAND2003 – 3769, Unlimited Release, February 2003, 1-15.

[24] Supanee Arthasartsri and He Ren (2009). "Validation and Verification Methodologies in A380 Aircraft Reliability Program", IEEE, 978-1-4244-4905-7, 2009, 1356-1363.

[25] Schwer L. E., Mair H. U., et al. (1998) "Guide for Verification and Validation in Computational Solid Mechanics[EB/OL] ". The American Society of Mechanical Engineers, 1998, http://cstools.asme.org/.

[26] Melike Nikbay and Muhammet N. Kuru (2013). "Reliability Based Multidisciplinary Optimization of Aeroelastic Systems with Structural and Aerodynamic Uncertainties", *Journal of Aircraft,* Vol. 50, No. 3, 708-714.

[27] Zimmerman, N. H. and Weissenburer, J. T. (1964). "Prediction of Flutter Onset Speed Based on Flight Testing at Subcritical Speeds", *Journal of Aircraft,* Vol. 1, No. 4, 190–202.

[28] Bingyuan Yang and Weili Song (2001). "The Application of Sub-critical Technology in Wind-tunnel Flutter Experiment for Rudder Model", the 7th aeroelastic Chinese conference, 34-39.

[29] MSC Software Corporation (MSC). "Aeroelastic Analysis User's Guide ", 14.

Chapter 3

An Optimization Procedure of Model's Base Construction in Multimodel Representation of Complex Nonlinear Systems

Bennasr Hichem and M'Sahli Faouzi

Abstract

The multimodel approach is a research subject developed for modeling, analysis and control of complex systems. This approach supposes the definition of a set of simple models forming a model's library. The number of models and the contribution of their validities is the main issues to consider in the multimodel approach. In this chapter, a new theoretical technique has been developed for this purpose based on a combination of probabilistic approaches with different objective function. First, the number of model is constructed using neural network and fuzzy logic. Indeed, the number of models is determined using frequency-sensitive competitive learning algorithm (FSCL) and the operating clusters are identified using Fuzzy K- means algorithm. Second, the Models' base number is reduced. Focusing on the use of both two type of validity calculation for each model and a stochastic SVD technique is used to evaluate their contribution and permits the reduction of the Models' base number. The combination of FSCL algorithms, K-means and the SVD technique for the proposed concept is considered as a deterministic approach discussed in this chapter has the potential to be applied to complex nonlinear systems with dynamic rapid. The recommended approach is implemented, reviewed and compared to academic benchmark and semi-batch reactor, the results in Models' base reduction is very important witch gives a good performance in modeling.

Keywords: nolinear systems, multimodel, optimization of submodel's reactor

1. Introduction

There are several representations of industrial processes: linear, nonlinear or other technique using fuzzy logic and/or neural networks. Nonlinear models are found in a large part and are used to properly represent the dynamics of real processes. However, their comlexcity proves a real obstacle for control or when designing an observer or in a diagnostic strategy. So, the multimodel approach is a powerful approach developed in the aim to overcome problems related to modeling and control of industrial processes which are often complex, nonlinear and/or nonstationary.

The multimodel approach supposes the representation of the nonlinear model by a set of linear models (as designed in future by submodels or model's base) thus forming what is called a model library or model's base. The interaction of each models of this model's base through a certain normalized validity calculation forming the global nonlinear system in its all area of operating.

The different models of the base could be of different structures and orders and no model can represent the system in its whole operating domain. Therefore, the multimodel approach aims at lowering the system complexity by studying its behavior under specific condition. **Figure 1** illustrates the concept of a process formed by a multimodel approach. This mechanism evaluates the contribution of each model in the description system's behavior.

The decision unit estimates by means of the numerical validity of each model the selection of the most relevant models at each time. The contribution of the different model's is made by a decision output unit that compute the multimodel output.

Several researchers have been interested in multimodel analysis and control and many applications have been proposed in different contexts. The multimodel approach has knowledgeable a sure interest since the publication of the work of [1]. The idea of the multimodel approach is to describe the complex nonlinear systems by a set of local models (linear or affine) characterizing the operation of the system in different areas of operation. In spite of its success in many fields, the multi-model approach remains confronted with several difficulties and some design problems such as the determination of the models' base and the adequate validity computation. Several works have treated these two points in the literature. We can refer to the works presented in [2–8]. Indeed once the model library is built it will remain intact. The number of model found is, first responsible to adequately represent the nonlinear model of the system and second, it will remain static when it is used in control or diagnosis. In this context there are very few works that have tried to reduce the models' base once it is built and this opens up a new axis of research added to the design of the multimodel approach in the representation of nonlinear systems. We can cite in this context the works of Gasso in [9–11], where the reduction of the models' base is based on an iterative procedure which include tree operation that is: elimination of less important local submodel, merging of the neighbouringsubmodel that describe the same behavior of the nonlinear system and finally a parameters optimization of the resultant structure is made on. In [12–15], The nonlinear system is modeled by a set of linear models and with the of a Gap metric in the model based control technique we can decide how many models are sufficient for control. In other context, when the process can be represented by a fuzzy model, the complexity of rule base can be minimized through two procedure: illumination of the less important rule and merging of neighboring rules that can

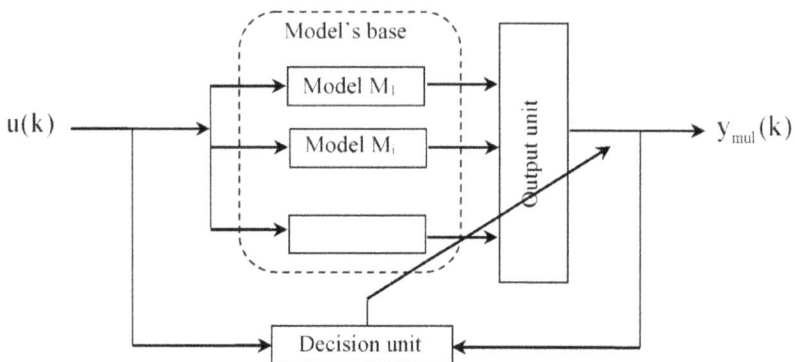

Figure 1.
Multimodel representation.

describe the same behavior of the nonlinear system [16]. The Chiu's classification method [17] is used in order to obtain the optimal systematic determination of model's base [18]. The latest work is a new approach permit to optimize the number of submodels with respect to the submodel complexity is presented. In [19] the optimal number of submodels is done based on both a reinforced combinatorial particle swarm optimization and a hybrid K-means.

In this chapter book we study a very relevant modeling approach especially where do not have an adequate model of the process. We will use the data set relative to the identification of the process in its area of operating. On the other hand the input and output measurement. The multimodel approach in this case becomes a very efficient method to overcome the problem of modeling.

The outline of this work is as follows: in Section 2, we will present the procedure used in the multimodel representation of complex nonlinear systems. Indeed, the number of models is determined using frequency-sensitive competitive learning algorithm and the operating clusters are identified using Fuzzy K- means algorithm. The structure of each local model and the validity computation are also presented to each submodel of the model's base (number of model built). Section 3 presents the recommended approach to reduce the number of submodels. Then, focusing on the use of both two type of validity calculation for each submodel and an SVD technique is used to evaluate the adequate number that can be retained in the model's base.

2. Multimodel system identification

In this section a two classification algorithms are used to allow the determination of the number, structure and parameters of the different models of the base. So, for the determination of model's base, we propose both Frequency Sensitive Competitive Learning (FSCL) and Fuzzy K-means to obtain respectively the number of models and the operating regime of each one. In the second step we propose the validity computation for each submodel in the model's base. The validity of each model is designed in such a way that it involves two types of validity, namely a simple and reinforced validity. The contribution of each submodel is done by each validity value and thus appears via an optimization technique. On the basis of the last results, one can in fact, minimize the number of submodels for those that have no effect on multimodel modeling. This can be explaining by a measure of similarity between each submodel and the global one. This technique is highlighted via an SVD technique.

2.1 Determination of the number models

Most existing clustering algorithms [20, 21], when we have not an idea about the number they cannot handle the selection of the appropriate number of cluster. However, when it is used for clustering, the FSCL algorithm automatically allocates the suitable number of units for an input data set. So, for the determination of model's base, we propose Frequency Sensitive Competitive Learning (FSCL). Frequency Sensitive Competitive Learning (FSCL) [22] is a competitive algorithm with N neuron that is trained by a dataset of P data vectors $x(t)$. In the FSCL, the competitive computing units are penalized in proportion to the frequency of their winning. Giving an input x_i each time, FSCL determines the winner by:

$$u_i = \begin{cases} 1 & if \ \ i = c \\ 0 & otherwise, \end{cases} \tag{1}$$

Such that

$$\gamma_c \|x - w_c\|^2 = \min_j \gamma_j \|x - w_j\|^2 \qquad (2)$$

where.

N: Is initial estimation of the number of clusters in the given data.

$u_i 1 \leq i \leq k$: The output units of dimension k.

$w_i \ 1 \leq i \leq k$: Weight vectors each of dimension k.

$x_i \ 1 \leq i \leq d$: The d-dimensional input vector from data set P.

w_c: D-dimensional weight vector corresponding to the winner.

$\| * \|$: Euclidean distance;

c: index of the unit which wins the competition;

γ_j: Conscience factor used to reduce the winning rate of the frequent winners defined as follows [23]:

$$\gamma_j = n_j \Big/ \sum_{i=1}^k n_i \qquad (3)$$

Where n_j refers to the cumulative number of occurrences the node j has won the competition.

After selecting the winner, FSCL updates the winner as follow:

$$w_j(t+1) = w_j(t) + \alpha_g(t)(x - w_j(t)) \qquad (4)$$

Where α_g is the learning rate defined as follow:

$$\alpha_g = \alpha_g^i \left(\frac{\alpha_g^f}{\alpha_g^i} \right)^{t/t_{max}} \qquad (5)$$

The convergence of the appropriate FSCL algorithm to a local minimum is studied in [24]. Using this technique with its parameters as defined so that, the maximum number of iteration designed by t_{max} and with its initial and final learning rate designed respectively α_g^i, α_g^f, convergence of the algorithm visualize in the final step that there are some of the data clusters are more densely populated than others which ideally result that there are some wining units are more often in those clusters than other. It must be considered that when the number of unit (neuron) is larger than the real number of cluster in the input data-set, the extra units are gradually driven faraway from the distribution of the data-set. If the number of cluster c is not equal to the true value, FSCL will lead to an incorrect clustering result. So, it needs to pre-assign the number of clusters c. For all the studied examples we have varied the cluster number, the learning rate for each case study and take the opportunity to determine the appropriate one in such away that $\alpha_g \in [0.1 - 0.5]$.

2.2 Fuzzy k-means clustering

In order to establish the operating cluster, the use of Fuzzy k-means is considered. This last algorithm was defined by [25] and improved by [26]. The determination of different cluster centers and dada set x_i assigned to each clusters for each centers is done by the use of the minimization objective function mentioned by:

$$J_m = \sum_{j=1}^{K} \sum_{i=1}^{N} \mu_{ij}^m \|x_i - c_j\|^2, 1 \leq m \leq \infty; \tag{6}$$

Where:

μ_{ij}: represents degree of membership of x_i to cluster j and stands for the local model's activation degree for that observation $\sum_{j=1}^{K} \mu_{ij} = 1$;

N: is the number of data points,

K: Number of cluster or local models;

x_i: i^{th} data point;

m: (real number greater than 1) is the "fuzzy exponent" that influences the membership values and represents the overlapping shape between clusters.

c_j: Center of cluster j.

The algorithm consists of the following steps:

1. Initialize the membership matrix $U = [\mu_{ij}]$ with random value between 0 and 1.

2. Calculate fuzzy centers using:

$$c_j = \frac{\sum_{i=1}^{N} \mu_{ij}^m x_i}{\sum_{i=1}^{N} \mu_{ij}^m}. \tag{7}$$

3. Compute a new matrix U using

$$\mu_{ij} = \left[\sum_{r=1}^{K} \left(\frac{\|x_i - c_j\|}{\|x_i - c_r\|} \right)^{2/(m-1)} \right]^{-1} \tag{8}$$

4. The iteration will stop if:

$$\|U(k-1) - U(k)\| < \xi. \tag{9}$$

Where ξ is a termination criterion between 0 and 1.

2.3 Local linear models

By identifying the cluster number and the datset for each cluster, the next step focuses primarily on obtaining cluster sub-models. This last step requires two phases: the first for the structure of submodel and the second deal with its parameters identification. For the different obtained vectors representative of cluster, a parametric estimation uses the Recursive Least-Square method is retained. The ARX model being chosen for each cluster whose equation is given by:

$$y(k) = -\sum_{i=1}^{n_a} a_i y(k-i) + \sum_{j=1}^{n_b} b_j u(k-j) \tag{10}$$

Where a_i and b_j are parameters of the ith submodel. n_a and n_b are the lags respectively in input and output. The order of each model is determined by the instrumental determinants' ratio-test [27]. For every order value d, the instrumental determinants 'ratio $RDI(d)$ is computed and the retained order d isthe value for

which the ratio $RDI(d)$ quickly increases for the first time. This method consists in building an information matrix Q_d giving by:

$$Q_d = \frac{1}{n_d} \sum_{k=1}^{n_d} \begin{bmatrix} u(k) \\ u(k+1) \\ u(k-d+1) \\ u(k+d) \end{bmatrix} \begin{bmatrix} y(k+1) \\ u(k+1) \\ y(k+d) \\ u(k+d) \end{bmatrix}^T \tag{11}$$

Where n_d is the observations' number and the instrumental determinants' ratio RDI (d) is given by the following relation:

$$RDI(d) = \left| \frac{\det(Q_d)}{\det(Q_{d+1})} \right| \tag{12}$$

2.4 Computation of validity

In the proposed approach the validity computation of each submodel is proposed by this expression:

$$v_i^{mul} = \alpha_i v_i^{simp} + \beta_i v_i^{renf} \tag{13}$$

Where v_i^{simp} and v_i^{renf} are respectively simple and reinforced validity. The simple validity is defined by:

$$v_i^{simp}(k) = \frac{1 - r_i^{norm}(k)}{N-1}, i = 1, N_m \tag{14}$$

And the reinforced validity is expressed by

$$v_i^{renf}(k) = v_i(k) \prod_{\substack{j=1 \\ j \neq i}}^{N} (1 - v_j(k)) \tag{15}$$

The r_i^{norm} is the normalized residue given by:

$$r_i^{norm}(k) = \frac{r_i(k)}{\sum_{i=1}^{N} r_i(k)}, i = 1, N_m \tag{16}$$

The residue is expressed as the distance between the process output y and the considered local output y_i of the submodel M_i which has the following formula:

$$r_i = |y - y_i|, i \in [1, N_m] \tag{17}$$

Where N_m is the number of submodel in the models' base.
α_i, β_i: are respectively the adequate values which can be calculated as:

$$\underset{\alpha_i, \beta_i}{Min} \left[\left(\left(\sum_{i=1}^{N_m} \left(\alpha_i v_i^{simpl} + \beta_i v_i^{renf} \right) y_i \right) - y \right) \right] \tag{18}$$

$$\sum_{i=1}^{n} \alpha_i = 1, \quad \sum_{i=1}^{n} \beta_i = 1.$$

A Least Square Estimation is used in order to prove the value of α_i and β_i. In fact, the multimodel output y_{mul} is calculated by a fusion of the models' outputs y_i weighted by their respective multi-validity indexes v_i^{mul} which is illustrated by the following expression:

$$y_{mul}(k) = \sum_{i=1}^{N_m} v_i^{mul}(k) y_i(k) = \theta^T \varphi(k) \tag{19}$$

Where we introduce the following regressor vector:

$$\varphi^T(k) = [y_1(k-1), \quad y_i(k-1),], i = 1, N \tag{20}$$

And the parameters vector:

$$\theta^T(k) = [\alpha_1, \alpha_i, \beta_1, \beta_i], i = 1, \; N \tag{21}$$

The recursive least squares estimate of θ is:

$$\begin{cases} \theta(k) = \theta(k-1) + P(k)\varphi(k)e(k) \\ P(k) = P(k-1) - \dfrac{P(k-1)\varphi(k)\varphi^T(k)P(k-1)}{1 + \varphi^T(k)P(k-1)\varphi(k)} \\ e(k) = y(k) - \theta^T(k-1)\varphi(k) \end{cases} \tag{22}$$

Where P denote the covariance matrix and e is an error of modeling.

3. Reduction of the model's base

Based on the new validity calculation of each submodel, the need of each one in the model's base is identified via the numerical values α and β. Thus, the impact of each submodel above the global model is also determined by the numerical values α and β. Each submodel can contribute in one way or another via its new validity and one can thus determine the adequate number of submodels where their contribution in modeling is important compared to the other submodels. A comparison is made on to have a good performance by comparing each submodel to the global model. The submodel whose contribution is important will be retained. To solve this problem, the following matrix X defined by Eq. (23) have to be determined first and an analysis using singular value decomposition analysis should be carried out.

$$X = \begin{bmatrix} \left(\alpha_1 v_1^{simp} + \beta_1 v_1^{renf}\right) y_1 \circ y & \left(\alpha_2 v_2^{simp} + \beta_2 v_2^{renf}\right) y_2 \circ y \\ & \left(\alpha_i v_i^{simp} + \beta_i v_i^{renf}\right) y_i \circ y \end{bmatrix} \tag{23}$$

SVD analysis is a numerical algorithm that decomposes an $N_m \times m$ matrix into three unique component matrices:

$$X = U\Sigma V^T \tag{24}$$

Where U is left singular vectors with $N_m \times N_m$ matrix, V is right singular vectors with $N_m \times m$ matrix and Σ is singular value with $m \times m$ matrix. Among these component matrices, U vectors provide the information of retained submodel in an orthogonal form. U_1 and U_2 which is the first two rows in the U vectors indicate the two most combinations of required submodel. Another approach that can facilitate this analysis is modified version of principal component analysis (PCA). In modified PCA analysis, the differences between U_1 and U_2 are also calculated by:

$$Z_i = |U_{1i}| - |U_{2i}| \tag{25}$$

Where Z_i denotes the absolute value of difference between the best two U vectors. The result of this analysis is easier to interpret as only one function is plotter versus submodel.

4. Model validity tests

In order to compare the modeling performances in both cases with all the number of submodels in the library and in the reduction submodel case, there are three different performance criteria considered: The normalized mean-square error (NMSE), the best fit (FIT) and the variance accounted for (VAF) criterion defined respectively as follows:
The normalized root mean squared error (NRMSE) is defined as:

$$NRMSE = \frac{1}{\max{(y)}} \sqrt{\frac{1}{M} \sum_k (y - y_s)} \tag{26}$$

The best fit is defined as:

$$FIT = \left(1 - \frac{\|y(k) - y_S\|}{\|y(k) - y_{mean}\|}\right).100\% \tag{27}$$

The variance accounted for is defined as:

$$VAF = \max\left\{1 - \frac{\text{var}(y(k) - y_S)}{\text{var}(y(k))}, 0\right\}.100\% \tag{28}$$

Where y denotes the measured output and y_S is the estimated output of the multimodel.

5. Simulation examples

Numerical simulations are presented to compare the proposed approach to the conventional multimodel design with full submodels number and investigate the impact to model highly nonlinear systems.

5.1 Example 1: second order continuous system

Taken from [28], the first simulation example considered is the system whose evolution is described by the following equations:

$$\ddot{y}(t) + \dot{y}(t) + y(t) + y(t)^3 = u(t) \tag{29}$$

The system has been sampled at a period of 0.2 s. A 1500 sampled data output y are used for the identification of the system. The adequate number of clusters is determined using the FSCL algorithm. **Figure 2** gives the results. With twelve neurons used in the output layer six centers move away from the observation data. We can conclude that the number of cluster is equal to six. An identification procedure is than applied in order to determinate the efficient transfer function of each local model for the library. So, the required data set for each data is used in a parametric and structure identification where the RDI index gives that the order of each model is equal to two. An SVD technique is applied in order to reduce the model's base. **Figure 3** shows the elements of U vector plotted versus submodels. The maximum absolute value of U_1 plot suggests submodels give the first submodels. While maximum absolute value of U_2 plot gives the secondary submodels. In the case study submodels 2 and 4 can be retained to represent the whole real process. The difference Z between the best two U vectors is also given prove the same retained submodels. Based in these results the following input sequence is considered to validate the proposed process modeling:

$$u(k) = 0.2 \sin\left(\pi k / 80\right) + 0.5 \sin\left(3\pi k / 100\right) \tag{30}$$

The results given in **Figure 4** demonstrate that the novel approach tracks the real output with a very small error.

Numerical performance comparison values are given in **Table 1**. Illustrate that the novel approach has good properties in modeling compared to the whole process with 6 submodels. With two submodels the modeling approach is accuracy with small NRMSE.

5.2 Example 2

The non linear plant considered given by [11] as a second simulation is described by the following nonlinear equation:

$$y(t) = 0.4u(t-1)^3 \exp\left(-0.5|y(t-1)|\right) \tag{31}$$

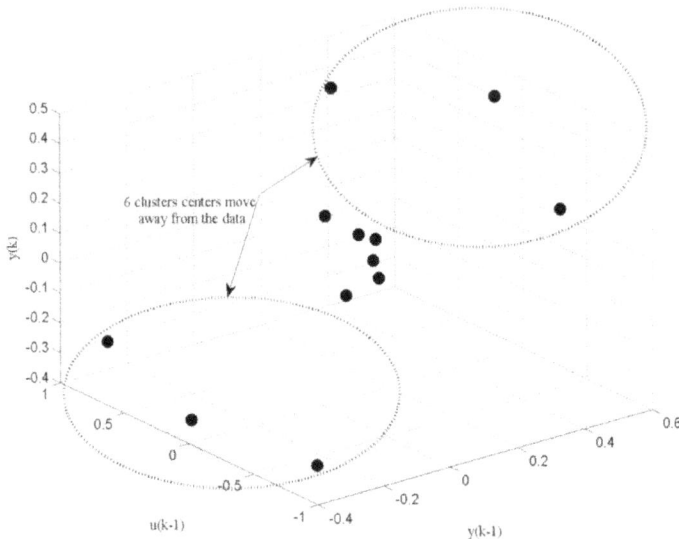

Figure 2.
Determination of the number of cluster via FSCL (second order continuous system).

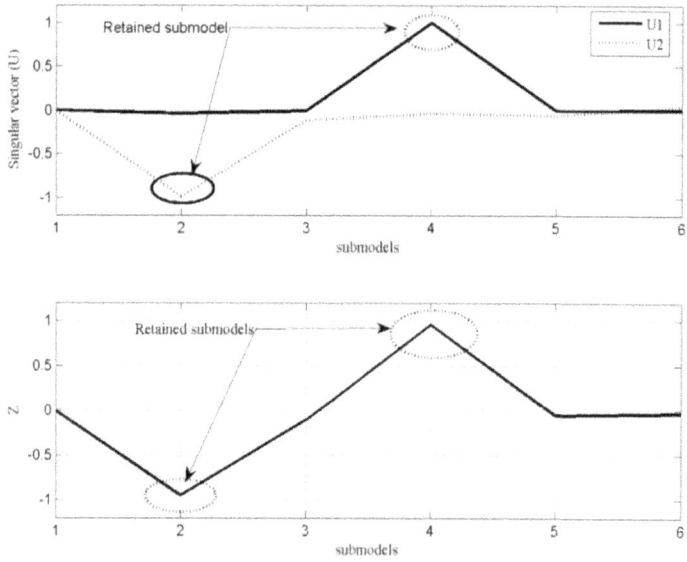

Figure 3.
Determination of the optimal number of submodels (second order continuous system).

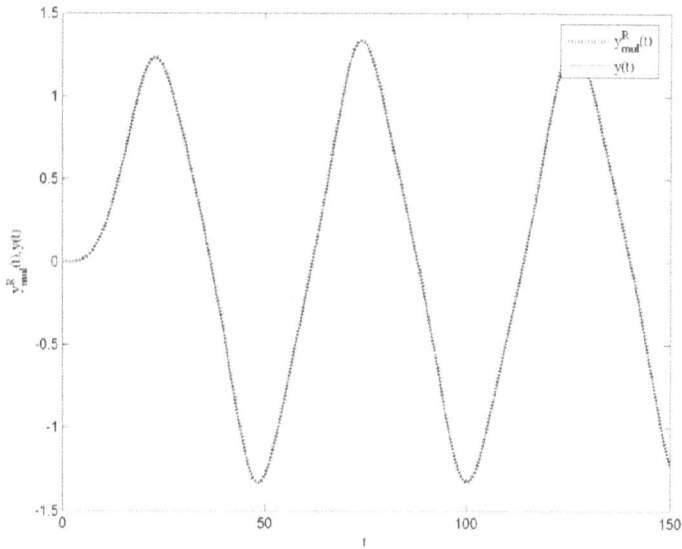

Figure 4.
Real and multimodel outputs proposed reduction approach. The solid line is the plant output y, and dashed line is the reduction multimodel y_{mul}^{R} (second order continuous system).

Number of submodel	VAF (%)	FIT (%)	NRMSE
6 submodels	100.0000	99.9985	7.5795e-004
2 submodels	100.0000	99.9951	0.0014

Table 1.
Performance comparison (second order continuous system).

The input u of the system is formed by the concatenation of piecewise constant signals with variable amplitude and duration. A set of 2500 data points is used to build the model. The FSCL algorithm is used to search the adequate number of clusters. With the following parameters of $\alpha_g^i = 0.45$ and $\alpha_g^f = 0.041$ we consider a 500 training iteration of the 2500 data. By the use of 36 neurons, fourteen clusters centers move away from the data which results that the number of cluster is equal to 22 (**Figure 5**). This result is also confirmed by [11]. According to Eq. (31), the non-linearity of the system are due to the variables $u(t-1)$ and $y(t-1)$. Therefore the feature variables considered are: $u(t-1)$ and $y(t-1)$ in the local models forming a first order model. The procedure of submodels elimination is also considered. The analysis of the singular vector U_1 and U_2 or the difference between the absolute value of U_1 and U_2 given in **Figure 6** show that the number of submodels can be reduced only to three submodels. In fact, the maximum absolute value of U_1 appears on submodels 5 and 16. Moreover, the maximum absolute value of U_2 appears on submodels 4 and 5. The submodels 4, 5 and 16 are those retained in future to validate the proposed process modeling with the following input sequence:

$$u(t) = \sin(\pi t / 100) \tag{32}$$

We have recorded in **Figure 7**, the evolution of the real process given by $y(k)$ and the evolution of the multimodel output reduction submodels given by $y_{mul}^R(t)$. The envisaged approach always promise best results of modeling. **Table 2** compares the performances of the structures with respectively 22 and 3 local submodels. Weremark that the deletion of 19 local submodels does not affect significantly the performances of the results in multiple model process representation. The new approach isable to identify and accurate the nonlinear process. The multimodel approach and the new approach both achieve the performance of VAF =99.9999 and VAF = 99.9854. So, clearly there is no significant difference between the performances of modeling.

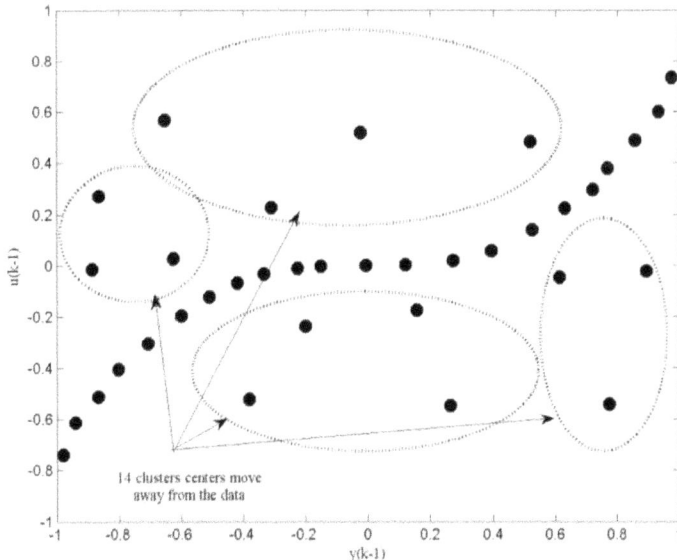

Figure 5.
Determination of the number of cluster (FSCL c = 36) and clustering results (c = 22) (second example).

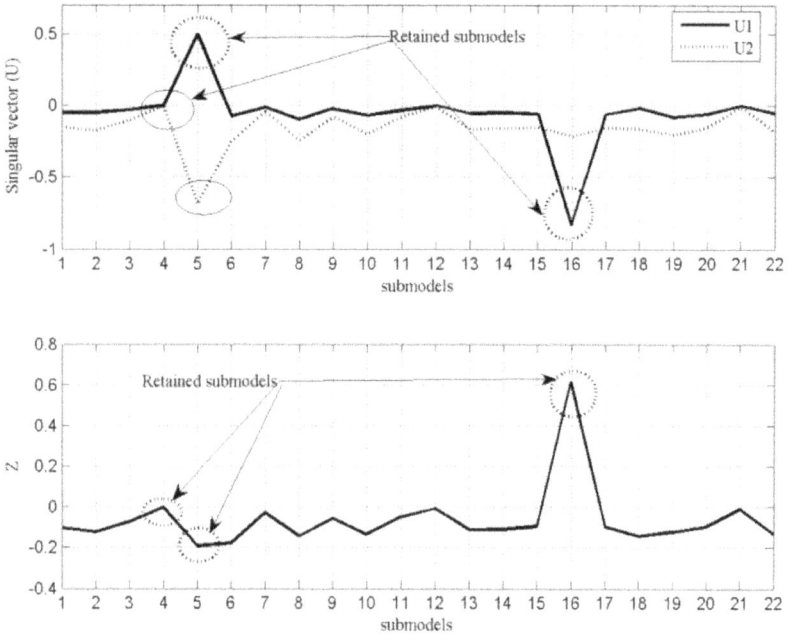

Figure 6.
Determination of the optimal number of submodels (second example).

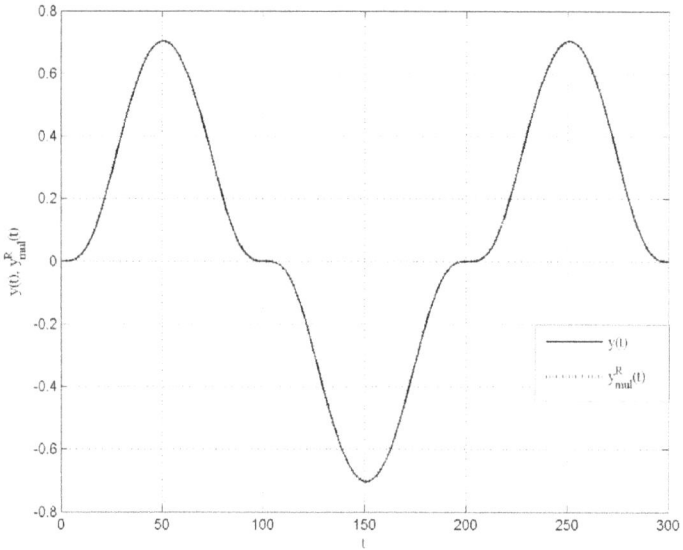

Figure 7.
Real and multimodel outputs proposed reduction approach. The solid line is the plant output y, and dashed line is the reduction multimodel y_{mul}^R (second example).

5.3 Process validation

5.3.1 Biological reactor

In order to highlight the interest and contributions of our modeling approach, based on the same system that is the bio-reactor, we compared our results with

Number of submodel	VAF (%)	FIT (%)	NRMSE
22 submodels	99.9999	99.8798	0.0078
3 submodels	99.9854	98.7907	0.0086

Table 2.
Performance comparison (second example).

those given by [3, 29]. Being a good academic example of nonlinear system, the biological reactor has been treated in some works for the purpose of illustration in different approaches of modeling and controlling non-linear systems [30, 31]. The nonlinear model is a bioreactor where its expression is given by the Contois [32] model in its discrete form as developed in the following expression:

$$x_{k+1}^{(1)} = x_k^{(1)} + 0.5\frac{x_k^{(1)}x_k^{(2)}}{x_k^{(1)}x_k^{(2)}} - 0.5u_kx_k^{(1)},$$

$$x_{k+1}^{(2)} = x_k^{(2)} - 0.5\frac{x_k^{(1)}x_k^{(2)}}{x_k^{(1)}x_k^{(2)}} - 0.5u_kx_k^{(2)} + 0.05u_k, \tag{33}$$

$$y_k = x_k^{(1)},$$

In this equation y denote the output and the control input denoted u. Based on equation given above, the system is excited by a signal of the form of 4-seconds-long stairs augmented with random amplitude between $0 \le u \le 0.7$. So that a collection of 3018 experimental data set are used. The first part of data with a number of 2416 is used in the identification procedure to give the adequate number of submodels with his structure and order. The second type of data is of 602 training point used to validate the modeling strategy. The FSCL algorithm is used to search the adequate number of clusters. Considering a 500 training iteration of the 2416 data by the use of 15 neurons with following parameters of $\alpha_g^i = 0.45$ and $\alpha_g^f = 0.041$. **Figure 8** presents the cluster centers repartition. Six clusters centers move away from the data which results that the number of cluster is equal to nine. Thus

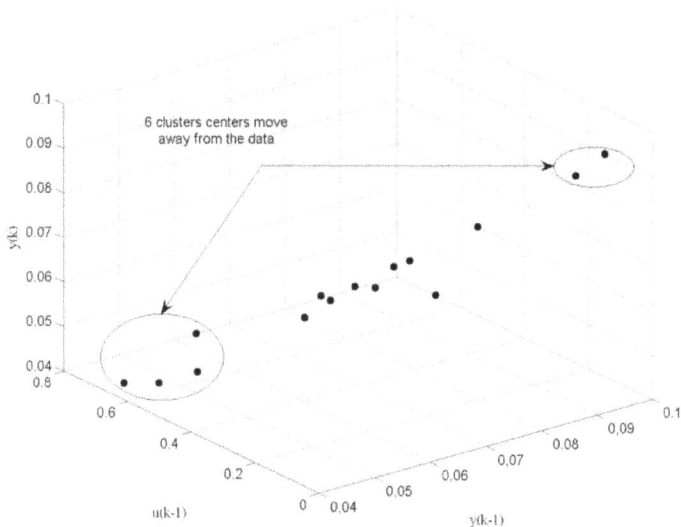

Figure 8.
Determination of the number of cluster via FSCL (biological reactor application).

with only 9 linear models against 10 model obtained by [4] and against 196 models obtained by the modeling application proposed in [29], we have succeeded firstly in designing a new multimodel structure.

Taken into account the analysis of the evolution of the singular vectors U_1 and U_2 or the difference between the absolute value of U_1 and U_2 given by **Figure 9** we can conclude that submodels 2, 6 and 7 are sufficient to represent the process reactor. In fact, the maximum absolute value of U_1 is signaled on the submodels 6 and 7. Against, the maximum absolute value of U_2 is signaled on the submodels 2 and 6. This difference in size of the model's base helps to highlight our approach which guarantees a satisfied representation with a smaller number of models compared to a same studied system. Knowing that a large number of model base risk of constituting a handicap in terms of command calculation and/or analysis. This result shows that the proposed concept for reduction in multiple-linear modeling identical in accuracy to the modeling paradigm using the classical multimodel approach. The modeling results of the proposed approach compared to the full model's base is shown in **Figure 10** and the performance comparison is given in **Table 3**. The two curves can hardly be distinguished from each other and there is no significant difference between the performances. The results described in this section prove the efficiency and the precision of the proposed modeling strategy and show that the method works well with various processes. Hence, there is a potential for improved quality and flexibility of final product if the cost of the model development can be reduced. In fact, for example, in the high on-line computational need to solve an optimal control actions in nonlinear model predictive control, which results in a non-convex optimization, can be compared with the new proposed concept of modeling. We can reduce the on-line computation in the NMPC scheme by transforming the NLMPC problem into a LMPC and a quadratic programming can be used to handle constraints. Limiting this paper only to modeling, this last observation will be studied in future works based on the results given in [33].

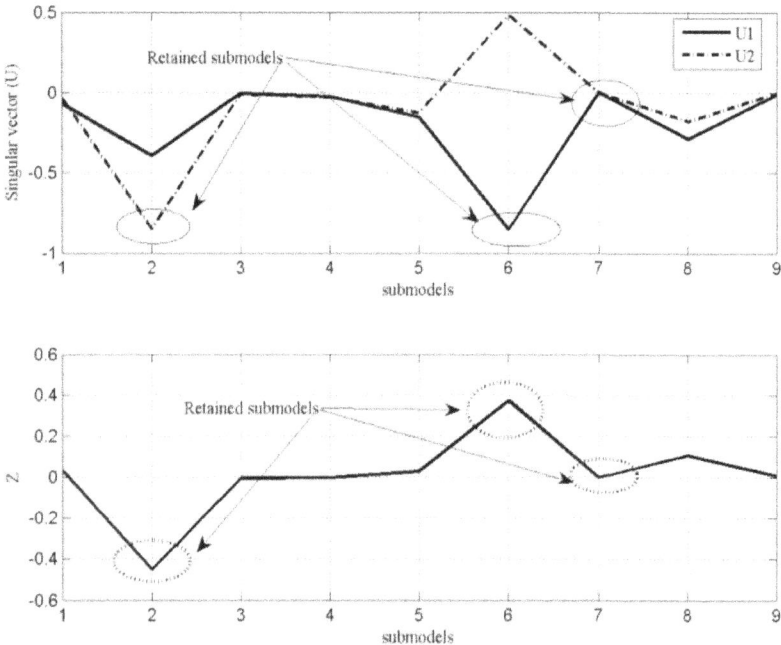

Figure 9.
Determination of the optimal number of submodels (biological reactor).

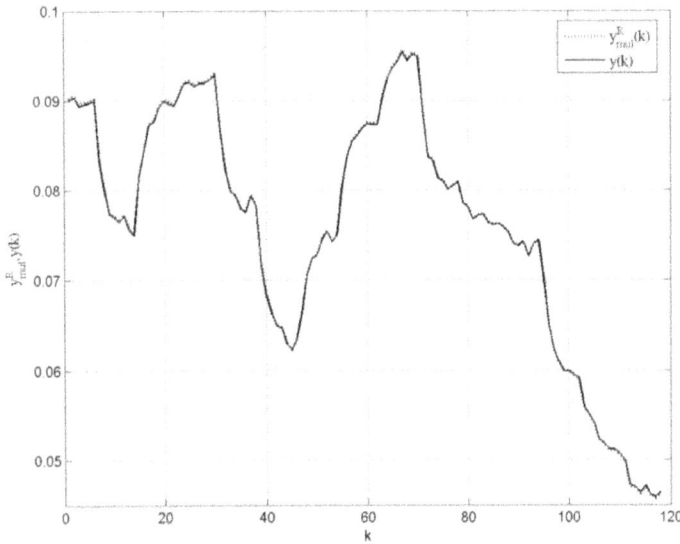

Figure 10.
Real and multimodel outputs proposed reduction approach. The solid line is the plant output y, and dashed line is the reduction multimodel y_{mul}^R (biological reactor).

Number of submodel	VAF (%)	FIT (%)	NRMSE
9 submodels	99.9999	99.8798	0.00478
3 submodels	99.9854	98.7907	0.0086

Table 3.
Performance comparison (biological reactor).

5.3.2 Liquidlevel process

The model is obtained through identification of a laboratory-scale liquid-level system [34] is used to illustrate the advantages of the proposed modeling method. The model of the plant is described by the following NARX model:

$$
\begin{aligned}
y(k) &= 0.9722y(k-1) + 0.3578u(k-1) - 0.1295u(k-2) \\
&\quad -0.3103y(k-1)u(k-1) - 0.04228y(k-2)^2 + 0.1663y(k-2)u(k-2) \\
&\quad -0.03259y(k-1)^2y(k-2) - 0.3513y(k-1)^2u(k-2) \\
&\quad +0.3084y(k-1)y(k-2)u(k-2) + 0.1087y(k-2)u(k-1)u(k-2);
\end{aligned}
\tag{34}
$$

The application of the FSCL algorithm is highlighted on the set of data collected from the process in order to determine the number of models of the library. Using 8 neurons with following parameters of $\alpha_g^i = 0.42$ and $\alpha_g^f = 0.01$, after 500 iterations of the 1600 data set, the algorithm leads to a concentration of 4 clusters centers and 4 centers have moved away from the dataset, which leads to conclude that the considered process can be modeled by 4 submodels (**Figure 11**).

Based on clustering results, the RDI value for each cluster lead to a value 2 and a least square estimation is used to develop the different submodels of the Model's base. In order to reduce the number of submodels, the SVD technique is applied on the formed matrix given by Eq. (23), so the plotted singular vectors U_1 and U_2 or the absolute value between U_1 and U_2 indicate that the two submodels 2 and 3 are

45

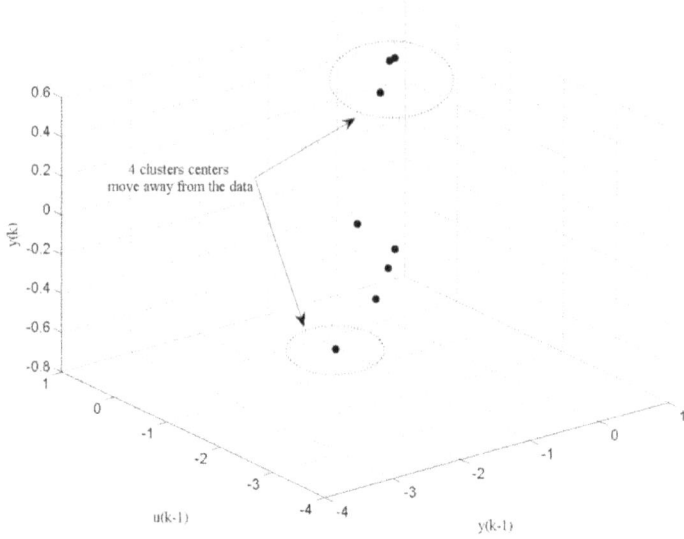

Figure 11.
Determination of cluster number (liquid level process application).

retained and sufficient to represent the nonlinear plant (**Figure 12**). In order to validate the proposed process modeling, the following input sequence is considered:

$$u(k) = 0.25 \sin\left(\pi k / 100\right) + 0.5 \sin\left(\pi k / 30\right) \qquad (35)$$

The results of multimodel reduction approach y_{mul}^{R} is plotted in **Figure 13**, demonstrate that the novel approach tracks the real output with a very small error.

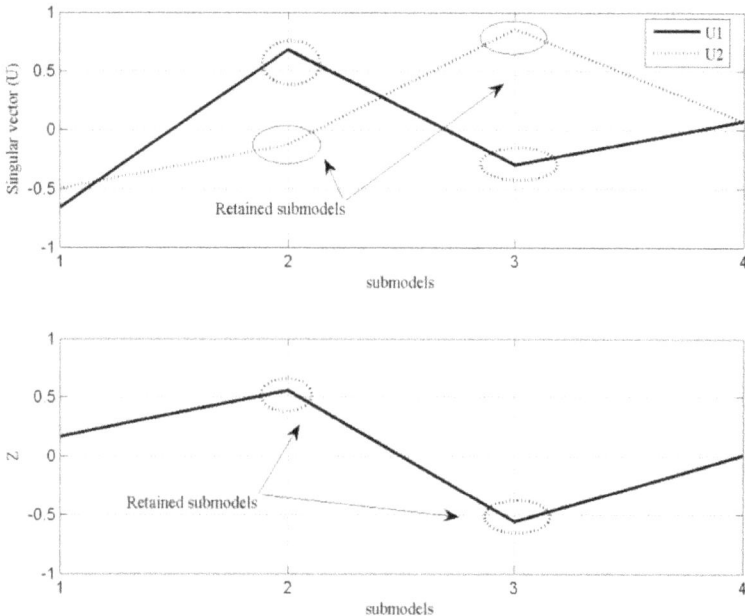

Figure 12.
Determination of the optimal number of submodels (liquid level control).

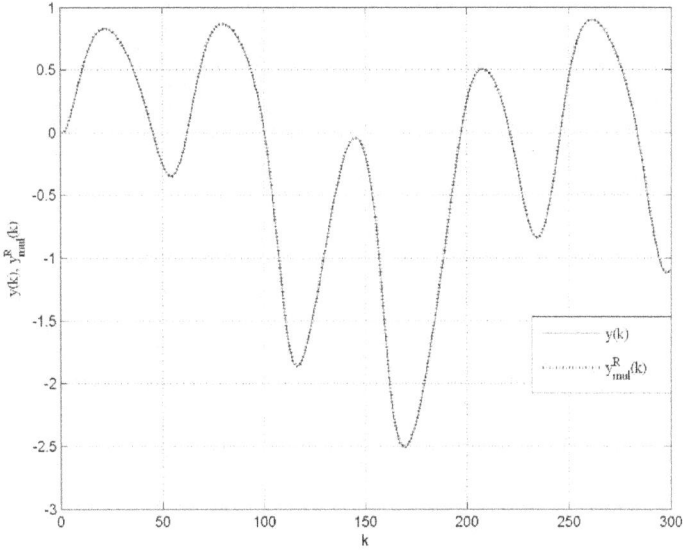

Figure 13.
Real and multimodel outputs reduction approach. The solid line is the plant output y and the dashed line is the reduction multimodel y_{mul}^R (liquid level plant).

This is confirmed by inspecting respectively the value of NRMSE, FAV and the FIT of modeling given in **Table 4**.

5.4 Experimental validation

The pilot unit that we are going to study is a process installed in the laboratory of process control at the Engineering school of Gabes (Tunisia) (**Figure 14**) it consists mainly on:

- A 2 l jacketed reactor equipped with a drain valve to empty its contents.

- A stirrer connected to an adjustable speed motor 0–3000 tr/min.

- A tube condenser. Cooling is provided by tap water;

- Two pumps P1 and P2: P1 ensures the supply of alcohol while P2 ensures the circulation of the heat transfer fluid;

- Two reservoirs R1 and R2: the reservoir R1 is used to store the alcohol which will be used as a reagent while the reservoir R2 allows the collection of the condensa at the outlet of the condenser;

- A heater resistor E1 provides variable power from 0 to 3500w;

Number of submodel	VAF (%)	FIT (%)	NRMSE
4 submodels	100.0000	99.9747	0.0049
2 submodels	99.9987	99.6333	0.0130

Table 4.
Performance comparison (liquid level control).

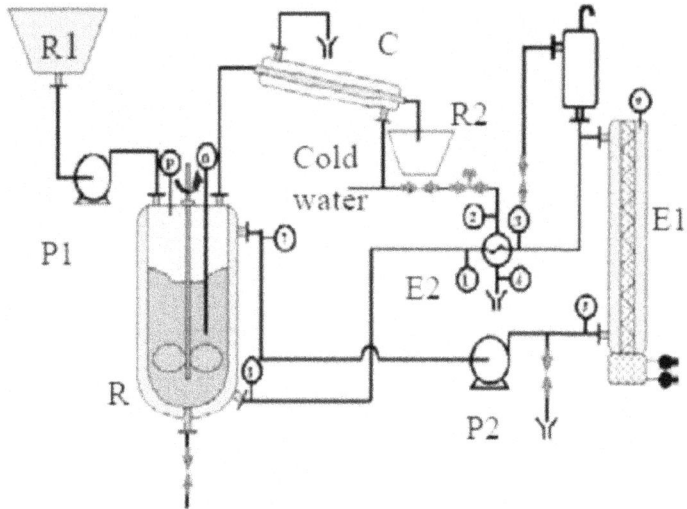

Figure 14.
Process reactor.

- E2 exchanger cools the heat transfer fluid.

The reaction carried in this semi batch reactor is an esterification reaction, which is given by the following scheme:

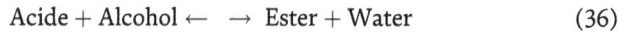

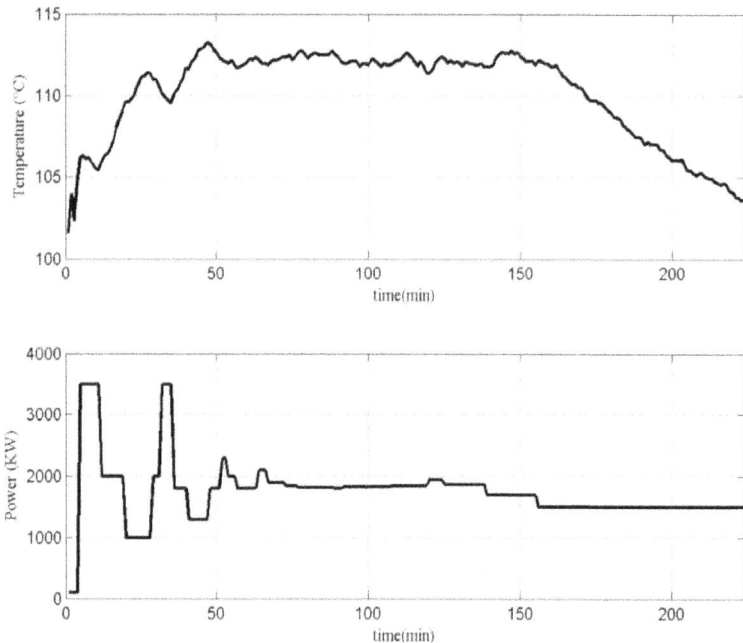

$$Acide + Alcohol \longleftrightarrow Ester + Water \tag{36}$$

Figure 15.
Reactor dataset for system identification.

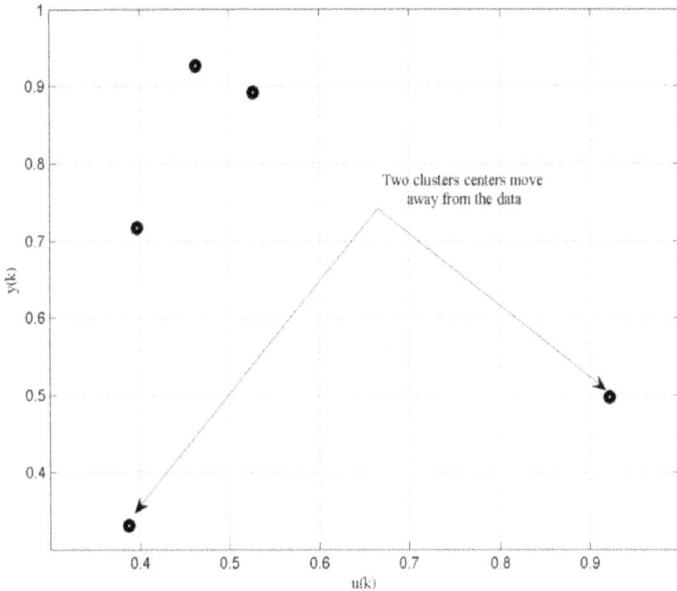

Figure 16.
The FSCL determination of cluster number in process reactor application.

For identification experiments, it has proved in previous works [35, 36]. Then the reaction was heated as quickly as possible with the maximum power (3 *KW*) up to a temperature of 110 °C, that is to say close to desired temperature. The collected data used for the identification of the process are plotted in **Figure 15**. Where the process sampling time is 180 s.

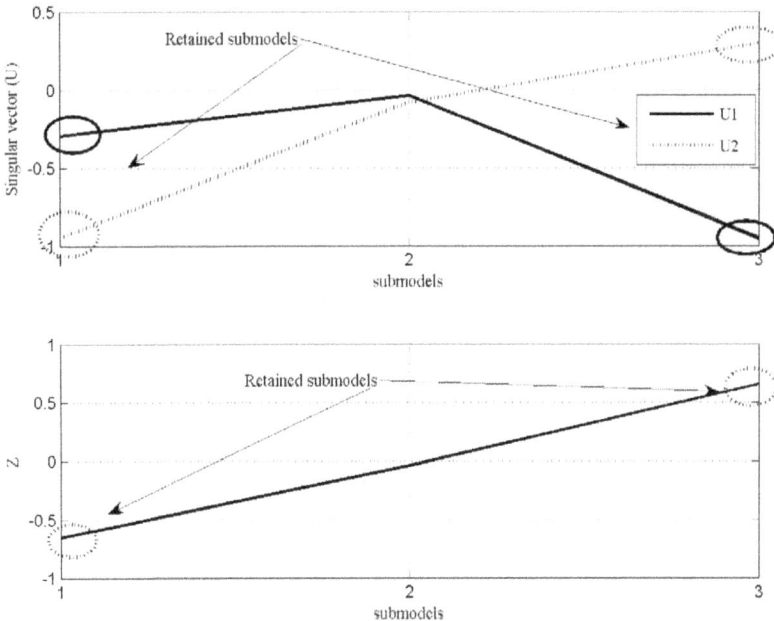

Figure 17.
Determination of the optimal number of submodels (process reactor).

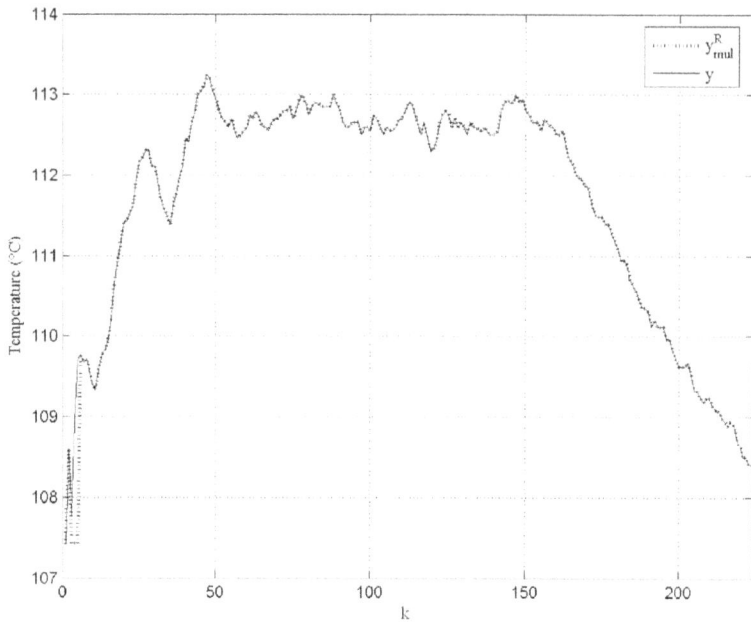

Figure 18.
Evolutions of the reduction multimodel output y^R_{mul} and the reactor output $y(k)$.

Number of submodel	VAF (%)	FIT (%)	NRMSE
3 submodels	98.3043	86.9121	0.0318
2 submodels	98.3043	86.9121	0.0318

Table 5.
Performance comparison (process reactor application).

Based on the FSCL algorithm, we have considered five neurons in the output layer with following parameters of $\alpha^i_g = 0.35$ and $\alpha^f_g = 0.01$. Because that two centers have moved apart from the data as illustrated in **Figure 16** we can terminate the adequate number of cluster which is equal to three. For ach cluster, the data set is used to determine the appropriate local model, after having carried out a structure and parametric identification as well as the determination of its order using the RDI procedure.

The SVD technique is applied in order to reduce the number of submodels. The plotted singular vectors U_1 and U_2 and the absolute value between U_1 and U_2 in **Figure 17** indicate that the two submodels1 and 3 are retained and sufficient to represent the nonlinear plant. The fusion of each model by the new technique of validity computation leads to the results given by **Figure 18**. The results of multimodel reduction approach y^R_{mul} demonstrate that the novel approach tracks the real output with a very small error. This is confirmed by inspecting respectively the value of NRMSE, FAV and FIT given in **Table 5**.

6. Conclusion

A new practice method to reduce the number of model in the model's base is designed for multimodel strategy. The proposed technique is an automatic procedure for decreasing the number of submodel in the model's base. From an initial number of submodel determined using frequency-sensitive competitive learning algorithm (FSCL) and the K- means algorithm, the reduction model procedure repose on the use of an adequate validity computation for each submodel and an analysis of an SVD technique is made on to select the adequate number of submodel. The proposed design has been applied to numerical examples shows its effectiveness compared to conventional approachesThe novel approach is also tested for a real process presents comparable results than those of the literature. A real time application is made on in order to model a process reactor using the same technique shows very remarkable performance. In the context of our approach, we do not seek to study the structure distribution of clusters. In future work, we will study the influence of the clustering algorithm such as the kohonen network or the kmeans algorithms and other competitive learning. The influence of the simple or reinforced validity is also a subject for future work in multimodel reduction procedure. The multiagent model predictive control can be very useful for future works in order to reduce the time consumption with this structure of modeling.

Author details

Bennasr Hichem[1]* and M'Sahli Faouzi[2]

1 Higher Institute of Technological Studies of Sfax (ISET), El Bustan, Tunisia

2 Ecole Nationale d'Ingénieurs de Monastir (ENIM), Monastir, Tunisia

*Address all correspondence to: hichem2001tn@yahoo.fr

IntechOpen

References

[1] Johansen T.A. and Foss B.A, (1997),"Operating regime based process modeling and identification", Computers and Chemical Engineering, vol. 21, pp.159–176.

[2] Bennasr H. and M'Sahli Faouzi, "Multimodel Representation of Complex Nonlinear Systems: A Multifaceted Approach for Real-Time Application," Mathematical Problems in Engineering, vol. 2018, Article ID 1829396, pp.: 15, 2018.

[3] Elfelly N., J.Y. Dieulot, Benrejeb M., Pierre B. (2010)." A new approach for multimodel identification of complex systems based on both neural and fuzzy clustering algorithms". Engineering Applications of Artificial Intelligence 23 pp. 1064–1071

[4] Elfelly N., J-Y Dieulot, M. Benrejeb, P. Borne, (2012),"A Multimodel Approach for Complex Systems Modeling based on Classification Algorithms". Int J. Comput. Commn., ISSN 1841–9836 Vol.7 (2012), No. 4 (November), pp. 645–660

[5] Raja B.M., Hichem B.N and M'Sahli F.(2011), "A multimodel approach for nonlinear system based on neural network validity", International Journal of Intelligent Computing and Cybernetics Vol3, N°3, pp: 331–352.

[6] Talmoudi S., Ben Abdennour R., Abderrahim K. and Ksouri M, (2003) "A New technique of validities'computation for multi-model approach", Wseas Transactions on circuits and systems, vol. 2, Issue 4, (2003), pp: 680–685.

[7] Talmoudi S., Abderrahim K., Ben Abdennour R. and Ksouri M., (2004) "Exprimental Validation of a Systematic Determination Approach of a Models'Base", Wseas Transactions on circuits and systems, vol. 1, Issue 2, pp: 410–416

[8] Talmoudi S., K. Abderrahim, R. Ben Abdennour and M. Ksouri (2008), "Multimodel Approach using Neural Networks for Complex Systems Modeling and Identification." Nonlinear Dynamics and Systems Theory, 8(3),pp. 299–316

[9] Gasso K., Mourot G., Boukhris A. and Ragot J. (1999) "Fuzzy rule base optimisation in TSK modelling" *Proc* of *LFA'99.* pp 233-240. Valenciennes, France.

[10] Gasso K., Mourot G., Ragot J. (2000) "Fuzzy rule base optimisation: a pruning and merging approach". IEEE International Conference on Systems, Man, and Cybernetics, Nashville, Tennessee, USA, October 8–11.

[11] Gasso K., Mourot G., Ragot J. (2001) "Structure identification in multiple model representation elimination and merging of local models". Proceedings of the 40th Conference on Decision and Control, Orlando

[12] Galan, O & Jose A. Romagnoli, Yaman Arkun, Palazoglu, A. (2000a). On the use of gap metric for model selection in multilinear model-based control. Proceedings of the American Control Conference. Vol 6. pp.3742–3746.

[13] Galan, O. & Jose A. Romagnoli, Yaman Arkun, Palazoglu Ahmet. (2000b). Use of gap metric for model selection in multi-model based control design: An experimental case study of pH control. Computer Aided Chemical Engineering. 8.

[14] Galan, O. & Jose A. Romagnoli, Palazoğlu A, Yaman A. (2003). Gap Metric Concept and Implications for Multilinear Model-Based Controller

Design. Industrial & Engineering Chemistry Research - IND ENG CHEM RES. 42

[15] Galan, O., Jose A. Romagnoli, Palazoğlu A, Yaman A. (2004). Experimental Verification of Gap Metric as a Tool for Model Selection in Multi-Linear Model-Based Control. IFAC Proceedings Volumes. 37.pp. 257–261.

[16] Boukhris A., Mourot G., Ragot J (2000). "System identification using multi-model approach: model complexity reduction". 12th IFAC Symposium on System Identification, SYSID'2000, Santa-Barbara, California, USA, June 21–23.

[17] Chiu S.L. Fuzzy model identification based on cluster estimation. Journal of Intelligent and Fuzzy Systems, vol.2,no. 3, pp. 267–278, 1994.

[18] Majda L., Anis Messaoud, Ben Abdennour R. (2014).Optimal Systematic Determination of Models' Base for Multimodel Representation: Real Time Application. International Journal of Automation and Computing, 2014,V11(6): 644–652

[19] Adeniran, A., & El Ferik, S. (2017). A reinforced combinatorial particle swarm optimization based multimodel identification of nonlinear systems. ArtificialIntelligence for Engineering Design, Analysis and Manufacturing, 31 (3),pp:327–358.

[20] Jain A.K and Dubes R.C, Algorithms for Clustering Data, Prentice-Hall Inc., Upper Saddle River, NJ, 1988.

[21] Mirkin B., Mathematical Classification and Clustering, Kluwer Academic Press, Boston-Dordrecht, 1996.

[22] Ahalt, S.C. Krisnamurthy A.K., Chen, P. and Melton D.E. (1990),"Competitive learning

algorithms for vector quantization", Neural Networks 3, pp: 277–290.

[23] Nair M and Zheng C.L and Lynn F. and Stuart R. and Gribskov M. (2003),"Rival penalized competitive: a topology-determining algorithm for analyzing gene expression data", Computational Biology and Chemistry, 27,(2003) 565–574.

[24] Galanopoulos, A.S. Moses R. L., and Ahalt S.C. (1997). "Diffusion approximation of frequency sensitive competitive learning". *IEEE Trans. Neural Networks*, 8(5).pp.1026–1030.

[25] Dunn, J.C.(1973), "A fuzzy relative of the ISODATA process and its use in detecting compact well- separated clusters". Journal of Cybernetics 3,pp. 32–57.

[26] Bezdek, J.(1981), "Pattern Recognition with Fuzzy Objective Function Algorithms". Plenum Press, New York.

[27] Ben Abdennour, R, Borne, P., Ksouri, M., M'sahli, F.(2001), " Identification et commande numérique des procédé s industriels". Editions Technip, Paris, France.

[28] Gasso K., Gilles Mourot and Jose Ragot, (2002) "Structure Identification of models with output error local models".15th Triennial World Congress, Barcelona, Spain, IFAC 2002

[29] Cho J. and Principe J.C and D. Erdogmus and Motter M.A. (2007), "Quasi-sliding mode control strategy based on multiple-linear models", Neurocomputing, 70,pp. 960–974,.

[30] Bastin G. and Dochain D. (1980), "On-line estimation and adaptive control ofbioreactors", Elsevier, Amsterdam.

[31] Gauthier J.P and Hammouri H and Othman S. (1992), A simple observer for

nonlinear systems applications to bioreactors, IEEE Transactions on Automatic Control, 37,pp. 875–880.

[32] Contois D., (1959)"Kinetics of bacterial growth relationship between population density and specific growth rate of continuous cultures," *Journal of Genetic Macrobiol*, 21, pp.40–50.

[33] Bennasr H and M'Sahli Faouzi, (2008)," A Multi-agent Approach To TS-Fuzzy Modeling and Predictive Control of Constrained Nonlinear System", International Journal of Intelligent Computing and Cybernetics. Vol, 3, p.398–424.

[34] Ieroham S. Baruch; Rafael Beltran Lopez; Jose-Luis Olivares Guzman; Jose Martin Flores (2008) A fuzzy-neural multi-model for nonlinear systems identification and control. Fuzzy Sets and Systems, ISSN: 0165–0114,Vol: 159, Issue: 20, Page: 2650–2667.

[35] M'Sahli F, R.B. Abdennour, and M. Ksouri,"Non-linear Model Based predictive control using a generalised Hammerstein model and its application to a semi-batch reactor", International Journal of Advanced Manufacturing Technology, No20, pp. 844–852, 2002.

[36] M'Sahli F, R.B. Abdennour and M. Ksouri "Identification and predictive control of nonlinear process using a parametric volterra model", International journal of computational engineering science, Vol.2, No.4, pp.633–651,2001.

Chapter 4

Approximation Algorithm for Scheduling a Chain of Tasks for Motion Estimation on Heterogeneous Systems MPSoC

Afef Salhi, Fahmi Ghozzi and Ahmed Fakhfakh

Abstract

Co-design embedded system are very important step in digital vehicle and airplane. The multicore and multiprocessor SoC (MPSoC) started a new computing era. It is becoming increasingly used because it can provide designers much more opportunities to meet specific performances. Designing embedded systems includes two main phases: (i) HW/SW Partitioning performed from high-level (eclipse C/C++ or python (machine learning and deep learning)) functional and architecture models (with virtual prototype and real prototype). And (ii) Software Design performed with significantly more detailed models with scheduling and partitioning tasks algorithm DAG Directed Acyclic Graph and GGEN Generation Graph Estimation Nodes (there are automatic DAG algorithm). Partitioning decisions are made according to performance assumptions that should be validated on the more refined software models for ME block and GGEN algorithm. In this paper, we focus to optimize a execution time and amelioration for quality of video with a scheduling and partitioning tasks in video codec. We show how they can be modeled the video sequence test with the size of video in height and width (three models of scheduling tasks in four processor). This modeling with DAG and GGEN are partitioning at different platform in OVP (partitioning, SW design). We can know the optimization of consumption energy and execution time in SoC and MPSoC platform.

Keywords: Motion Estimation, Video Codec, Video and image processing, DAG, GGEN, Localization, Scheduling, Partitioning, H 264/AVC, H 265, MPSoC, SoC, OVP, Mapping, Co-design, Xilinx-SoC-Platform, multiple streams processing

1. Introduction

Multi-core and Multi-processor architectures (SoC and MPSoC) started a new computing era. They are becoming increasingly used as they can provide designers with new opportunities to meet desired requirements in embdedd system for different application and domain. Multi-media and telecommunication streaming applications are now widely used in several domains such as visio-conference, networking, video cripttage and compression, surveillance, medical services, military imaging and telecommunication applications. These applications are characterized with stringent delay time of tasks. Model of scheduling tasks in 4 CPUs for Motion Estimation "ME" and DAG algorithm is illustrated in **Figure 1**.

Motion estimation module is very important complex module of a video codec. MPSoCs have the most suitable architecture to meet real time high-definition "HD" encoding requirements. This real time HD coding is based on a block matching algorithm "BMA" which locates matching blocks in a sequence of digital video frames. This technique is used to discover temporal redundancy in the video sequence, increasing the effectiveness of inter-frame video compression. The scheduling of tasks is an important step to accelerate the motion estimation process. The objective is to minimize the execution time of this algorithm, by distributing computing the tasks that describe the algorithm to the various cores of an MPSoC. In this case, we can see various methods to jobs scheduling and partitioning tasks for video codec applications, but are not adapted to our motion estimation block problem.

In this project, we choose the second method to schedule our algorithm based on DAG and GGEN for many reasons. This method minimizes the complexity of NP-complete problems in tasks. This methods are automatic, generic and periodic algorithm. It refine a granularity of tasks in ME block. The true parallelism tasks in SoC and MPSoC system, we start in 4 CPUs as in [1, 2], after we can works 8, 16, … , 1024 CPUs. There are some works and scheduling in tasks, it is periodic and acyclic. This methods give us an optimal solution for complex scheduling and

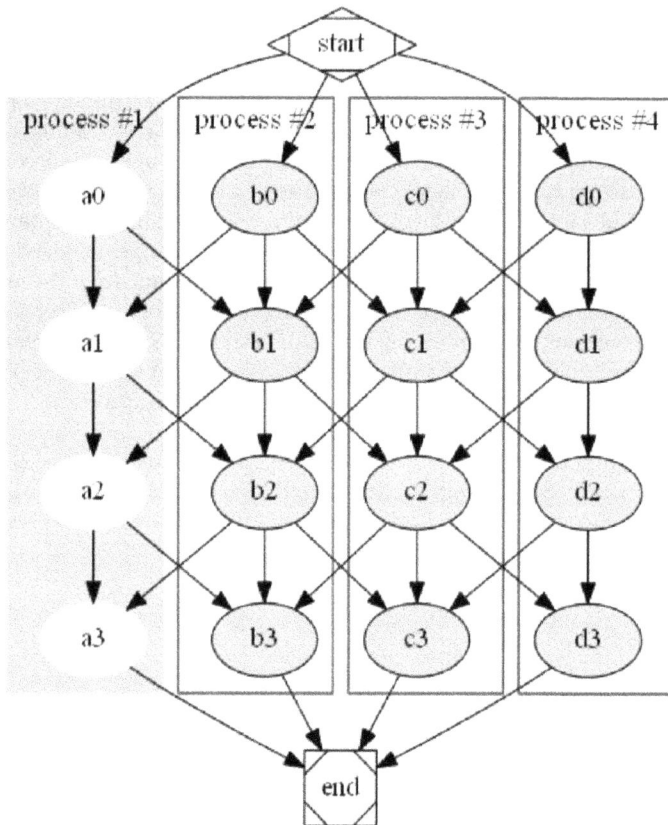

Figure 1.
Model of scheduling tasks in CPUs.

partitioning tasks on MPSoC system, as in [3, 4]. In this case, we choose parallelizing tasks and instructions in codec video blocks, there is very interessting step for scheduling and partitioning tasks in embedded platforms. We can cite: FPGA, GPU, DSP target, as in [5, 6]. The standard H 264/AVC is a new video codec configured and released by ITU-T and ISO/IEC [7, 8]. This standard give a very important results in bit rate, quality of image and others criteria in video codec block comparing with others standards [2, 7–9].

This paper is scheduled as follows: In section II, we detail the applied algorithm for block motion estimation with scheduling and partitioning tasks method in SoC and MPSoCs systems, this works are designed with acyclic algorithm (DAG-TPG) and co-designed in OVP platform. After, we present the ME module and importance in video codec standards. In section III, we can describe a new scheduling tasks approach and co-design in platform OVP (partitioning). Then, we synthesis with results, section IV. We finished by our conclusion.

2. ME in codec video H 264 and scheduling tasks approach

In this paper, high-quality video encoding imposes unprecedented performance requirements on real time mobile devices. To address the competing requirements of high performance and real time, embedded mobile multimedia device manufactures have recently adopted MPSoC (multiprocessor system-on-chip). Despite the advancements in new technology digital mobile device, computer system and I-Pad, the execution time, energy consumption and quality of image in SoC and MPSoC systems, we needed parallelism and scheduling tasks applied for H 264/AVC video codec. This approach of scheduling can eliminate problems in ME blocks and remains artifacts in image. We can start to minimize a time delays and time execution in ME blocks in video codec. Secondly, video codec blocks very important in structure and function, needed predictive coding blocks. These structures and functions, can be defined and modeled as with acyclic approach (semi-automatic and automatic) directed acyclic graphs (DAG) and generated graphs GGEN. With this approach, we can give a good and important solution for criteria evaluation in video codec, we can site execution time, time delays in tasks, and others [3, 4, 9]. We can finished to describe the delays tasks in some CPUs (SoC and MPSoCs target), there are considered real-tim applications [3, 4, 9]. In other case, video frames and video sequence should meet their deadlines and their estimation, the quality of image is very important. Then, many execution time in SoC and MPSoC targets with scheduling tasks approachs, there are exploit execution time and others criteria evaluations in video codec, we can see [9–12].

In **Table 1**, we classify these representative solutions based on their utilized optimization horizons, application models, complexity models, scheduling granularities, and considered sources of execution time. The scheduling tasks approach was chosen due to its efficiency and implementation simplicity for video codec H 264. The main idea of the scheduling tasks in ME is to decompose a sequence video into a frame and a frame into a Macro-block, so we scheduled and partitioned a tasks in parity order, we can see our method [9–12].

2.1 Scheduling and partitioning algorithm tasks

In previous works, scheduling and partitioning algorithms in MPSoCs systems have been formulated heuristically due to the inherent complexity of the multi-machine scheduling problem. Such algorithms constitute a very important step in multimedia applications. There is an NP-complete step in co-design. A schedule is

Scheduling and partitioning tasks	Platform	Algorithms	Application	Optimal	Type of granularity
[13]	SoC	DAG	Signal	Yes	Law
[3, 4]	SoC /MPSoC	DAG-GGEN /SDF-CSDF	Treatment	Yes	High
[14]	SoC/MPSoC	DAG	H 264	Yes	Law
Our sollution	MPSoC	GGEN	H 265	Yes	High
[15]	SoC	DAG	H 264	No	High
[9–12]	SoC/MPSoC	DAG/GGEN	H 264	Yes	High
[16]	MPSoC	DAG	H 264	Yes	Law

Table 1.
Comparaison our solution with different approachs scheduling tasks.

an assignment of each task to a machine. For scheduling, the load M_i defines the requirement of total processing jobs assigned, and the scheduling length is the load on the busiest machine. We want to find a minimum length for the schedule, as in [3, 13]. There are three distinct approaches to these scheduling problems: the theory of queues for networks scheduling, the deterministic scheduling and the software engineering scheduling. The study of approximation algorithms for NP-complete scheduling and partitioning problems for MPSoC systems has started with the work of Graham in 1996, who has analyzed a simple algorithm. When designing an approximate algorithm to minimize such NP-complete problems, we can evaluate its performance in a different way. One would use an algorithm that approximates with a guarantee of the performance of the deviation of the optimum value of the worst case as in [3, 4]. However, most scheduling and partitioning algorithms make assumptions on the relationship between the task dependencies. We may then classify scheduling algorithms. Scheduling, data partitioning and parallelism identification are three key points for an efficient application deployment. There are several approaches to manage scheduled tasks for real-time applications. The most important approaches are cyclic and acyclic approaches. Cyclic approaches have been used for Static Data Flow "SDF", Cyclo-Static Data Flow "CSDF" and Petri networks [4, 13]. Acyclic approaches have used DAG, Unified Directed Acyclic Graph "UDAG, Weighted Directed Acyclic Graph "WDAG", etc.

In this paper, we consider the problem of partitioning and scheduling tasks with a Task Precedence Graph "TPG" which is given as a DAG. A solution based on the DAG representation has the advantage to be generic, simple and usable with a refined granularity. A node in the DAG is a task which is a set of instructions to be executed sequentially without preemption in the same processor. Modeling, terminologies and all the mathematical formalism of the DAG algorithm is presented in [3, 4].

2.2 DAG algorithm

DAG algorithm is based on the asynchronous message passing paradigm. The parallel architectures are increasingly popular, but scheduling is very difficult because the data and the program must be partitioned and distributed to processors [3, 17, 18]. The methodology for MPSoC in embedded applications is the Adequacy-Algorithm-Architecture "AAA". The following issues are of major importance for distributed memory architectures:

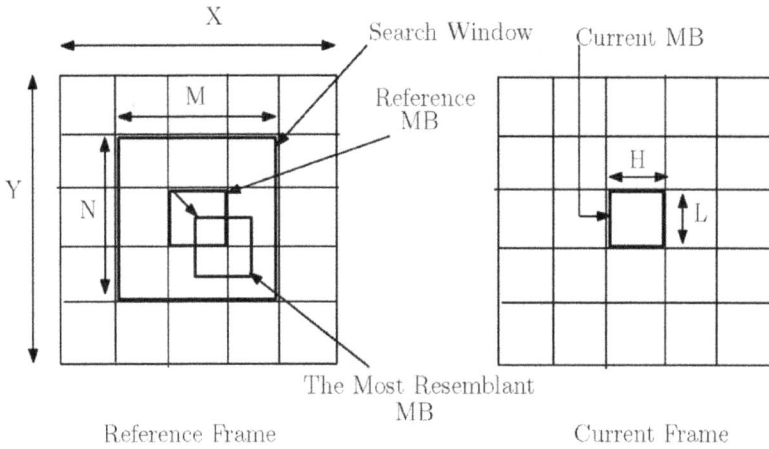

Figure 2.
Principle of SAD function.

- Scheduling, data partitioning and parallelism identification.

- Data mapping and program architecture.

- Scheduling and partitioning the execution of the task.

2.3 Motion estimation "ME": application

ME block is very important in H 264 and H 265 video codec. In various standards, ME needs very complex tasks and instructions, and takes the largest part of video codec [2, 10–12, 19].

2.3.1 Principle of the ME block

Principle of ME is the following: for a MB in the current frame, we define a search window in the reference frame. There are several evaluation criteria, such as MSE, SAD, BBM, MAD, NCF. We seek in this window the best MB using the Sum of Absolute Difference "SAD" distortion criterion given by Eq. (1) as in [2, 19–21], we desibe this function SAD in **Figure 2**. To optimize the complexity level of the ME module, several fast search algorithms have been defined in the literature as DS, FS, TSS, HDS, PMVFAST, LDPS and the block-matching algorithm "BMA". Inter-coding consists in finding a similar block that is aware of a reference frame block. This process is performed by a BMA. General principle of BMA is to exploit the temporal redundancies between consecutive frames.

$$SAD(x,y) = \begin{cases} \sum(i=0, j=15)\sum(i=0, j=15) \\ |R(i,j) - F(x+i, y+j)| \end{cases} \tag{1}$$

3. Scheduling based on DAG formalism: approach

3.1 Block matching algorithm "BMA"

Inter-coding consists in finding a block similar to the current block of a reference frame. This process is performed by a block-matching algorithm. The general principle

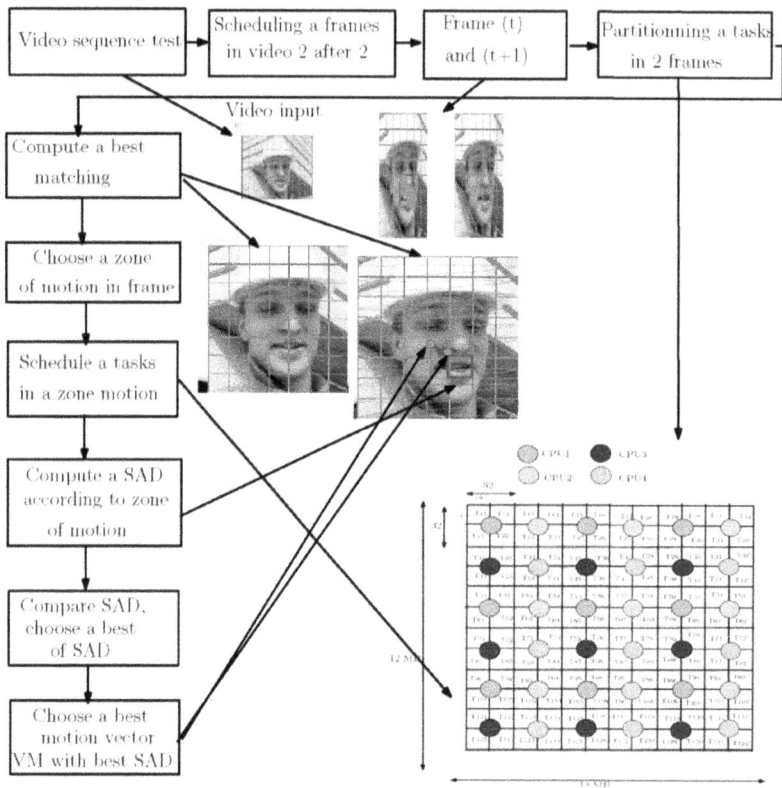

Figure 3.
Principal of block matching algorithm "BMA".

of the BMA is to exploit existing temporal redundancies between consecutive frames. This method involves searching for each point of the frame of interest It, the point of the frame $It + 1$ USD which maximizes a correlation score. The search is performed in a search block [7, 10–12, 19]. We describe a principal of this method in **Figure 3**.

Object of interest is determined when the number of corresponding blocks in the previous and current frame is higher than the value of a certain threshold. The threshold value is obtained experimentally [22]. We define the principle from SAD function in Eq. (1), where $(R(i; j))$ denotes the pixels of the reference MB and $(F(x + i; y + j))$ denotes the pixels of the current MB. **Figure 4** below shows the flow chart of the ME block of H 265 video codec. We use a padding method in order to enforce a whole number of packets. The scheduling methodology is defined in the **Figure 5**, which illustrates the flow chart of the MB $(16 * 16)$ for H 265 the video codec. The new idea for the flow chart of H 265 is to work with the padding technique rather than the affinity of the granularity $(1/2)$, $(1/4)$ and the $(1/8)$ pixel method. The padding added to the end of a packet in order to enforce a whole number of packets.

3.2 DAG applied to ME

In this section, we describe our new approach for the scheduling tasks DAG. Firstly, we should to make the method generic, we can validated for any frame size. Then, we chose to work on block parity in scheduling tasks for ME blocks. Four combinations are possible: odd odd, odd even, even odd, even even. The different

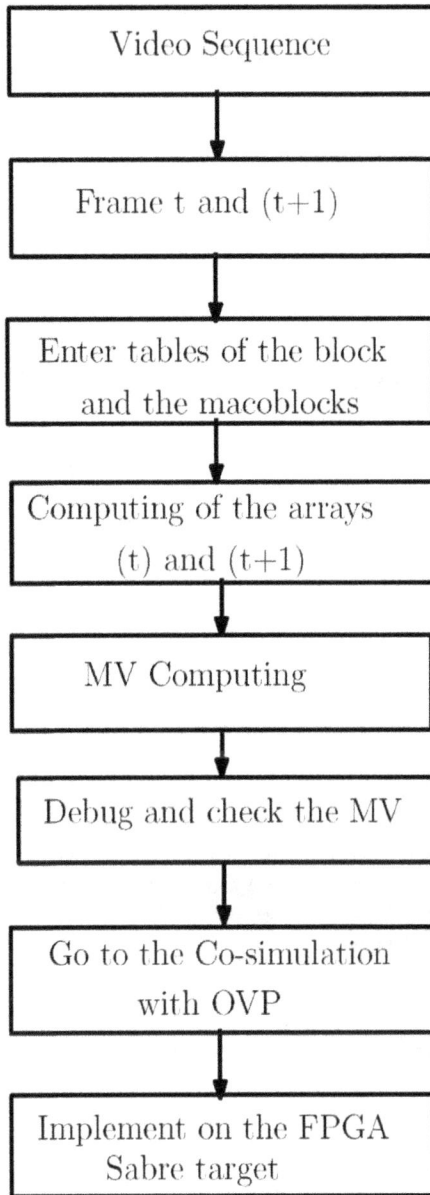

Figure 4.
Flow chart of ME algorithm.

sequence test in our works are modeled in three models for the scheduling tasks methods. The generic frame size is ("X = N", "Y = M") where "X" represents the pixels for lines and "Y" denotes the column pixels. After scheduling a three models, we can see a problem in border of image or frame. This problem, we can apply padding method to add rows and columns to the frames by adding empty pixels. We notice that "N" has to take an even value that is divisible by 4. **Figure 6** shows the work-flow of an entire frame with size (N*M).

For each node in the task graph, one must compute the start and end times of execution cycles by using the weight of each node. The "Gantt of chart" represents

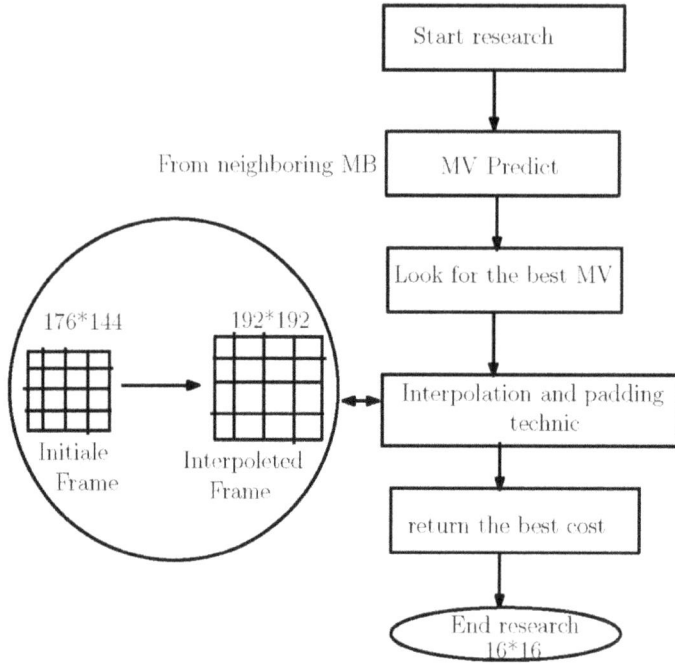

Figure 5.
Flow chart of the 16 ∗ 16 for ME block in H 264 video codec.

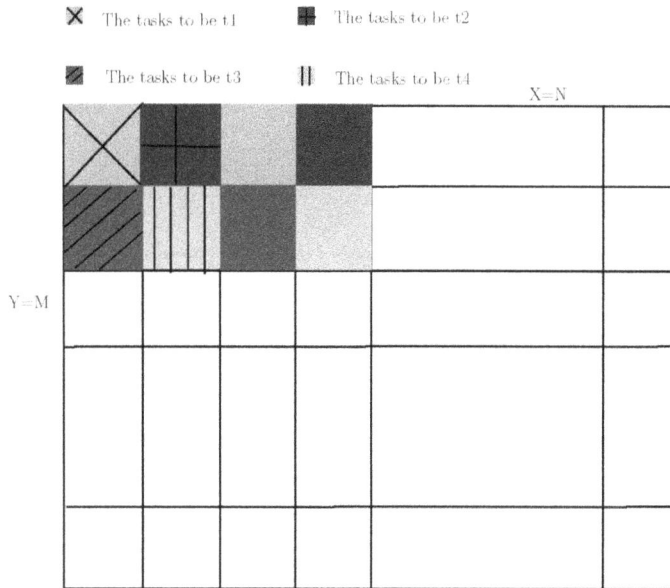

Figure 6.
Scheduling of an frame of size (N ∗ M).

the task scheduling in processors in terms of time and gives the order of the tasks of the ME block. This Gantt of chart is illustrated in **Figure 7**. Clustering is the placement of the various tasks of our application on different clusters. It depends on

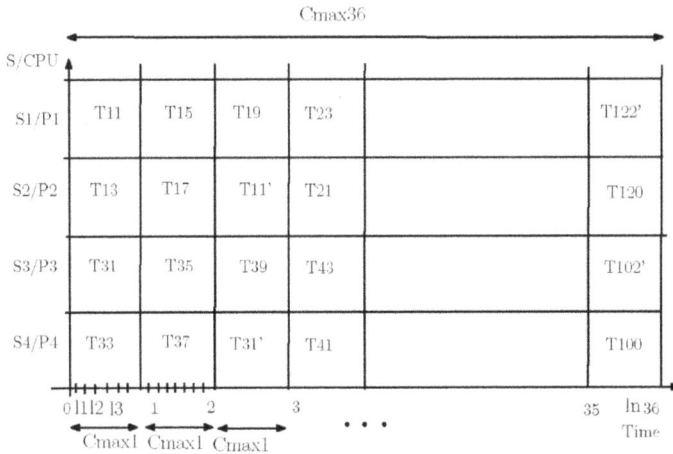

Figure 7.
Gantt of chart from ME block.

the number of processors in the implementation platform. In our case, we chose a platform with four processors. We need four groups, four instants of time and then we have four outputs.

3.3 Parameters setting

For a set P of processors, each node must be assigned to a single processor. For all the the addressable memory, we give a portion of the same space to each case. When setting up parameters, we have to take in consideration the time constraints and the constraint of targets HW/SW. The classical assumptions made on the target MPSoC are:

- The tasks are non-preemptive.

- Local communication is costless.

Despite, the following equations show the scheduling tasks of a classic frame. First, we treat the odd tasks and the odd nodes in the TPG graph, $ts(n_{ii})$ and $w(n_{ii})$ presenting the time and weight in odd nodes.

$$tf(n_{ii}) = ts(n_{ii}) + w(n_{ii}) \qquad (2)$$

At the end, we compute the equations of even tasks, where $ts(n_{jj})$ is the time of the node (jj) and $w(n_{jj})$ is the weight in this node.

$$tf(n_{jj}) = ts(n_{jj}) + w(n_{jj}) \qquad (3)$$

For the time end calculation "tf" in the node $(n_i; n_j)$. The sequence tasks from the ME blocks are defined in **Figure 8**.

In general, the nodes and tasks in TPG graph are made of nodes $n_{ii}, n_{ij}, n_{ji}, n_{jj}$. They are illustrated in Eqs. (4)–(7).

$$N_{ii} = (n_{ii} + n_{ij}) \qquad (4)$$

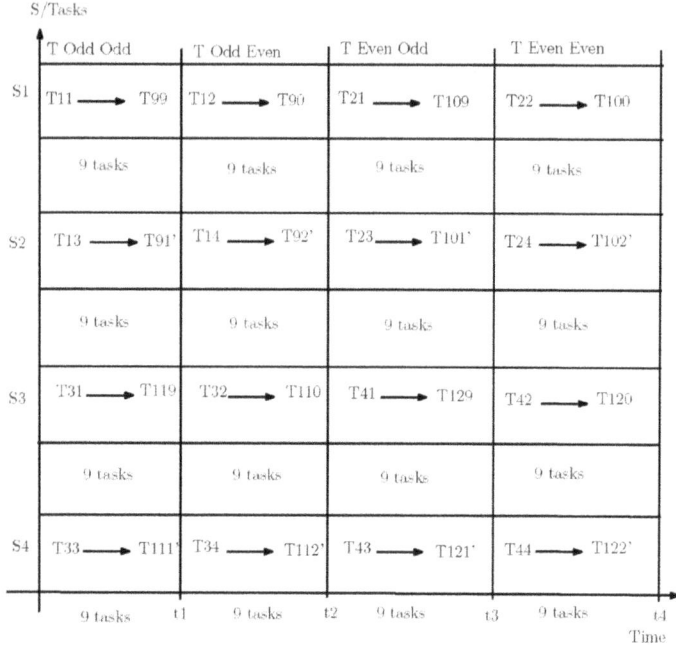

Figure 8.
Gantt of chart for tasks sequences by CPU.

$$N_{ij} = \left(n_{ij} + n_{ji}\right) \tag{5}$$

$$N_{ji} = \left(n_{ji} + n_{ii}\right) \tag{6}$$

$$N_{jj} = \left(n_{jj} + n_{ji}\right) \tag{7}$$

Execution time for the node of the test sequence is presented by Eq. (8) below.

$$t_f\left(n_{ii}, n_{ij}\right) = \begin{cases} t_{f(n_{ii})} \\ n_{11}, \ldots, n_{ii} \text{ in the same processor} \end{cases} \tag{8}$$

For the nodes or the tasks executed on different processors, the execution period is computed in the output frame for the current reconstruction frame. The nodes are spread by parities over each processor. In the first level, we have 16 nodes. In Level II, we add nodes to the neighbor to reconstruct the macro-blocks. The same is done for all levels, the goal being to reconstruct the frame. In our case, we have four groups since both architectures test four processors. We compute the execution period in Eq. (9) below.

$$t_f\left(n_i, n_j\right) = \begin{cases} t_{f_{n_i}} + C\left(n_i, n_j\right) \\ Otherwise\ n_i, n_j \text{ in different processor} \end{cases} \tag{9}$$

Hence, the set of end times is computed in the three models for scheduling and partitioning tasks. This algorithm is applied to ME blocks for the test video sequence in H 265 video codec: $t_{f(n_{ii}, n_{ii})}, t_{f(n_{ij}, n_{ij})}, t_{f(n_{jj}, n_{ij})}, t_{f(n_{ji}, n_{ji})}$. It is applied to all nodes in the graph found for the task with the ME block using test sequence "Akiyo". Two processors p_i and p_j are isomorphic if read times are equal. Then the weight odd nodes or tasks $w(n_i)$ are equal to the weight nodes or tasks $w(n_j)$. Thus, the set of nodes is $n_{ii}, n_{ij}, n_{jj}, n_{ji}$ and the set of tasks is $T_{ii}, T_{ij}, T_{jj}, T_{ji}$. Succeeding in scheduling and partitioning the tasks with DAG is equivalent to the following conditions::

- $\text{Pred}(n_i) = \text{pred}(n_j)$.

- $W(n_i) = W(n_j)$.

- $\text{Succ}(n_i) = \text{Succ}(n_j)$.

In our case, the platforms contain four processors, therefore we seek other processors satisfying the conditions. Basing on the assumptions above, we can deduce other parameters. The scheduling length for tasks and partitioning problems of processor $(P+1)$ is always less than or equal to the processor P.

$$SL\left(S_{opt}(P+1)\right) < = SL\left(S_{opt}(P)\right) = max\left(t_f(n_{xx})\right) \qquad (10)$$

Where n_{xx} is the set of nodes $n_{ii}, n_{ij}, n_{ji}, n_{jj}$. This equation is applied to the three models, where "SL(S)" contains the scheduling graph length "Sopt(P)" which has the optimal schedule of P processors. For our application, we have four processors. To apply the hypothesis, we must know the maximum execution time for all nodes ("X" nodes = "X" tasks).

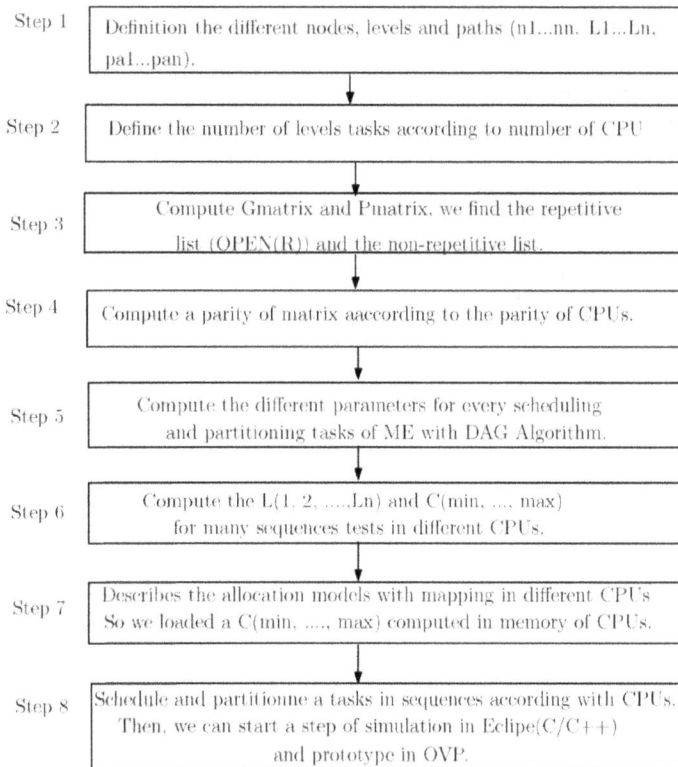

Step 1	Definition the different nodes, levels and paths (n1...nn, L1...Ln, pa1...pan).
Step 2	Define the number of levels tasks according to number of CPU
Step 3	Compute Gmatrix and Pmatrix, we find the repetitive list (OPEN(R)) and the non-repetitive list.
Step 4	Compute a parity of matrix aaccording to the parity of CPUs.
Step 5	Compute the different parameters for every scheduling and partitioning tasks of ME with DAG Algorithm.
Step 6	Compute the L(1, 2, ...Ln) and C(min, ..., max) for many sequences tests in different CPUs.
Step 7	Describes the allocation models with mapping in different CPUs So we loaded a C(min, ..., max) computed in memory of CPUs.
Step 8	Schedule and partitionne a tasks in sequences according with CPUs. Then, we can start a step of simulation in Eclipe(C/C++) and prototype in OVP.

Figure 9.
Flow-chart of steps DAG algorithm with ME.

Steps Val of	Model of sequences	Nbre of frames	Nebre levels	Val of paths	Nbre of tasks	Nbre of L (1, ..., n)
1	Q-cif	100	8	2/../288	144	36/36 + c
2	Q-cif	495	8	2/../288	144	36/36 + c
3	Cif	150	8	2/../480	240	144/144 + c
4	Cif	280	8	2/../480	240	144/144 + c
5	Sif	300	8	2/../480	240	144/144 + c
6	Q-cif	995	8	2/../480	240	144/144 + c
7	HD	25	16	2/../3840	1920	4096/4096 + c
8	HD	25	16	2/../3840	1920	4096/4096 + c

Table 2.
Parameters for steps of tests videos sequences.

3.4 Applied DAG algorithm with ME in video codec

Flow chart of the global steps is illustrated in **Figure 9**.
So, we can resume the scheduling steps are as follows in **Table 2**.

3.4.1 Proposed architecture and performance of the DAG algorithm applied to ME blocks in OVP

How to compute the execution time?
To calculate the processing time of an instruction, since the message transfer time depends sent to the calculation of instructions executed by a processor flows. Each instruction requires multiple clock cycles, the instruction is executed in as many cycles steps. Sequential microprocessors run the following statement when they finish first. In the case of instruction parallelism, the microprocessor can process several of these steps at the same time for several different instructions because different internal resources are mobilized. In other words, the processor executes instructions in parallel and sequencially at various stages of completion. This execution queue is called a pipeline. This mechanism has been implemented for the first time in the 1960 by IBM. **Figure 10** describe the canonical example of this type of pipeline, under the form of a RISC processor, in five stages. Sequencing instructions in a processor with a 5-stage pipeline needs 9 cycles to execute 5 instructions. At t = 5, all floors have solicited the pipeline, and 5 operations occur simultaneously.

How to compute the transfer time?
Assume that the data transfer time is proportional to the size of the data exchanged (between processors), this transfer time is the time of sending data within

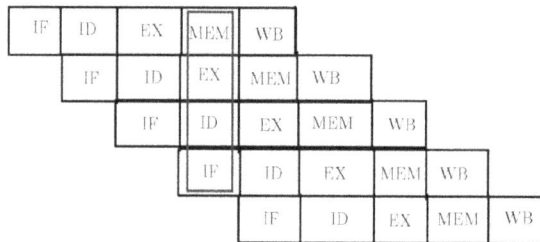

Figure 10.
The canonical example of this type of pipe is that of a RISC processor (5 stages).

tasks in between processors. Then "Tt" is the transfer time and "Td" is the data size. k($p1; p2$) is the transfer duration between p1 and p2. Tt is defined by Eq. 11:

$$Tt = td \times K(p1, p2) \tag{11}$$

The execution time of an instruction and message transfer between processors, and tasks between processors processed by the ME block are presented in **Table 3**.

Virtual prototyping with OVP environment

We work with a project virtual prototype in OVP simulator. The strategies from a project and simulations in OVP are illustrated in [23]. We describes the co-simulation in **Figure 11**. This figure is composed the implementation and prototype in OVP (with thread method in CPUs).

Affinity scheduling granularity of the ME algorithm

The following diagram shows the flow chart of mode (16 * 16) for the video coding in H 264/AVC. The difference with the flowchart of H 264/AVC standard is the interpolation technique, used instead of the affinity method granularity 1/2, 1/4 and 1/8 pixels. Using the interpolation technique minimizes the number of jobs and the number of level task graphs. Thereafter, the execution time of block ME is optimized for the standard video codec HEVC/H 265 and the standard H.264/AVC old report time. Thus, we have one level graph TPG-DAG. Both graphs were improving accuracy and frame quality. In this section, we consider time communications between different tasks, as independent tasks in the same processor itself or in different processors as shown in **Figure 12**. We define the communication delays between tasks and scheduling tasks lengths. **Figure 12** presents the partitioning and scheduling algorithm for the ME block in the H 265/AVC video codec and the different communications and mappings for the different processors.

Table 4 shows the test video sequences with theoretical results. We can see that the theoretical execution time is close to the practical execution time observed in

L	Nbr T(1)	Nbr T(2)	Nbr T(3)	Tdt (1)	Tdt (2)	Tdt (3)	k (p1,p2)
L1	144	576	16384	3600	14400	409600	2/./135462
L2	72	288	8192	1800	7200	204800	2/./67736
L3	36	144	4096	900	3600	102400	2/./33868
L4	18	72	2048	450	1800	51200	2/./26844
L5	9	36	1024	225	900	25600	2/./18422
L6	5	18	512	112.5	450	12800	2/./9216
L7	2	9	256	56.25	225	6400	2/./4608
L8	1	5	128	28.12	112.5	3200	2/./2304
L9	0	2	64	14.6	56.25	1600	2/./1152
L10	0	1	32	7.3	28.12	800	2/./576
L11	0	0	16	3.8	14.6	400	2/./256
L12	0	0	8	1.9	7.3	200	2/./128
L13	0	0	4	0.95	3.8	100	2/./64
L14	0	0	2	0.475	1.9	50	2/./32
L15	0	0	1	0.24	0.95	25	2/./16

Table 3.
The results of scheduling and partitioning levels of the model for the sequence test 1.

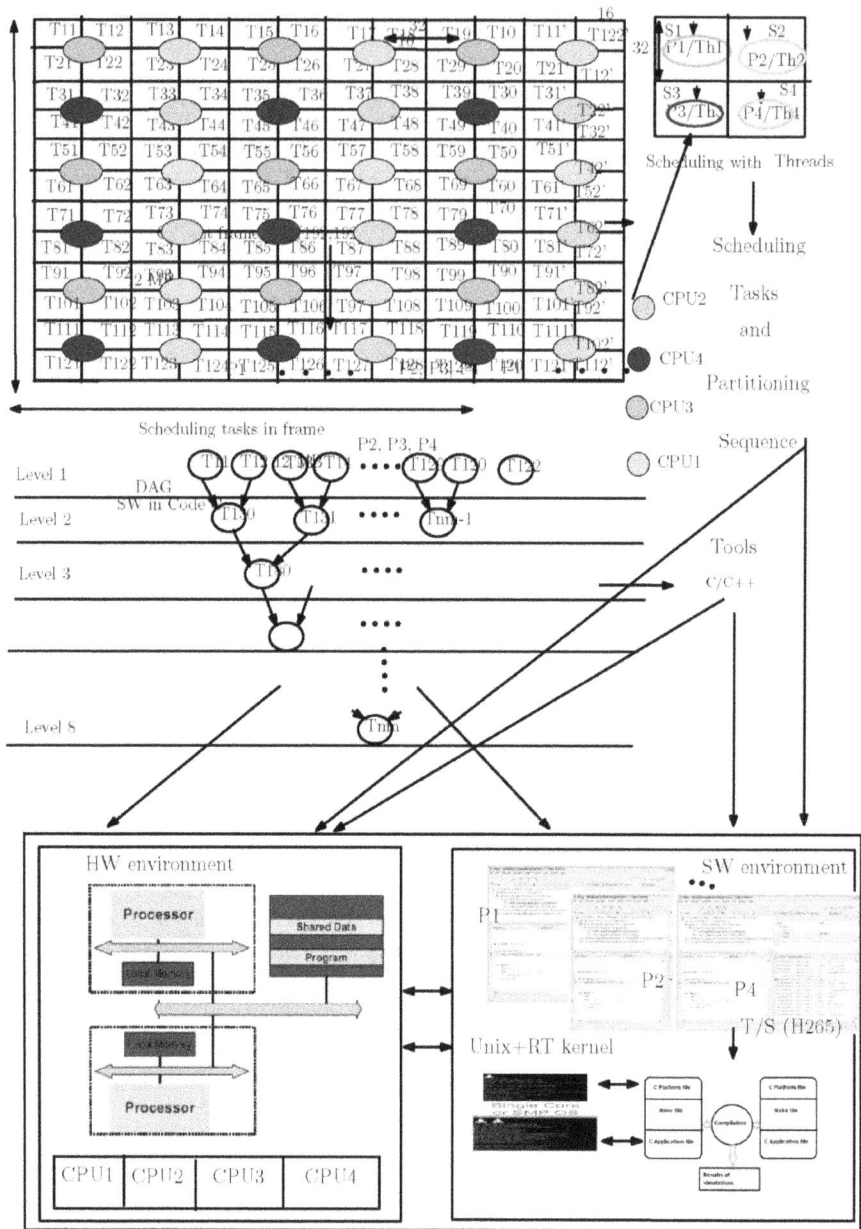

Figure 11.
Prototype and implementation in OVP.

co-simulations of OVP. Also, we see that our scheduling and partitioning algorithm is optimal within the approximation and modeling formalisms considered.

3.5 The results of the ME block applied to scheduling tasks approach

Some experimental curves of execution time for the ME block applied to the DAG algorithm are presented in **Figure 13**, obtained in OVP for SoC and MPSoC based ARM Cortex A9MP. The important metric in our works is execution time.

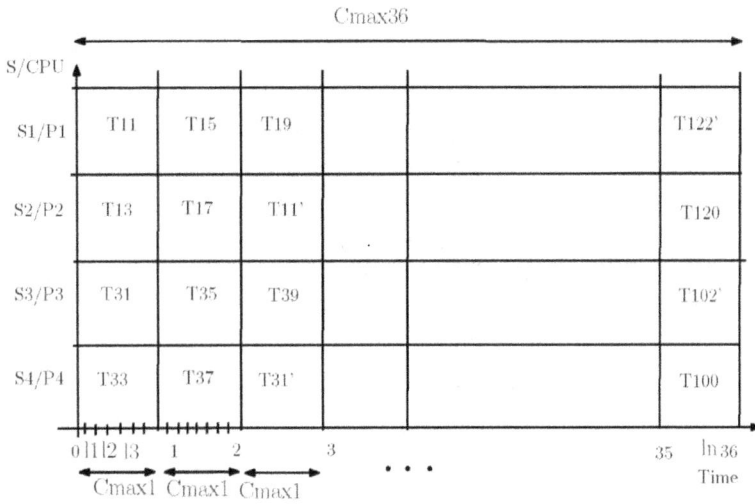

Figure 12.
Independence tasks in sequences with CPU.

Seq	Tt (cycls)	EX-T in Fr(cycls)	EX-T in Seq(cycls)	EX-T in seq(s)
Qcif	25	3600	1080000	7.2
Qcif	25	3600	540000	5.4
Qcif	25	3600	1616400	10.76
Cif	25	14400	4320000	8.8
Sif	25	14400	1612800	10.75
Cif	25	14400	1440000	9.6
Sif	25	14400	12960000	8.64
Qcif	25	14400	1778400	11.85
HD [9]	25	409600	117964800	7.9
HD [10]	25	409600	117964800	7.9

Table 4.
Modeling test sequences using the DAG algorithm.

After, execution time simulation, we chose co-simulation in OVP. In this virtual platform, we can try variety SoC and MPSoC targets in OVP. The execution time is ameliorated compared with other results in literature with respect to the number or frames in test video sequences. For more details we can see [9], in our works we presents a different values in three models in scheduling and partitioning tasks co-designed in platform OVP (SW/HW). Our scheduling algorithm DAG applied to ME block in video codec is optimal because $(0 < = C < = 0 : 5)$. T_1 is the execution time for one processor and T_2 is the execution time for many processors. The gain for the application is computed in Eq. 12.

$$Gain = [(|T1 - T2| \div T1) \times 100] \qquad (12)$$

Also, our parameters for DAG and GGEN algorithms applied to ME blocks in codec video is very important if you compared with others approachs in scheduling tasks. In **Figure 13**, we give the formulations for scheduling dependent tasks into

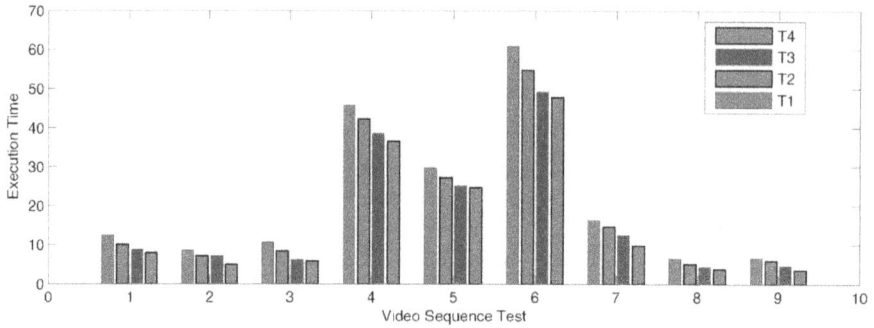

Figure 13.
Results execution time in SoC and MPSoCs targets in OVP with DAG algorithm.

homogeneous multiprocessor architectures of an arbitrary DAG and GGEN, taking into account communication delays. The time execution is in seconds. T_1 presents the execution time for simulations with C/C++. Then T_2 is the execution time for simulations with SoC platform in OVP. T_3 and T_4 present the execution times when scheduling and partitioning tasks with the MPSoC system in OVP. P_1 is the platform Versatile ARM CortexA9MP*4, P_2 is the platform Ukernel arm Cortex A9MP*4.

We observe our results in simulation for execution time are very more high compared with our results in co-simulation in Virtual Platform OVP, we can see the important minimization. **Figure 14** shows the results obtained for the gain of the entire test frame with the DAG algorithm. From **Figure 14**, we can see that using our technique reduces the execution time of the ME block. The experimental results illustrates the substantial enhancement of execution time. The scheduling and partitioning tasks algorithm DAG is give a true parallism with 4 processors in SoC and MPSoCs targets. For those comparisons, the metric value is the latency time of executions "t" for the SoC system. We selected SOC and tow platforms MPSoC (the models of MPSoC are: Platform including ARM Cortex-A9MPx4 to run ARM MPCore Sample Code and Versatile Express booting Linux on Cortex-A9MP Single, Dual and Quad Core in OVP). We remark that the execution time decreases when changing the platform, as in SoC and MPSoC systems. We deduce that the

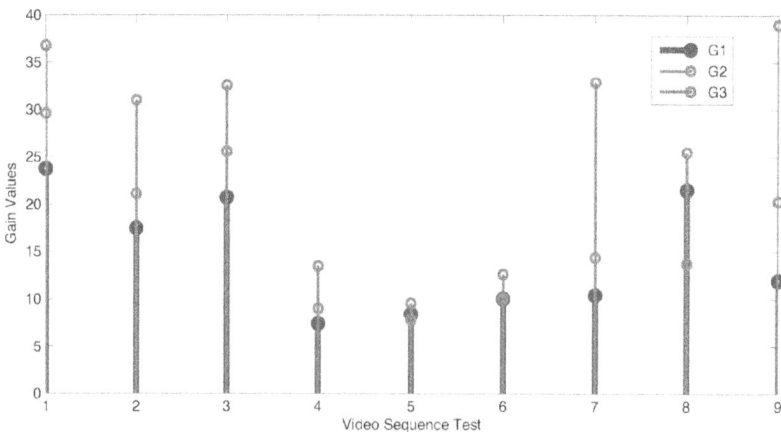

Figure 14.
The results of gain for the entire test frame with the DAG algorithm.

Sequences	F(sc)/Frame	F(sc)/Video
Akiyo	525533184	157659955200
Car(Avnet)[HD]	42998169600	2579890176000
Avion(Avnet)[HD]	42998169600	2149908480000
Forman	1751777280	525533184000
Miss-America	525533184	78829977600
Claire	525533184	259613392896
Mobile	1751777280	196199055360
Bus1	1751777280	175177728000
Skin1	1751777280	1576599522000
HD$_{seq}$ [9]	42998169600	1031956070400
HD$_{seq}$ [10]	42998169600	1031956070400

Table 5.
The results of the "F(sc)" function of the ME block for the different test sequences.

scheduling methodology has beneficial results for the execution times in OVP for the various test video sequences.

We show in this part the evaluation of the criteria of our application treated in our research work. We calculate the complexity of the H 264 video codec for ME block in the Eq. (14). **Table 5** prensent a results for the function F(sc) of ME blocks. The function (Fsc) is the function for calculating the complexity of the EM block; such as "(w, h)" frame size of the test sequence, "p" is the maximum authorized displacement, "N" is the size of MB processed, "h" is the factor and w is the width.

$$F_{sc}(Im) = \left((h \times w)(2 \times p + 1)^2 \times N^2 \right) avecp = 4 \qquad (13)$$

$$F_{sc}(Im) = \left[\left((h \times w)(2 \times p + 1)^2 \times N^2 \right) \right] \times (NF) \qquad (14)$$

We notice that the sequences that we use in our research are very complex, so we need a very optimal, precise, generic and automatic approach. With this we get good results in all H 264/AVC video codec.

4. Conclusion

In this paper, we presented a new scheduling and partitioning tasks algorithm DAG applied to ME for MPSoC platforms in OVP. The main contributions of our approach are the following ones: the semi-automatic scheduling and partitioning tasks, the performance with respect to granularity, the high quality, the accuracy and the short time execution for ME blocks in H 265 video codec. Stemming from complexity and profiling analysis, the DAG algorithm with ME blocks for three architectures platforms are presented in OVP (Soc and MPSoC system). The proto-type SoC and MPSoCs system in OVP results highlighted that the processors have interesting performances, complexity, and execution times compared to other published solutions. This is visible in the tables and figures. The co-design HW/SW high level is also presented in SoC and MPSoC systems in OVP with IP of the DAG algorithm applied to ME blocks in H 265 video codec. Our scheduling and partitioning algorithm DAG is able to handle the execution time efficiently with very limited resources, dropping the appropriate tasks in order to reduce the deadline time and the execution time in ME blocks. Our scheduling and partitioning algorithm DAG is fully semi-automatic to the characteristics of each application and it does not require any offline profiling data. The DAG model solution as a low-complexity applied to ME blocks for H 265 compared with other approaches. Besides, our results for the H 265 video codec have demonstrated that our proposed low-complexity solution for the general DAG model reduces the execution time by up to 70 %.

The execution times computed theoretically are almost the same that are found in OVP. We have also shown how our solution efficiently adapts with respect to the DAG type, and scales well with the number of cores and the number of deadlines considered in the buffer. The design of the bus and the interface of SoC and MPSoC platforms based on Arm Cortex A 9 MP is also described, allowing direct integration of the IP cores on-chip communication used in MPSoC for H 265 video codec. In a future work, we will try to minimize other metrics in the H 265 video codec. We will use an automatic scheduling and partitioning algorithm DAG. We will also implement this work in a real target, based on the ARM Cortex A 9 MP, which is composed of four processors and named Embest SABER Lite, Target from Development SABER Lite-i.MX 6 Quad.

Acknowledgements

This research was financially supported by the "Laboratory of Technology for Smarts systems in Digital Research Center of Sfax "CRNS"", We would like to thank all the team for Pr. Ahmed FAKHFAKH, phd and master students, for his ambition and support in this project.

Author details

Afef Salhi[1*], Fahmi Ghozzi[1,2] and Ahmed Fakhfakh[1,2]

1 Digital Research Center of SFAX (CRNS), Laboratory of Signals, Systems, Artificial Intelligence and Networks (SM@RTS), Sfax, Tunisia

2 ENET'COM, University of Sfax, Sfax, Tunisia

*Address all correspondence to: salhiafefge@gmail.com; afef.salhi.ge@enis.tn

IntechOpen

References

[1] L. Rainer, S. Frank, M. Grant, K. Tim, P. Roman and V. Martin, *Virtual Platforms: Breaking New Grounds* DATE12/2012 EDAA.

[2] B. Elias, S. Hassan and N. Smail, *H.264 Color Components Video Decoding Parallelization on Multi-Core Processors* 2010 13th euro-micro conference on digital system design: Architectures, Methods and Tools.

[3] Ch. philipe, G.C. Edward, K.L Jr Jan and L. Zhen, *Scheduling Theory and its Applications* 1995 by Jhon Wiley ans Sons Ltd, England.

[4] M.A. Kordon, *A Graph-Based Analysis of the Cyclic Scheduling Problem with Time Constraints: Schedulability and Periodicity of the Earliest Schedule*, 3rd ed. Springer Science+Business Media, February 2010.

[5] B. Sebastien, K. Jabran, B. Ccile, K.B. Muhammad,*Effectiveness of power strategies for video applications: a practical study* J Real-Time Image Proc, IF 2, n.10, 2014, pp.1-10.

[6] E. Wajdi, D. Julien, M. Johel, A. Mohamed, *An efficient low-cost FPGA implementation of a configurable motion estimation for H.264 video coding* J Real-Time Image Proc, IF 2, n.10, 2012, pp.1-12.

[7] W. Thomas, G. Sullivan, G. Bjntegaard, A. Luthra,*Overview of the H.264/AVC Video Coding Standard* IEEE Tran. on Circuits and systems for video tech, Vol 13, n.7, 2003, pp.560-576.

[8] JVT and ITU-T, *Draft ITU-T recommendation and final draft international standard of joint video specification (ITU-T Rec. H 264–ISO/IEC 14496-10 AVC.)*

[9] A. Salhi, F. Ghozzi, and A. fakhfakh, *Toward a Methodology for Object Tracking System in computer Vision* GAMMMART

TUNISIA, TJASSST'2017, collogue Japonais-Tunisia, section "manegment and innovation", pp:185-189.

[10] A. Salhi, F. Ghozzi, and A. fakhfakh, and M.A. Kordon, *Video Codec Applications Scheduling and Optimization Based on DAG and GGEN Algorithm* 2018 2nd European Conference on Electrical Engineering and Computer Science (EECS), Bern-Suisse, November-2018, 2018-IEEE, pp: 236-241.

[11] A. Salhi, F. Ghozzi, and A. fakhfakh, *Scheduling Tasks in MPSoC System for Motion Estimation in H26x Video Codec* International Journal of Science and Engineering Investigations, Volume 7, Issue 74, April 2018,ISSN: 2251-8843, pp:72-77.

[12] A. Salhi, F. Ghozzi, and A. fakhfakh, *Modeling and Scheduling with DAG used for tasks in ME with OVP*, International Conference on Smart Applications, Communications and Networking, SmartNets 2019, Sharm El Sheik, Egypt, December 17-19, 2019, IEEE 2019, ISBN 978-1-7281-4275-3, pp:1-6.

[13] C. Hanen, *Study of a NP-hard cyclic scheduling problem: The recurrent job-shop*, 3rd ed. European Journal of Operational Research 72 (1994) 82-101 North-Holland.

[14] K. Kanoun, and M. van der SchaaK, *Big-Data Streaming Applications Scheduling with Online Learning and Concept Drift Detection*, 3rd ed. 978-3-9815370-4-8/DATE15/c 2015 EDAA, DATE15/IEEE, pp: 1547–1550.

[15] K. Kanoun, M. Nicholas, A. David, and M. van der Schaar, *Online Energy-Efficient Task-Graph Scheduling for Multicore Platforms*, 3rd ed. IEEE TRANSACTIONS ON COMPUTER-AIDED DESIGN OF INTEGRATED CIRCUITS AND

SYSTEMS, VOL. 33, NO. 8, AUGUST 2014, pp:1194–1204.

[16] R. Ben Atittallah, S. Niar, A. Grainier, *Estimating Energy Consumption for an MPSoC Architectural Exploration*, 3rd ed. International Conference on Architecture of Computing Systems, 298-310, 2011.

[17] M. Ruggiero, D. Bertozzi, L. Benini, M. Milano, and A. Andrei, *Reducing the abstraction and optimality gaps in the allocation and scheduling for variable voltage/frequency MPSoC platforms*, 3rd ed. IEEE Transactions Computer-Aided Design Integr. Circuits Syst, vol. 28, no. 3, pp. 378391, Mar. 2009.

[18] J. Luo and N. K. Jha, *Power-efficient scheduling for heterogeneous distributed real-time embedded systems*, 3rd ed. IEEE Trans. Computer-Aided Design Integr. Circuits Syst, vol. 26, no. 6, pp. 11611170, Jun. 2007.

[19] S. Heiko, D. Marpe and W. Thomas, *Overview of the Scalable Video Coding Extension of the H.264/AVC Standard*. IEEE Transactions On Circuits And Systems For Video Technology, VOL. 17, NO. 9, 2007.

[20] J. Rainer G.J. Sullivain, *High efficiency video coding: The next frontier in video compression*. IEEE Signal Processing Magazine, 2013.

[21] M. Eric, I. Lahoucine, N.C Marcian, B. Imene, T. Alin, and N. Mohamed Wissem, *fpgas in Industrial Control Applications*, 3rd ed. IEEE Transactions On Industrial Informatics, 2011.

[22] B. Sugandi, H. Kim, J.K. Tan, and S. Ishikawa, *A Block Matching Technique for Object Tracking Based on Peripheral Increment Sign Correlation Image* Image and Vision Computing, 2008.

[23] Open Virtual Platform and Imperas, *http://www.ovpworld.org* Imperas-OVP, 2008.

Chapter 5

Analytical Solutions of Some Strong Nonlinear Oscillators

Alvaro Humberto Salas and
Samir Abd El-Hakim El-Tantawy

Abstract

Oscillators are omnipresent; most of them are inherently nonlinear. Though a nonlinear equation mostly does not yield an exact analytic solution for itself, plethora of elementary yet practical techniques exist for extracting important information about the solution of equation. The purpose of this chapter is to introduce some new techniques for the readers which are carefully illustrated using mainly the examples of Duffing's oscillator. Using the exact analytical solution to cubic Duffing and cubic-quinbic Duffing oscillators, we describe the way other conservative and some non conservative damped nonlinear oscillators may be studied using analytical techniques described here. We do not make use of perturbation techniques. However, some comparison with such methods are performed. We consider oscillators having the form $\ddot{x} + f(x) = 0$ as well as $\ddot{x} + 2\varepsilon\dot{x} + f(x) = F(t)$, where $x = x(t)$ and $f = f(x)$ and $F(t)$ are continuous functions. In the present chapter, sometimes we will use $f(-x) = -f(x)$ and take the approximation $f(x) \approx \sum_{j=1}^{N} p_j x^j$, where $j = 1, 3, 5, \cdots N$ only odd integer values and $x \in [-A, A]$. Moreover, we will take the approximation $f(x) \approx \sum_{j=0}^{N} p_j x^j$, where $j = 1, 2, 3, \cdots N$, and $x \in [-A, A]$. Arbitrary initial conditions are considered. The main idea is to approximate the function $f = f(x)$ by means of some suitable cubic or quintic polynomial. The analytical solutions are expressed in terms of the Jacobian and Weierstrass elliptic functions. Applications to plasma physics, electronic circuits, soliton theory, and engineering are provided.

Keywords: Nonlinear second-order equation, Duffing equation, Cubic-quintic Duffing equation, Helmholtz oscillator, Duffing-Helmholtz oscillator, Mixed parity oscillator, Damped Duffing equation, Damped Helmholtz equation, Forced Duffing equation, Nonlinear electrical circuit, Solitons

1. Introduction

Both the ordinary and partial differential equations have an important role in explaining many phenomena that occur in nature or in medical engineering, bio-technology, economic, ocean, plasma physics, etc. [1, 2]. Duffing equation is considered one of the most important differential equations due to its ability for demonstrating the scenario and mechanism of various nonlinear phenomena that occur in nonlinear dynamic systems [3–11]. It is one of the most common models for analyzing and modeling many nonlinear phenomena in various fields of science such as the mechanical engineering [12], electrical engineering [13], plasma physics [14, 15], etc. Mathematically, the Duffing oscillator is a second-order ordinary differential equation with a nonlinear restoring force of odd power

IntechOpen

$$\begin{cases} \ddot{x} + f(x) = 0, \\ f(x) = \sum_{i=1}^{\infty} K_i x^{2i-1}, \end{cases} \tag{1}$$

where $f(-x) = -f(x)$ is a continuous function on some interval $[-A, A]$ with $f(0) = 0$, K_i is a physical coefficient related to the physical problem under study, and $i = 1, 2, 3, \cdots \infty$. It is clear from Eq. (1) that there is no any friction/dissipation (this force arises either as a result of taking viscosity into account or the collisions between the oscillator and any other particle, etc.), and this only occurs in standardized systems such as superfluid (fluid with zero viscosity which it flows without losing any part from its kinetic energy *sometimes* like Bose–Einstein condensation) or the systems isolated from all the external force that resist the motion of the oscillator. The undamped Duffing equation [9] is considered one of the effective and good models for explaining many nonlinear phenomena that are created and propagated in optical fiber, Ocean, water tank, the laboratory and space collisionless and warm plasma (we will demonstrate this point below). As well known in fluid mechanics and in the fluid theory of plasma physics; the basic fluid equations of any plasma model can be reduced to a diverse series of evolution equations that can describe all phenomena that create and propagate in these physical models. For example, we can mention some of the most famous evolution equations that have been used to explain several phenomena in plasma physics and other fields of sciences; the family of one dimensional $(1 - D)$ korteweg–de Vries equation (KdV) and it is higher-orders, including the KdV, KdV-Burgers (KdVB), modified KdV (mKdV), mKdV-Burgers (mKdVB), Gardner equation or called Extended KdV (EKdV), EKdV-Burgers (EKdVB), KdV-type equation with higher-order nonlinearity. All the above mentioned equations are partial differential equations and by using an appropriate transformation, we can convert them into ordinary differential equations of the second orders. If the frictional force is neglected, some of these equations can be converted into the undamped Duffing equation with $f(x) \approx P(x) = K_1 x + K_2 x^3$ like the mKdV equation, the KdV equation can be transformed to the undamped Helmholtz equation with $f(x) \approx P(x) = K_1 x + K_2 x^2$ [16], the Gardner equation can be converted into the undamped H-D equation for $f(x) \approx P(x) = K_1 x + K_2 x^2 + K_3 x^3$ [17, 18], and so on the other mentioned equations.

However, these undamped models (without friction/dissipation) do not exist much in reality except under harsh conditions. In order to describe and simulate the natural phenomena that arise in many realistic physical models and dynamic systems, the friction/dissipation forces must be taken into account, as is the case in many plasma models and electronic systems. Accordingly, the following damped (non-conservative) Duffing equation will be devoted for this purpose

$$\ddot{x} + 2\varepsilon \dot{x} + f(x) = 0. \tag{2}$$

If the frictional force does not neglect, so that all PDEs that have "Burgers $\equiv \partial_x^2(\cdot)$" term like KdVB-, mKdVB-, EKdVB-, KPB-, mKPB-, EKPB-, ZKB-, mZKB, EZKB-Eq. [1, 2], etc. can be transformed to damped Duffing equation $(\ddot{x} + 2\varepsilon \dot{x} + K_1 x + K_2 x^3 = 0)$, damped Helmholtz equation $(\ddot{x} + 2\varepsilon \dot{x} + K_1 x + K_2 x^2 = 0)$, and damped Duffing-Helmholtz equation $(\ddot{x} + 2\varepsilon \dot{x} + K_1 x + K_2 x^2 + K_2 x^3 = 0)$. Eq. (2) without [19] and with [7, 20, 21] including damping term $(2\varepsilon \dot{x})$ for $f(x) \approx P(x) = K_1 x + K_2 x^3$ has been investigated and solved analytically and numerically by many authors using different approaches in order to understand its physical characters [22–28].

Many authors investigated the (un)damped Duffing equation, (un)damped Helmholtz Eq. [16, 29–31], and undamped H-D equation. On the contrary, there is a few numbers of published papers about damped Duffing-Helmholtz equation [32, 33]. For example, Zúñiga [32] derived a semi-analytical solution to the damped Duffing-Helmholtz equation in the form of Jacobian elliptic functions, but he putted some restrictions on the coefficient of the linear term, and then obtained a solution that gives good results compared to numerical solutions. Also, it is noticed that Zúñiga solution [32] is very sensitive to the initial conditions. Gusso and Pimentel [33] obtained obtain improved approximate analytical solution to the forced and damped Duffing-Helmholtz in the form of a truncated Fourier series utilizing the harmonic balance method.

In this chapter, we display some novel semi-analytical (approximate analytical) solutions to the strong higher-order nonlinear damped oscillators of the following initial value problem (i.v.p)

$$\begin{cases} \ddot{x} + 2\varepsilon\,\dot{x} + px + qx^3 + rx^5 = \mathcal{F}(t), \\ \quad x(0) = x_0\,\&\,x'(0) = \dot{x}_0, \end{cases} \tag{3}$$

and its family ($\varepsilon = 0$ or $r = 0$ or $\varepsilon = r = 0$).

Our new semi-analytical solution to Eq. (3) is derived in terms of Weierstrass and Jacobian elliptic functions. Also, we will solve Eq. (3) numerically using Runge–Kutta 4th (RK4) and make a comparison between both the semi-analytical and numerical solutions. Moreover, as some realistic physical application to the problem (3) and its family will be investigated.

2. Duffing equation

Let us consider the standard (undamping) Duffing equation in the absence both friction ($2\varepsilon\,\dot{x}$) and excitation ($\mathcal{F}(t)$) forces [34, 35]

$$\ddot{x} + px + qx^3 = 0, x = x(t), \tag{4}$$

which is subjected to the following initial conditions

$$x(0) = x_0\,\&\,x'(0) = \dot{x}_0. \tag{5}$$

The general solution of Eq. (4) maybe written in terms of any of the twelve Jacobian elliptic functions.

For example, let us assume

$$x(t) = c_1\mathrm{cn}\left(\sqrt{\omega}t + c_2, m\right). \tag{6}$$

By inserting solution (6) in Eq. (4), we get

$$\ddot{x} + px + qx^3 = \left(c_1^3 q - 2c_1 m\omega\right)\mathrm{cn}^3 + (2c_1 m\omega + c_1 p - c_1\omega)\mathrm{cn}, \tag{7}$$

where $\mathrm{cn} = \mathrm{cn}\left(\sqrt{\omega}t + c_2, m\right)$.

Equating to zero the coefficients of $\mathrm{cn}^{\,j}$ gives an algebraic system whose solution gives

$$\omega = \sqrt{p + qc_1^2} \text{ and } m = \frac{qc_1^2}{2\left(p + qc_1^2\right)}. \tag{8}$$

Thus, the general solution of Eq. (4) reads

$$x(t) = \text{cn}\left(\sqrt{p + qc_1^2}\,t + c_2,\; \frac{qc_1^2}{2(p + qc_1^2)}\right).$$
(9)

The values of the constants c_1 and c_2 could be determined from the initial conditions given in Eq. (5).

Definition 1. The number $\Delta = (p + qx_0^2)^2 + 2q\dot{x}_0^2$ is called the discriminant of the i.v.p. (4) -(5). Below three cases will be discussed depending on the sign of the discriminant Δ.

2.1 First case: $\Delta > 0$

For $\Delta > 0$, the solution of the i.v.p. (4)-(5) is given by

$$x(t) = \sqrt{\frac{\sqrt{\Delta} - p}{q}}\,\text{cn}\left(\sqrt[4]{\Delta}\,t - \text{sign}(\dot{x}_0)\text{cn}^{-1}\left(\sqrt{\frac{q}{\sqrt{\Delta} - p}}x_0, \frac{1}{2} - \frac{p}{2\sqrt{\Delta}}\right), \frac{1}{2} - \frac{p}{2\sqrt{\Delta}}\right).$$
(10)

Making use of the additional formula

$$\text{cn}(x + y, m) = \frac{\text{cn}(x, m)\text{cn}(y, m) + \text{sn}(x, m)\text{dn}(x, m)\text{sn}(y, m)\text{dn}(y, m)}{1 - m\,\text{sn}(x, m)\text{sn}(y, m)},$$
(11)

the solution (10) could be expressed as

$$x(t) = \frac{x_0\text{cn}(\sqrt{\omega}t|m) + \frac{\dot{x}_0}{\sqrt{\omega}}\text{dn}(\sqrt{\omega}t|m)\text{sn}(\sqrt{\omega}t|m)}{1 + \frac{p + qx_0^2 - \omega}{2\sqrt{\Delta}}\text{sn}(\sqrt[4]{\Delta}t|m)^2},$$
(12)

where

$$m = \frac{1}{2}\left(1 + \frac{p}{\sqrt{\Delta}}\right) \text{ and } \omega = \sqrt[4]{\Delta}.$$
(13)

Solution (12) is a periodic solution with period

$$T = 4\left|\frac{K(m)}{\sqrt{\omega}}\right|.$$
(14)

Example 1.
Let us consider the i.v.p.

$$\begin{cases} x''(t) + x(t) + x^3(t) = 0, \\ x(0) = 1\,\&\,x'(0) = -1. \end{cases}$$
(15)

Using formula (10), the exact solution of the i.v.p. (15) reads

$$x(t) = -\sqrt{\sqrt{6} - 1}\,\text{cn}\left(\sqrt[4]{6}t + \text{cn}^{-1}\left(-\frac{1}{\sqrt{\sqrt{6} - 1}}, \frac{1}{2} - \frac{1}{2\sqrt{6}}\right), \frac{1}{12}(6 - \sqrt{6})\right).$$
(16)

According to the relation (12)-(13), the exact solution of the i.v.p. (15) is also written as

$$x(t) = \frac{2\sqrt[4]{6}\,\mathrm{dn}\left(\sqrt[4]{6}t\,\middle|\,\frac{1}{12}\left(6-\sqrt{6}\right)\right)\mathrm{sn}\left(\sqrt[4]{6}t\,\middle|\,\frac{1}{12}\left(6-\sqrt{6}\right)\right) - 2\sqrt{6}\,\mathrm{cn}\left(\sqrt[4]{6}t\,\middle|\,\frac{1}{12}\left(6-\sqrt{6}\right)\right)}{\left(\sqrt{6}-2\right)\mathrm{sn}\left(\sqrt[4]{6}t\,\middle|\,\frac{1}{12}\left(6-\sqrt{6}\right)\right)^2 - 2\sqrt{6}},$$

$$(17)$$

and its periodicity is given by

$$T = \frac{4K\left(\frac{1}{12}\left(6-\sqrt{6}\right)\right)}{\sqrt[4]{6}} \approx 3.27458.$$

In **Figure 1**, the comparison between the exact analytical solution (17) and the approximate numerical RK4 solution is presented. Full compatibility between the two analytical and numerical solutions is observed.

2.2 Second case: $\Delta < 0$

For $\Delta < 0$, in this case $q < 0$ and then, $\delta = \frac{p^2 - \Delta}{-q} > 0$, $\delta \overset{\text{def}}{=} \left(2p + qx_0^2\right)x_0^2 + 2\dot{x}_0^2 > 0$. Let us introduce the solution in the following form

$$x(t) = A - \frac{2A}{1 + y(t)}, \tag{18}$$

where $y = y(t)$ is a solution of some Duffing equation

$$y''(t) + my(t) + ny^3(t), \tag{19}$$

with initial conditions

$$y(0) = y_0 = \frac{2A\dot{x}_0}{(A - x_0)^2}$$

$$y'(0) = \dot{y}_0 = \frac{A + x_0}{A - x_0}. \tag{20}$$

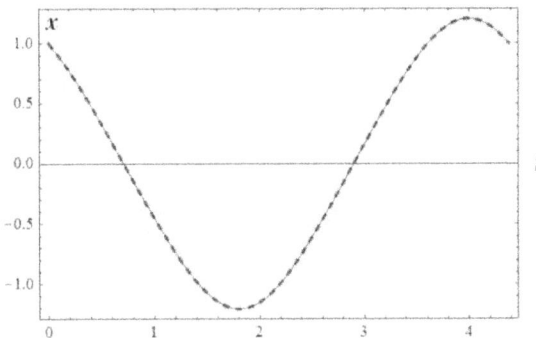

Figure 1.
A comparison between the analytical solution (17) and the approximate numerical RK4.

Inserting ansatz (18) into Eq. (4) and taking the below relation into account

$$y'(t)^2 = \dot{y}_0 + m y_0^2 + \frac{n}{2} y_0^4 - m y^2(t) - \frac{n}{2} y^4(t), \tag{21}$$

we get

$$
\begin{aligned}
&-A\left(A^2 q + 4 m y_0^2 + 2 n y_0^4 + p + 4\dot{y}_0^2\right) \\
&+A\left(3A^2 q - 2m - p\right) y(t) \\
&-A\left(3A^2 q - 2m - p\right) y(t)^2 \\
&+A\left(A^2 q - 2n + p\right) y(t)^3 = 0.
\end{aligned}
\tag{22}
$$

Equating the coefficients of $y^j(t)$ to zero, gives an algebraic system. A solution to this system gives

$$m = \frac{1}{2}\left(-p + 3A^2 q\right), n = \frac{1}{2}\left(p + A^2 q\right),$$

$$A = \sqrt[4]{\frac{(2p + q x_0^2) x_0^2 + 2\dot{x}_0^2}{-q}} = \sqrt[4]{\frac{\delta}{-q}}.$$

Note that the i.v.p. (19)-(20) has a positive discriminant and it is given by

$$\left(m + n y_0^2\right)^2 + 2n\,\dot{y}_0^2 = \frac{\delta(A - x_0)^4 \left(2A^4 \dot{x}_0^2 + \delta\left(A^2 + x_0^2\right)^2\right)}{4A^8 x_0^2}.$$

Then the problem reduces to the first case. Accordingly, the solution of the i.v.p. (4)-(5) maybe written in the form,

$$x(t) = A - \frac{2A}{1 + B\frac{b_0 \mathrm{cn}(\sqrt{\omega t}|m) + b_1 \mathrm{sn}(\sqrt{\omega t}|m)\mathrm{dn}(\sqrt{\omega t}|m)}{1 + b_2 \mathrm{sn}^2(\sqrt{\omega t}|m)}}, \tag{23}$$

where

$$m = \frac{B^2\left(A^2 q + p\right)}{2A^2\left(B^2 + 3\right)q + 2\left(B^2 - 1\right)p},$$

$$\omega = \frac{1}{2}\left(A^2\left(B^2 + 3\right)q + \left(B^2 - 1\right)p\right),$$

$$b_0 = \frac{A + x_0}{AB - Bx_0}, b_1 = \frac{2A\dot{x}_0}{B\sqrt{\omega}(A - x_0)^2},$$

$$b_2 = -\frac{2A x_0(x_0 - A)(p + q x_0^2) + \omega(A + x_0)(A - x_0)^2 + 4A\dot{x}_0^2}{2\omega(A - x_0)^2(A + x_0)}, \tag{24}$$

$$A = \sqrt[4]{\frac{(2p + q x_0^2) x_0^2 + 2\dot{x}_0^2}{-q}} = \sqrt[4]{\frac{\delta}{-q}},$$

$$B = \sqrt{\frac{A\left(2\sqrt{2}\sqrt{q(A^2 q - p)} - 3Aq\right) + p}{A^2 q + p}}.$$

The solution (23) is unbounded and its periodicity is given by

$$T = \left| \frac{4K(m)}{\sqrt{\omega}} \right| = \left| \frac{4K(1-m)}{m\sqrt{\omega}} \right|.$$ (25)

Example 2.

Let us assume the following i.v.p.

$$\begin{cases} x''(t) + x(t) - x^3(t) = 0, \\ x(0) = -1 \,\&\, x'(0) = -1. \end{cases}$$ (26)

The solution of the i.v.p. (26) according to the relation (23) reads

$$x(t) = 1.31607 - \frac{2.63215}{1 + \frac{0.13647\text{cn}(1.75396t|1.00353) - 0.27976\text{dn}(1.75396t|1.00353)\text{sn}(1.75396t|1.00353)}{1. - 1.00463\text{sn}(1.75396t|1.00353)^2}},$$ (27)

and the periodicity of this solution is given by

$$T = 9.57783.$$ (28)

Solution (27) is displayed in **Figure 2**.

2.3 Third case: $\Delta = 0$

If the discriminant vanishes ($\Delta = 0$), then $q < 0$ and the only solution of problem (4) with

$$x'(0)^2 = \dot{x}_0^2 = \frac{(p + qy_0^2)^2}{-2q},$$ (29)

reads

$$x(t) = \sqrt{-\frac{p}{q}} \tanh \left[\sqrt{\frac{p}{2}} t \pm \tanh^{-1} \left(x_0 \sqrt{-\frac{q}{p}} \right) \right].$$ (30)

which may be verified by direct computation.

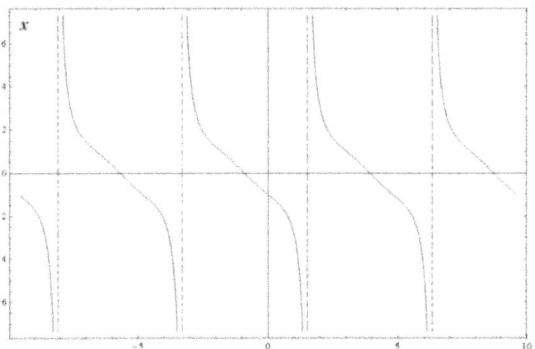

Figure 2.
The profile of solution (27) is plotted against t.

Remark 1. The solution of the i.v.p.

$$\begin{cases} \ddot{x} + px + qx^3 = 0, \\ x(0) = x_0 \ \& \ x'(0) = 0, \end{cases}$$

(31)

is given by

$$x(t) = x_0 \mathrm{cn}\left(\sqrt{p + qx_0^2}, \frac{qx_0^2}{2(p + qx_0^2)} \right).$$

(32)

Remark 2. For $p + \sqrt{p^2 + 2q}\, x\_0^{\,2} > 0$, then the solution of the i.v.p.

$$\begin{cases} \ddot{x} + px + qx^3 = 0, \\ x(0) = 0 \ \& \ x'(0) = \dot{x}_0. \end{cases}$$

(33)

is given by

$$x(t) = \frac{\sqrt{2}\dot{x}_0}{\sqrt{\sqrt{p^2 + 2q\dot{x}_0^2} + p}} \mathrm{sn}\left(\sqrt{\frac{p + \sqrt{p^2 + 2q\dot{x}_0^2}}{2}} t, \ -\frac{p^2 + q\dot{x}_0^2 - \sqrt{p^2 + 2q\dot{x}_0^2}\,p}{q\dot{x}_0^2} \right).$$

(34)

Remark 3. According to the following identity

$$\mathrm{cn}\left(\sqrt{\omega}t, m \right) = 1 - \frac{S_0}{1 + S_1\wp(t; g_2, g_3)},$$

(35)

with

$$S_0 = \frac{6}{(4m + 1)}, S_1 = \frac{12}{(4m + 1)\omega},$$

$$g_2 = \frac{1}{12}(16m^2 - 16m + 1)\omega^2,$$

$$g_3 = \frac{1}{216}(2m - 1)(32m^2 - 32m - 1)\omega^3,$$

the solution of the i.v.p. (4)-(5) could be written in terms of the Weierstrass elliptic function $\wp \equiv \wp(t; g_2, g_3)$. More precisely, if $\Delta > 0$ then

$$x(t) = A - \frac{A\left(\frac{4p}{3A^2q + p} + 2 \right)}{1 + \frac{12}{3A^2q + p}\wp(t + t_0; g_2, g_3)},$$

(36)

with

$$t_0 = \wp^{-1}\left(\frac{3A^3q + 3A^2qx_0 + 5Ap + px_0}{12(A - x_0)}; g_2, g_3 \right),$$

$$g_2 = \frac{1}{12}(-3A^4q^2 - 6A^2pq + p^2),$$

(37)

$$g_3 = \frac{p}{216}(9A^4q^2 + 18A^2pq + p^2),$$

and

$$A = \sqrt{\frac{-p \pm \sqrt{(p + qx_0^2)^2 + 2q\dot{x}_0^2}}{q}} = \pm\sqrt{\frac{-p \pm \sqrt{\Delta}}{q}}. \tag{38}$$

The solution (36) is periodic with period

$$T = 2\int_\rho^{+\infty} \frac{dx}{\sqrt{4x^3 - g_2 x - g_3}}, \tag{39}$$

where ρ is the greatest real root of the cubic $4x^3 - g_2 x - g_3 = 0$.

Remark 4. An approximate analytic solution of the i.v.p. (31) is given by

$$x(t) = \frac{x_0\sqrt{1 + \lambda}\cos{(wt)}}{\sqrt{1 + \lambda\cos^2{(wt)}}}, \tag{40}$$

where

$$w = \frac{1}{2}\sqrt{\frac{5(p + qx_0^2)\lambda^2 + (12p + 11qx_0^2)\lambda + 8p + 6qx_0^2}{3\lambda + 2}} \tag{41}$$

and λ is a root of the cubic

$$25(p + qx_0^2)\lambda^3 + (58p + 59qx_0^2)\lambda^2 + 2(16p + 21qx_0^2)\lambda + 8qx_0^2 = 0. \tag{42}$$

Example 3.
Let us consider the i.v.p.

$$\begin{cases} \ddot{x} + x + 10x^3 = 0, \\ x(0) = 4, x'(0) = 0. \end{cases} \tag{43}$$

The approximate solution in trigonometric form is given by

$$x_{app}(t) = \frac{3.34603\cos{(10.7542t)}}{\sqrt{1 - 0.300255\cos^2(10.7542t)}}. \tag{44}$$

The exact solution reads

$$x(t) = 4\text{cn}\left(\sqrt{161}t \Big| \frac{80}{161}\right), \tag{45}$$

with period

$$T = \frac{4K\left(\frac{80}{161}\right)}{\sqrt{161}}.$$

The error on the interval $0 \le t \le T$ equals 0.025.

The comparison between the approximate analytic solution (44) and the exact analytic solution (45) is illustrated in **Figure 3**.

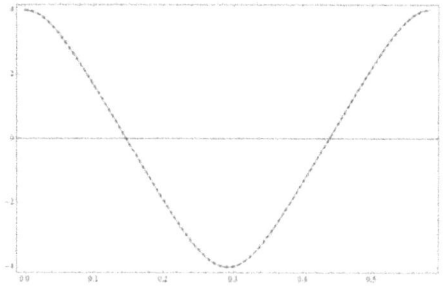

Figure 3.
A comparison between the approximate solution (44) and exact solution (45).

Remark 5. An approximate analytical solution of the i.v.p. (33) is given by

$$x_{app}(t) = \frac{\dot{x}_0 \, \sin\left(\sqrt{\omega}t\right)}{\sqrt{\omega}\sqrt{1 + \lambda \sin^2\left(\sqrt{\omega}t\right)}},$$ (46)

where

$$\omega = -\frac{\sqrt{\lambda^2\left(64p^2 - 160q\dot{x}_0^2\right) + 25p^2\lambda^4 + 80p^2\lambda^3 - 128q\dot{x}_0^2\lambda + p\lambda(5\lambda + 8)}}{16\lambda}$$ (47)

and λ is a solution of the quintic

$$125p^2\lambda^5 + 10\left(79p^2 + 125q\dot{x}_0^2\right)\lambda^4 + 40\left(43p^2 + 85q\dot{x}_0^2\right)\lambda^3 +$$
$$8\left(196p^2 + 389q\dot{x}_0^2\right)\lambda^2 + 64\left(8p^2 + 17q\dot{x}_0^2\right)\lambda + 128q\dot{x}_0^2 = 0$$ (48)

Example 4.
The approximate trigonometric solution of

$$\begin{cases} \ddot{x} + 3x + 5x^3 = 0, \\ x(0) = 0, x'(0) = 1. \end{cases}$$ (49)

reads

$$x_{app}(t) = \frac{0.499502 \sin\left(2.00199t\right)}{\sqrt{1 - 0.0817025 \sin^2(2.00199t)}}.$$ (50)

The exact solution is

$$x(t) = \sqrt{\frac{2}{3 + \sqrt{19}}} \mathrm{sn}\left(\sqrt{\frac{1}{2}\left(3 + \sqrt{19}\right)}t, \frac{1}{5}\left(-14 + 3\sqrt{19}\right)\right),$$ (51)

with period

$$T = 4\sqrt{\frac{1}{5}\left(\sqrt{19} - 3\right)}K\left(\frac{1}{5}\left(-14 + 3\sqrt{19}\right)\right) = 3.1383.$$ (52)

The error on the interval $0 \leq t \leq T$ equals 0.00018291.

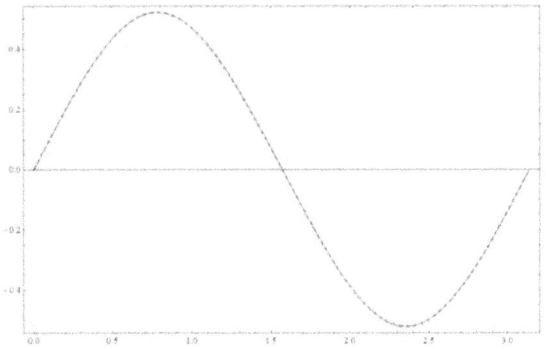

Figure 4.
A comparison between the approximate solution (50) and exact solution (51).

Figure 4 demonstrates the comparison between the approximate analytic solution (50) and the exact analytic solution.

3. An analytical solution of the undamped Duffing-Helmholtz Equation

The undamped Duffing-Helmholtz equation reads

$$\begin{cases} \ddot{x} + px + qx^2 + rx^3 = 0, \\ x(0) = x_0 \text{ and } x'(0) = \dot{x}_0. \end{cases} \tag{53}$$

We will give a solution to the i.v.p. (53) in terms of Weierstrass elliptic functions. For solving this problem the following ansatz is considered

$$x(t) = A + \frac{B}{1 + C\wp(t + t_0; g_2, g_3)}, \tag{54}$$

where $BC \neq 0$.

Substituting the ansatz (54) into the ordinary differential equation (ode) $\mathbb{R} \equiv \ddot{x} + px + qx^2 + rx^3 = 0$, gives

$$\frac{1}{2(1 + C\wp^3)} \sum_{j=0}^{3} K_j \wp^j = 0, \tag{55}$$

with

$K_3 = 2C^2 \left(A^3 Cr + A^2 Cq + ACp + 2B \right),$

$K_2 = 2C \left(3A^3 Cr + 3A^2 BCr + 3A^2 Cq + 2ABCq + 3ACp + BCp - 6B \right),$

$K_1 = C \left(6A^3 r + 12A^2 Br + 6A^2 q + 6AB^2 r + 8ABq + 6Ap + 2B^2 q - 3BCg_2 + 4Bp \right),$

$K_0 = A^3 r + 6A^2 Br + 2A^2 q + 6AB^2 r + 4ABq + 2Ap + 2B^3 r + 2B^2 q - 4BC^2 g_3 + BCg_2 + 2Bp.$

Equating the coefficients K_j to zero will give us an algebraic system. Solving this system, we finally get

$$B = -\frac{6A(A^2r + Aq + p)}{3A^2r + 2Aq + p}, C = \frac{12}{3A^2r + 2Aq + p},$$

$$g_2 = -\frac{1}{12}(3r^2A^4 + 4qrA^3 + 6prA^2 - p^2), \tag{56}$$

$$g_3 = \frac{1}{216}[(9pr^2 - 3q^2r)A^4 + (12pqr - 4q^3)A^3 + (18p^2r - 6pq^2)A^2 + p^3],$$

The values of t_0 and A could be determined from the initial conditions $x(0) = x_0$ and $x'(0) = \dot{x}_0$ and

$$\ddot{x}(0) + px(0) + qx^2(0) + rx^3(0) = 0. \tag{57}$$

We have

$$t_0 = \pm\wp^{-1}\left(\frac{x_0 - A - B}{C(A - x_0)}; g_2, g_3\right). \tag{58}$$

The number A is a solution to the quartic

$$3rA^4 + 4qA^2 + 6pA - (3rx_0^4 + 4qx_0^3 + 6px_0^2 + 6\dot{x}_0^2) = 0. \tag{59}$$

Example 5.
The solution of the i.v.p.

$$\begin{cases} \ddot{x} + x + 2x^2 + 3x^3 = 0, \\ x(0) = 1 \text{ and } x'(0) = 1, \end{cases} \tag{60}$$

according to the relation (54) is given by

$$x(t) = 1.07627 - \frac{2.72078}{1 + 0.762858\wp(t - 0.148317; -7.16667, 0.675926)}. \tag{61}$$

In **Figure 5**, the comparison with the approximate analytic solution (61) and the approximate numerical solution using RK4 is investigated.
The periodicity of solution (61) is given by

$$T = 2\int_{0.0938538}^{\infty} \frac{1}{\sqrt{4x^3 + 7.16667x - 0.675926}} \, dx = 3.12129.$$

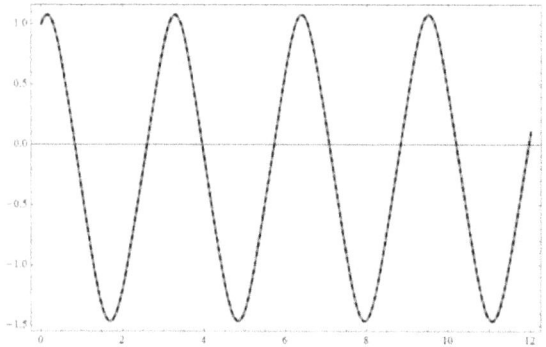

Figure 5.
A comparison between solution (61) and the approximate numerical solution using RK4.

4. The solution of the forced undamped Duffing-Helmholtz equation

Suppose that the physical system to be studied is under the influence of some constant external/excitation force, so the standard Duffing-Helmholtz equation can be reformulated to the following constant forced Duffing-Helmholtz i.v.p.

$$\begin{cases} \ddot{x} + px + qx^2 + rx^3 = F, \\ x(0) = x_0 \,\text{and}\, x'(0) = \dot{x}_0. \end{cases} \tag{62}$$

For solving the i.v.p. (62), the following assumption is introduced

$$x(t) = y(t) + \zeta, \tag{63}$$

where ζ is a solution to the cubic algebraic equation

$$r\zeta^3 + q\zeta^2 + p\zeta - F = 0. \tag{64}$$

Substituting Eq. (63) into the i.v.p. (62), we have

$$y''(t) + (p + 2q\zeta + 3r\zeta^2)y(t) + (q + 3r\zeta)y(t)^2 + ry(t)^3 = 0. \tag{65}$$

Note that the constant forced Duffing-Helmholtz Eq. (62) has been reduced to the standard Duffing-Helmholtz Eq. (65) with the following new initial conditions

$$y(0) = x_0 - \zeta \,\&\, y'(0) = \dot{x}_0. \tag{66}$$

Example 6.
Suppose that we have the following i.v.p. and we want to solve it

$$\begin{cases} \ddot{v} + 2v - 12v^2 + v^3 = 4, \\ v(0) = 1 \,\&\, v'(0) = 1. \end{cases} \tag{67}$$

It is clear that the i.v.p. (67) is a constant forced Duffing-Helmholtz equation. The solution of this problem is given by

$$v(t) = 15.8046 - \frac{15.7714}{1 + 0.0322539\wp(0.761045 - t; -9.41667, 47.287)}. \tag{68}$$

The comparison between the solution (68) and the RK4 solution is introduced in **Figure 6**.

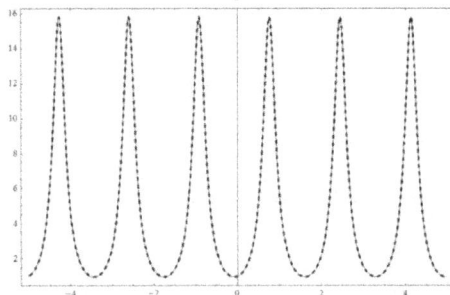

Figure 6.
A comparison between the solution (68) and the RK4 solution.

The periodicity of solution (68) is given by

$$T = 2\int_{1.93657}^{\infty} \frac{1}{\sqrt{4x^3 + 9.41667x - 47.287}}\, dx = 1.68202.$$

5. An approximate analytic solution of the forced damped Duffing-Helmholtz equation

Let us define the following i.v.p.

$$\begin{cases} \ddot{x} + 2\varepsilon\,\dot{x} + px + qx^2 + rx^3 = F, \\ x(0) = x_0 \,\&\, x'(0) = \dot{x}_0. \end{cases} \tag{69}$$

Suppose that

$$\lim_{x \to +\infty} x(t) = d, \varepsilon > 0, \tag{70}$$

then the first equation in system (69) can be written as

$$pd + qd^2 + rd^3 = F. \tag{71}$$

For solving the i.v.p. (69), the following ansatz is assumed

$$x(t) = \exp(-pt)y(\,f(t)), \tag{72}$$

with

$$f(t) = \frac{1 - \exp(-2(\varepsilon - p)t)}{2(\varepsilon - p)}, \tag{73}$$

where the function $y \equiv y(t)$ represents the exact solution to the following i.v.p.

$$\begin{cases} y''(t) + (3d^2r + 2dq - 2\varepsilon p + p + p^2)y(t) + (3dr + q)y(t)^2 + ry(t)^3 = 0, \\ y(0) = x_0 - d \,\&\, y'(0) = \dot{x}_0 + p(x_0 - d). \end{cases} \tag{74}$$

Let us define the following residual

$$R(t) \equiv \ddot{x}(t) + 2\varepsilon\,\dot{x}(t) + px(t) + qx^2(t) + rx^3(t) - F, \tag{75}$$

and by applying the condition $R'(0) = 0$, we obtain

$$4p^3 - 12\varepsilon p^2 + (3d^2r + 3dq + 3drx_0 + 8\varepsilon^2 + 4p + 5qx_0 + 6rx_0^2)p$$
$$-4\varepsilon(d^2r + dq + drx_0 + p + qx_0 + rx_0^2) = 0. \tag{76}$$

By solving this equation we can get the value of p.

Example 7.

Let

$$\begin{cases} \ddot{x} + 0.02\dot{x} + 5x + 2x^2 + x^3 = 1/2, \\ x(0) = 0.1 \,\&\, x'(0) = 0.1. \end{cases} \tag{77}$$

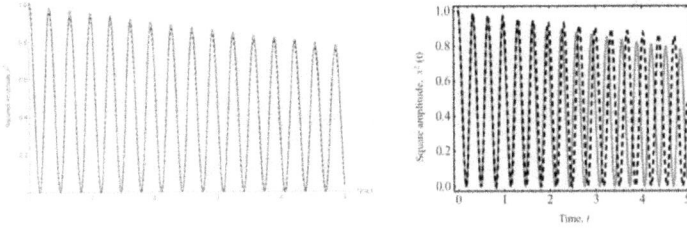

Figure 7.
A comparison between solution (78) and RK4 solution.

The approximate analytic solution of the i.v.p. (77) reads

$$x_{\text{app}}(t) = 0.0961263$$

$$+ e^{-0.0099959t}\left(-\frac{0.0429135}{1 + 2.13749\wp\left(0.631364 - 121944.\left(1 - e^{-8.2 \times 10^{-6}t}\right); 2.43588, 0.737005\right)} \right).$$

$$(78)$$

The distance error as compared to the RK4 numerical solution is given by

$$\max_{0 \le t \le 5} |x_{\text{app}}(t) - x_{\text{RK4}}(t)| = 0.000944148. \tag{79}$$

Also, the comparison between solution (78) and RK4 solution is presented in **Figure 7**.

Remark 5. For the damped and constant forced Helmholtz equation

$$\begin{cases} \ddot{x} + 2\varepsilon\,\dot{x} + px + qx^2 = F, \\ x(0) = x_0 \,\&\, x'(0) = \dot{x}_0. \end{cases} \tag{80}$$

The value of d can be determined from: $pd + qd^2 = F$. However, if this equation has no real solutions we can choose $d = 0$.

Remark 6. Letting $q = 0$, we obtain the damped and constant forced Duffing equation

$$\begin{cases} \ddot{x} + 2\varepsilon\,\dot{x} + px + rx^3 = F, \\ x(0) = x_0 \,\&\, x'(0) = \dot{x}_0. \end{cases} \tag{81}$$

In this case, the number d must be a root to the cubic $pd + rd^3 = F$.

6. Approximate analytic solution of the damped and trigonometric forced Duffing-Helmholtz equation

Let us define the following new i.v.p.

$$\begin{cases} \ddot{x} + 2\varepsilon\,\dot{x} + px + qx^2 + rx^3 = F\cos(\omega t), \\ x(0) = x_0 \,\&\, x'(0) = \dot{x}_0. \end{cases} \tag{82}$$

We suppose that $q^2 - 4pr < 0$, and the following residual is defined

$$R(t) \equiv \ddot{x}(t) + 2\varepsilon \dot{x}(t) + px(t) + qx^2(t) + rx^3(t) - F\cos(\omega t). \tag{83}$$

Let us define the solution of i.v.p. (82) as follows

$$x(t) = \exp(-\rho t)y(t) + c_1\cos(\omega t) + c_2\sin(\omega t), \tag{84}$$

where

$$9F^2r^2c_1^3 + 96\varepsilon^2Fr\omega^2c_1^2 +$$
$$4(64\varepsilon^4\omega^4 + 16\varepsilon^2\omega^6 + 3F^2pr - 3F^2r\omega^2 + 16\varepsilon^2p^2\omega^2 - 32\varepsilon^2p\omega^4)c_1 \tag{85}$$
$$-4F(-16\varepsilon^2\omega^4 + 3F^2r + 16\varepsilon^2p\omega^2) = 0.$$

$$-6144\varepsilon^3Fr\omega^3 + 432F^2r^3c_2^3 + 2304\varepsilon Fr^2\omega(p - \omega^2)c_2^2$$
$$+3072\varepsilon^2r\omega^2(4\varepsilon^2\omega^2 + p^2 - 2p\omega^2 + \omega^4)c_2 = 0. \tag{86}$$

The function $y \equiv y(t)$ is a solution to the i.v.p.

$$\begin{cases} y''(t) + 2\varepsilon y'(t) + \tilde{p}y(t) + qy(t)^2 + ry(t)^3 = 0, \\ y(0) = (x_0 - c_1) \,\&\, y'(0) = (\dot{x}_0 - \omega c_2). \end{cases} \tag{87}$$

where $\tilde{p} = \frac{1}{2}(2p + 3rc_1^2 + 3rc_2^2 - 4\varepsilon\rho + 2\rho^2)$.
The value of ρ can be determined from the following equation

$$4\rho^3 - 12\varepsilon\rho^2 + (4p + 8\varepsilon^2 - 5qc_1 + 12rc_1^2 + 6rc_2^2 + 5qx_0 - 12rc_1x_0 + 6rx_0^2)\rho \tag{88}$$
$$-2\varepsilon(2p - 2qc_1 + 5rc_1^2 + 3rc_2^2 + 2qx_0 - 4rc_1x_0 + 2rx_0^2) = 0.$$

Example 8.
Let

$$\begin{cases} \ddot{x} + 0.2\,\dot{x} + 13x + x^2 + x^3 = 0.25\cos(0.5t), \\ x(0) = 0 \,\&\, x'(0) = -0.2. \end{cases} \tag{89}$$

The approximate analytic solution of the i.v.p. (89) is given by

$$x_{app}(t) = e^{-0.100036t}\left(\frac{0.0587764 - \dfrac{0.350893}{1 + 0.914739\wp(0.529144 - 13766.9(1 - e^{0.0000726379t}); 14.0404, 10.1967)}}{}\right)$$
$$+0.000153771\sin(0.5t) + 0.0196062\cos(0.5t) \tag{90}$$

The distance error according to the RK4 numerical solution is calculated as

$$\max_{0 \le t \le 60}|x_{app}(t) - x_{RK4}(t)| = 0.000671928. \tag{91}$$

Moreover, solution (90) is compared with RK4 solution as shown in **Figure 8**.

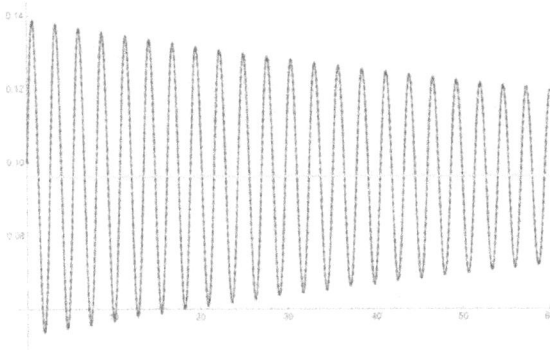

Figure 8.
A comparison between solution (90) and RK4 solution.

7. An analytic solution of cubic-quintic Duffing equation

Let us consider the following ordinary differential equation [36]

$$\ddot{x} + ax + \beta x^3 + \gamma x^5 = 0, x = x(t), \tag{92}$$

which is subjected to the following initial conditions

$$x(0) = x_0 \text{ and} x'(0) = \dot{x}_0. \tag{93}$$

Theorem 1.
a. Suppose that $x_0 \neq 0$, then the solution of the i.v.p. (92)-(93) is given by

$$x(t) = x_0 \frac{\sqrt{1 + \lambda v(t)}}{\sqrt{1 + \lambda v^2(t)}}, \tag{94}$$

where the function $v \equiv v(t)$ is the solution to the following Duffing equation

$$\begin{cases} \ddot{v} + pv + qv^3 = 0, \\ v(0) := v_0 = 1, \\ v'(0) := \dot{v}_0 = \dfrac{\dot{x}_0}{x_0}(1 + \lambda). \end{cases} \tag{95}$$

The values of the coefficients p and q are given by

$$p = -\frac{\lambda^2(4a + 3\beta x_0^2 + 2\gamma x_0^4) + \lambda(3\beta x_0^2 + 4\gamma x_0^4) + 2\gamma x_0^4}{2\lambda^2}, \tag{96}$$

$$q = -\frac{2(\lambda^2(a + \beta x_0^2 + \gamma x_0^4) + \lambda(\beta x_0^2 + 2\gamma x_0^4) + \gamma x_0^4)}{\lambda}, \tag{97}$$

and the value of the quantity λ is a solution of the cubic

$$6\dot{x}_0^2\lambda^3 - (6ax_0^2 + 6\beta x_0^4 + 6\gamma x_0^6)\lambda^2 - (3\beta x_0^4 + 6\gamma x_0^6)\lambda - 2\gamma x_0^6 = 0. \tag{98}$$

The solution to the the i.v.p. (95) is obtained from the formulas in the first section.

b. Suppose that $x_0 = 0$, in this case, the solution of the i.v.p. (92)-(93) is given by

$$x(t) = \frac{\sqrt{1 + \lambda v(t)}}{\sqrt{1 + \lambda v^2(t)}},$$ (99)

where the function $v \equiv v(t)$ is the solution of the following Duffing equation

$$\begin{cases} \ddot{v} + pv + qv^3 = 0, \\ v(0) := v_0 = 0, \\ v'(0) := \dot{v}_0 = \dfrac{\dot{x}_0}{\sqrt{1 + \lambda}}. \end{cases}$$ (100)

The values of the coefficients p and q are expressed as

$$p = -\frac{2\gamma + (3\beta + 4\gamma)\lambda + (4\alpha + 3\beta + 2\gamma)\lambda^2}{2\lambda^2},$$ (101)

$$q = -\frac{2(\gamma + (\beta + 2\gamma)\lambda + (\alpha + \beta + \gamma)\lambda^2)}{\lambda},$$ (102)

and the value of λ is a solution of the cubic

$$(6\alpha + 3\beta + 2\gamma - 6\dot{x}_0^2)\lambda^3 + (6\alpha + 6\beta + 6\gamma)\lambda^2 + (3\beta + 6\gamma)\lambda + 2\gamma = 0.$$ (103)

Note that the solution of the i.v.p. (100) could be obtained from the formulas in the first section.

- Proof: case (a)

Inserting ansatz (94) into Eq. (92) taking the following equation into consideration

$$\dot{v}^2 = \dot{v}_0^2 + pv_0^2 + \frac{q}{2}v_0^4 - pv^2 - \frac{q}{2}v^4,$$ (104)

and using Eq. (100), we have

$$\sum_{j=1}^{5} H_j v(t)^j = 0,$$ (105)

with

$$H_1 = \frac{\left(-6px_0^2\lambda - 3qx_0^2\lambda + 2x_0^2\alpha - 6\dot{x}_0^2\lambda^3 - 12\dot{x}_0^2\lambda^2 - 6\dot{x}_0^2\lambda - 2x_0^2\right)}{2x_0^2},$$

$$H_3 = -\left(-3p\lambda + q - x_0^2\beta\lambda - x_0^2\beta - 2\alpha\lambda + \lambda\right),$$

$$H_5 = \frac{1}{2}\left(q\lambda + 2x_0^2\beta\lambda^2 + 2x_0^2\beta\lambda + 2x_0^4\gamma\lambda^2 + 4x_0^4\gamma\lambda + 2x_0^4\gamma + 2\alpha\lambda^2\right),$$

where $j = 1, 3, 5$.

Equating the coefficients H_j to zero gives an algebraic system: $H_1 = 0$, $H_3 = 0$, and $H_5 = 0$. Solving $H_1 = 0$ and $H_3 = 0$ will give the values of p and q that are given in Eqs. (101)-(102). Finally, by inserting the values of p and q into $H_1 = 0$, we obtain the cubic Eq. (103). Likewise, the case (b) can be proved.

8. Damped Cubic-Quintic Oscillator

Let us define the following i.v.p.

$$\begin{cases} \ddot{x} + +2\varepsilon\dot{x} + px + qx^3 + rx^5 = 0, \\ x(0) = x_0 \,\&\, x'(0) = \dot{x}_0. \end{cases} \tag{106}$$

We seek approximate analytic solution in the ansatz form

$$x(t) = \exp{(-\rho t)}y(\,f(t)), \tag{107}$$

with

$$f(t) = \frac{1 - \exp{(-2(\varepsilon - \rho)t)}}{2(\varepsilon - \rho)}, \tag{108}$$

where $y \equiv y(t)$ is the exact solution to the i.v.p.

$$\begin{cases} \ddot{y} + (p - \varepsilon\rho + \rho^2)y + qy^3 + ry^5 = 0, \\ y(0) = x_0 \,\&\, y'(0) = \dot{x}_0 + x_0\rho. \end{cases} \tag{109}$$

Define the residual

$$R(t) = \ddot{x} + 2\varepsilon\dot{x} + px + qx^3 + rx^5, \tag{110}$$

then, the condition $R'(0) = 0$ gives

$$2\rho^3 - 6\varepsilon\rho^2 + (2p + 3qx_0^2 + 4rx_0 + 4\varepsilon^2)\rho - 2p\varepsilon - 2qx_0^2\varepsilon - 2rx_0^4\varepsilon = 0. \tag{111}$$

Some real roots of Eq. (111) give the value of ρ. For $x_0 = 0$, the default value of ρ could be chosen as $\rho = (2/3)\varepsilon$.

Example 9.

Let

$$\begin{cases} \ddot{x} + 0.05\dot{x} + 96.6289x - 3.5x^3 - 0.8x^5 = 0, \\ x(0) = 1 \,\&\, x'(0) = 0. \end{cases} \tag{112}$$

The approximate analytical solution of the i.v.p. is given by

$$x_{\text{app}}(t)$$

$$= \frac{1.00575e^{-0.0257101t}(\text{cn}(\,f(t)|m) - 0.00261773\text{dn}(\,f(t)|m)\text{sn}(\,f(t)|m))}{\left(1 + 5.14 \times 10^{-9}\text{sn}(\,f(t)|m)^2\right)\sqrt{1 + \frac{0.0107805(\text{cn}(\,f(t)|m) - 0.00261773\text{dn}(\,f(t)|m)\text{sn}(\,f(t)|m))^2}{\left(1 + 5.14 \times 10^{-9}\text{sn}(\,f(t)|m)^2\right)^2} + \frac{0.000750033(\text{cn}(\,f(t)|m) - 0.00261773\text{dn}(\,f(t)|m)\text{sn}(\,f(t)|m))^4}{\left(1 + 5.14 \times 10^{-9}\text{sn}(\,f(t)|m)^2\right)^4}}}, \tag{113}$$

where

$$f(t) = 6807.39 - 6807.39e^{0.00142017t} \,\&\, m = -0.000750607. \tag{114}$$

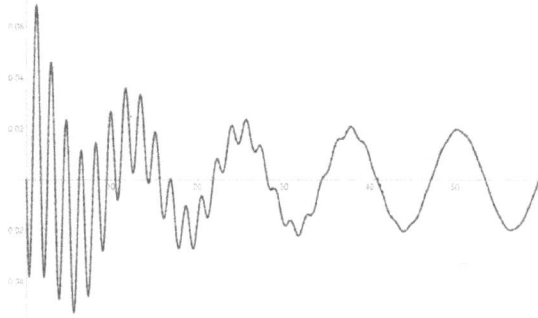

Figure 9.
A comparison between RK4 solution and (a) Solution (113) and (b) Zuñiga solution [17].

The distance error as compared to the RK4 numerical solution reads

$$\max_{0 \le t \le 5} \left| x_{\text{app}}(t) - x_{\text{RK4}}(t) \right| = 0.0561216. \tag{115}$$

In **Figure 9**, we make a comparison between our solution, RK4 solution, and Zuñiga solution given in Ref. [17]. It is clear that the accuracy of our solution is better than the solution of Zuñiga [17].

9. Realistic physical applications

The above solutions could be applied to various fields of physics and engineering such as they could be used for describing the behavior of oscillations in RLC electronic circuits, plasma physics etc. In the below section, the above solution will be devoted for studying oscillations in various plasma models.

9.1 Nonlinear oscillations in RLC series circuits with external source

In the RLC series circuits consisting of a linear resistor with resistance R in Ohm unit, a linear inductor with inductance L in Henry unit, and nonlinear capacitor with capacitance C in Farady unit as well as external applied voltage E in voltage unit, the Kirchhoff's voltage law (KVL) could be written as

$$L\partial_t i'(t) + i(t)R + sq + aq^2 = E, \tag{116}$$

where the relation between the current the charge is given by $i = \partial_t q \equiv \dot{q}$, $i' \equiv \partial_t i$, the coefficients (a, s) are related to the nonlinear capacitor, and E represents the voltage of the battery which is constant. By reorganizing Eq. (116), the following constant forced and damped Helmholtz equation could be obtained as

$$\ddot{q} + 2\gamma\dot{q} + \alpha q + \beta q^2 = F, \tag{117}$$

with $\gamma = R/(2L)$, $\alpha = 1/(LC)$, $\beta = 1/(Cq_0L)$, and $F = E/L$ where $q_0 = q(t = 0)$ is the initial charge value at $t = 0$, $\ddot{q} \equiv \partial_t^2 q$, and $\dot{q} \equiv \partial_t q$.

The solution of Eq. (117) can be devoted for interpreting and analyzing the oscillations that can generated in the RLC circuit.

9.2 Duffing-Helmholtz equation for modeling the oscillations in a plasma

For studying the plasma oscillations using fluid theory, the basic equations of plasma particles using the reductive perturbation method (RPM) will be reduced to some evolution equations such as KdV equation and its family [37–41]. Let us consider a collisionless and unmagnetized electronegative complex plasma, consisting of inertialess cold positive and negative ion species, inertia non-Maxwellian electrons in addition to stationary negative dust impurities [42]. Thus, the quasi-neutrality condition reads: $n_2^{(0)} + n_e^{(0)} = n_1^{(0)}$ where $n_{s,e}^{(0)}$ donates the unperturbed number density of the plasma particles (here, the index "s" = "1" and "2" point out the positive ion and negative ion, and "e" refers to the electron, respectively). It is assumed that the plasma oscillations take place only in x–directional which means that the fluid equations of the plasma particles become perturbed only in x–directional. If the effect of the ionic kinematic viscosities η_s for both positive (η_1) and negative (η_2) ions are included in the present investigation, as a source of damping/dissipation, in this case we will get a new evolution equation governs the dynamics of damping pulses. The dynamics of plasma oscillations are governed by the following fluid equations: $\partial_x n_s + \partial_t(n_s u_s) = 0$, $\partial_t u_s + u_s \partial_x u_s +$ $(\delta/Q_s)\partial_x \phi - \eta_s \partial_x^2 u_s = 0$, and $\partial_x^2 \phi - n_d^{(0)} - n_2 - n_e + n_1 = 0$, where $n_e = \mu(1 - \beta\phi + \beta\phi^2) \exp \phi$. Here, n_s donates the normalized number density of positive and negative ions, and u_s represents the normalized fluid velocity of positive and negative ions, and ϕ is the normalized electrostatic wave potential. The mass ratio is defined as: $Q_s = m_1/m_s$ (note that $Q_1 = m_1/m_1 = 1$ and $Q_2 \equiv Q = m_1/m_2$), where m_s is the ionic mass, $\delta = 1(-1)$ for positive (negative) ion, and β illustrates nonthermality parameter. The quasi-neutrality condition in the normalized form reads: $\mu = 1 - \alpha$, where $\alpha = n_2^{(0)}/n_1^{(0)}$ gives the the negative ion concentration and $\mu = n_e^{(0)}/n_1^{(0)}$ is the electron concentration.

Now, the RPM is introduced to reduce the fluid plasma equations to the evolution equation. According to the RPM, the independent quantities (x, t) are stretched as: $\xi = \varepsilon(x - V_{ph}t)$, $\tau = \varepsilon^3 t$, and $\eta_s = \varepsilon\eta_s^{(0)}$ where V_{ph} is the wave phase velocity of the ion-acoustic waves and ε is a real and small parameter $(0 < \varepsilon < 1)$. The dependent perturbed quantities $\Pi(x, t) \equiv (n_1, n_2, u_1, u_2, \phi)^T$ are expanded as: $\Pi = \Pi^{(0)} + \sum_{j=1}^{\infty} \varepsilon^j \Pi(\xi, \tau)^{(j)}$, where $\Pi^{(0)} = (1, \alpha, 0, 0, 0)^T$ and T represents the transpose of the matrix. Inserting both the stretching and expansions of the independent and dependent quantities into the basic fluid equations and after boring but straightforward calculations, the Gardner-Burgers/EKdVB equation is obtained

$$\partial_\tau \varphi + (P_1 \varphi + P_2 \varphi^2)\partial_\xi \varphi + P_3 \partial_\xi^3 \varphi + P_4 \partial_\xi^2 \varphi = 0, \qquad (118)$$

with the coefficients of the quadratic nonlinear, cubic nonlinear, dispersion, and dissipation terms P_1, P_2, P_3, and P_4, respectively,

$$P_1 = \frac{3}{2}P_3\left[\frac{1}{V_{ph}^4} - \frac{3\alpha}{V_{ph}^4 Q^2} - \frac{2h_2}{3}\right], P_2 = \frac{3}{4}P_3\left[\frac{5}{V_{ph}^6} - \frac{5\alpha}{V_{ph}^6 Q^3} - 2h_3\right],$$

$$P_3 = \frac{V_{ph}^3 Q}{2(Q + \alpha)}, P_4 = -P_3\left[\frac{\eta_1^{(0)}}{V_{ph}^4} + \frac{\alpha\eta_2^{(0)}}{QV_{ph}^3}\right] \& V_{ph} = \sqrt{\frac{(Q + \alpha)}{Qh_1}},$$

where $\varphi \equiv \phi^{(1)}$, $h_1 = \mu(1 - \beta)$, $h_2 = \mu/2$, and $h_3 = \mu(1 + 3\beta)/6$.

It is shown that the coefficients P_1, P_2, P_3, and P_4, are functions in the physical plasma parameters namely, negative ion concentration α, the mass ratio Q, and the electron nonthermal parameter β. It is known that at the critical plasma compositions say β_c or α_c (critical value of negative ion concentration), the coefficient P_1 vanishes and in this case Eq. (118) will be reduced to the following mKdVB equation which is used to describe the damped wave dynamics at critical plasma compositions

$$\partial_\tau \varphi + P_2 \varphi^2 \partial_\xi \varphi + P_3 \partial_\xi^3 \varphi + P_4 \partial_\xi^2 \varphi = 0, \tag{119}$$

To convert EKdVB Eq. (118) to the damped H-D Eq. (4), the traveling wave transformation $\varphi(\xi, \tau) \to \varphi(X)$ with $X = (\xi + \lambda\tau)$ should be inserted into Eq. (118) and integrate once over η, and by applying the boundary conditions: $(\varphi, \varphi', \varphi'') \to 0$ as $|X| \to \infty$, the constant forced damped following constant forced damped Duffing-Helmholtz equation is obtained

$$\varphi'' + 2\varepsilon\varphi' + p\varphi + q\varphi^2 + r\varphi^3 + D = 0, \tag{120}$$

where λ represents the reference frame speed, φ' and φ'' denote the first and second ordinary derivative of regarding X, $\varepsilon = P_4/(2P_3)$, $p = \lambda/P_3$, $q = P_1/(2P_3)$, $r = P_2/(3P_3)$, and $D = C/P_3$.

Note that the coefficient q may be positive or negative according to the values of plasma parameters and for studying oscillations using (120), solution (72) can be devoted for this purpose. In the absence of the ionic kinematic viscosity ($P_4 = 0$ or $\varepsilon = 0$), then Eq. (120) reduces to the constant forced undamping Duffing-Helmholtz equation and in this case the solution (63) can be applied for investigating the undamped oscillations in the present plasma model. Also, for $q = 0$, Eq. (120) reduces to the constant forced damped Helmholtz equation. Moreover, the constant forced damped Duffing equation can be obtained for $p = 0$.

10. Conclusion

The analytical and semi-analytical solutions for nonlinear oscillator integrable and non-integrable equations have been investiagted. First, the standard integrable Duffing equation has been analiyzed and its solutions have been obtained depending on the sign of its discriminant Δ. Accordingly, three cases ($\Delta > 0, \Delta < 0$, and $\Delta = 0$) have been discussed in details and the solutions of each case has be obtained. Second, the analytical and semi-analytical solutions of the integrable Duffing-Helmholtz equation and its non-integrable family including the damped Duffing-Helmholtz equation, forced undamped Duffing-Helmholtz equation, forced damped Duffing-Helmholtz equation, and the damped and trigonometric forced Duffing-Helmholtz equation have been obtained and discussed in details. Third, the solutions to the intgrable cubic-quintic Duffing equation and the non-intgrable damped cubic-quintic Duffing equation have been investigated. Moreover, some realistic applications reaslted to the RLC circuits and physics of plasmas have been introduced and discussed depending on the solutions of the mentioned evolution equations.

Author details

Alvaro Humberto Salas[1*] and Samir Abd El-Hakim El-Tantawy[2,3]

1 Universidad Nacional de Colombia, Fizmako Research Group, Colombia

2 Department of Physics, Faculty of Science, Port Said University, Port Said, Egypt

3 Research Center for Physics (RCP), Department of Physics, Faculty of Science and Arts, Al-Mikhwah, Al-Baha University, Al-Baha, Saudi Arabia

*Address all correspondence to: ahsalass@unal.edu.co

IntechOpen

References

[1] A. M. Wazwaz, *Partial Differential Equations and Solitary Waves Theory*, Higher Education Press, Beijing, USA, (2009).

[2] A. M. Wazwaz, *Partial Differential Equations: Methods and Applications*, Lisse: Balkema, cop. (2002).

[3] Alvaro H. Salas, S. A. El-Tantawy, and Noufe H. Aljahdaly, Mathematical Problems in Engineering 2021, 1-8, 2021, Article ID 8875589, https://doi.org/10.1155/2021/8875589

[4] Alvaro H. Salas and S. A. El-Tantawy, Eur. Phys. J. Plus 135, 833-17, 2020.

[5] Alvaro H. Salas and J. E. Castillo, Visión electrónica, DOI: https://doi.org/10.14483/22484728.7861.

[6] O. N. F. Nelson, Z. Yu, B. P. Dorian, and Y. Wang, J. Appl. Math. 6, 2718 (2018).

[7] K. Johannessen, Eur. J. Phys. 36, 065020 (2015) and references in therein.

[8] K. Johannesen, Int. J. Appl. Comput. Math. 3, 3805-3816, 2017.

[9] S. K. Lai and C. W. Lim, Int. J. Comput. Meth. Eng. Sci. Mech. 7, 201 (2006).

[10] I. Kovacic and M. J. Brennan, *The Duffing Equation: Nonlinear Oscillators and their Behaviour*, 1st ed. John Wiley & Sons, Ltd., 2011.

[11] P. S. Landa, *Nonlinear Oscillations and Waves in Dynamical Systems*, Springer, 1996.

[12] N. Srinil, and H. Zanganeh, Ocean Engineering, 53, 83 (2012).

[13] U. R. Singh, Int. J. Nonlinear Dynamics and Control 1, 87 (2017).

[14] S. A. El-Tantawy and P. Carbonaro, Phys. Lett. A 380, 1627 (2016).

[15] Noufe H. Aljahdaly and S. A. El-Tantawy, Chaos 30, 053117 (2020).

[16] J. P. Praveen and B. N. Rao, MAYFEB Journal of Mathematics 2, 7 (2016).

[17] A. E.-Zúñiga, C. A. Rodrguez, and O. M. Romero, Comput. Math. Appl. 60, 1409 (2010).

[18] Y. Geng, Chaos, Solitons and Fractals 81, 68 (2015).

[19] E. T. Whittaker and G. N. Watson, *A Course of Modern Analysis* 4th edn (Cambridge: Cambridge University Press), 1980).

[20] A. E.-Zúñiga, Nonlinear Dyn. 42, 175 (2005).

[21] S. Nourazar, A. Mirzabeigy, scientia Iranica, 20, 364 (2013).

[22] R. E. Mickens, J. Sound Vib. 244, 563, 2001.

[23] J. H. He, Eur. J. Phys. 29, 19 (2008).

[24] J. H. He, Comput. Methods in Appl. Mech. Eng. 178, 257 (1999).

[25] Y. Khan, M. Akbarzade, and A. Kargar, Scientia Iranica 19, 417 (2012).

[26] M. El-Shahed, Comm. Nonlinear Sci. Numer. Simulat. 13, 1714 (2008).

[27] Y. Khan, and Q. Wu, Computers& Mathematics with Applications 61, 1963 (2011).

[28] Y. Khan, and F. Austin, Zeitschrift für Naturforschung A 65a, 849 (2010).

[29] J. A. Almendral and M. A. F. Sanjuan, J. Phys. A: Math. Gen. 36, 695 (2003).

[30] J.-w. Zhu, Appl. Math. Model. 38, 5986 (2014).

[31] K. Tamilselvan, T. Kanna, and A. Govindarajan, chirped elliptic and solitary waves, Chaos 29, 063121 (2019).

[32] A. E.-Zúñiga, Appl. Math. Lett. 25, 2349 (2012).

[33] André Gussoa and Jéssica D. Pimentel, Appl. Math. Model. 61, 593 (2018).

[34] Alvaro H. Salas and S.Casanova, Mathematical Problems in Engineering 2020, Article ID 3985975, https://doi.org/10.1155/2020/3985975

[35] Alvaro H. Salas, Applicable Analysis, 2019, DOI: 10.1080/00036811.2019.1698729.

[36] Ma'mon Abu Hammad, Alvaro H. Salas, and S. A. El Tantawy, AIP Advances 10, 085001, 2020.

[37] N. H. Aljahdaly and S.A. El-Tantawy, Chaos 30, 053117 (2020).

[38] Bothayna S. Kashkari, S. A. El-Tantawy, A. H. Salas, and L. S. El-Sherif, Chaos, Solitons and Fractals 130, 109457 (2020).

[39] S. A. El-Tantawy, T. Aboelenen, and S. M. E. Ismaeel, Phys. Plasmas 26, 022115 (2019).

[40] S. A. El-Tantaw and A. M. Wazwaz, Phys. Plasmas 25 092105 (2018).

[41] S. A. El-Tantawy, Astrophys. Space Sci. 361 249 (2016).

[42] S. A. El-Tantawy, Astrophys. Space Sci. 361, 164 (2016).

Chapter 6

Optimal Heat Distribution Using Asymptotic Analysis Techniques

Zakaria Belhachmi, Amel Ben Abda, Belhassen Meftahi and Houcine Meftahi

Abstract

In this chapter, we consider the optimization problem of a heat distribution on a bounded domain Ω containing a heat source at an unknown location $\omega \subset \Omega$. More precisely, we are interested in the best location of ω allowing a suitable thermal environment. For this propose, we consider the minimization of the maximum temperature and its L^2 mean oscillations. We extend the notion of topological derivative to the case of local coated perturbation and we perform the asymptotic expansion of the considered shape functionals. In order to reconstruct the location of ω, we propose a one-shot algorithm based on the topological derivative. Finally, we present some numerical experiments in two dimensional case, showing the efficiency of the proposed method.

Keywords: Topological optimization, Asymptotic analysis, Coated inclusion, Heat conduction

1. Introduction

The concept of topological derivative is a powerful tool for solving shape optimization problems constrained by partial differential equations. The method has a great potential of applications in the field of non-destructive control. In this chapter, the topological derivative is applied in the context of optimization of a heat distribution. More precisely, we consider the problem of locating circular coated inclusions in order to get an appropriate layout with minimized maximum temperature distribution. This problem can be encountered in the design of current carrying multicables and in some devices in hybrid and electric cars. The mathematical problem is similar to the mixture of materials with different conduction properties extensively studied in the case of two materials; see for instance [1–6].

The topological derivative measure the sensitivity of a given shape functional with respect to the insertion of a small hole inside the domain. More precisely, we consider a domain $\Omega \subset \mathbb{R}^2$ and a cost functional $\mathcal{J}(\Omega) = j(\Omega, u_\Omega)$, were u_Ω is the state variable, i.e. a solution of a given partial differential equation in Ω. For $\varepsilon > 0$, let $\Omega_\varepsilon = \Omega \backslash \overline{(x_0 + \varepsilon D)}$ be the domain obtained by removing a small part $\overline{(x_0 + \varepsilon D)}$ from Ω, at a location $x_0 \in \Omega$, and D is a fixed bounded subset of $\mathbb{R}^2$ with $(0, 0) \in D$.

Then, the shape functional $\mathcal{J}(\Omega_\varepsilon)$ associated with the topologically perturbed domain, admits the following topological asymptotic expansion

$$\mathcal{J}(\Omega_\varepsilon) = \mathcal{J}(\Omega) + f(\varepsilon)g(x_0) + o(f(\varepsilon)), \tag{1}$$

IntechOpen

where $\mathcal{J}(\Omega$ is the shape functional associated to the unperturbed domain Ω and $f(\varepsilon)$ is a positive function such that $f(\varepsilon) \to 0$ as $\varepsilon \to 0$. The function $x_0 \to g(x_0)$ is called topological gradient of $\mathcal{J}$ at x_0. Note that the topological derivative is defined through the limit passage $\varepsilon \to 0$. However, according to (1), it can be used as a descent direction in an optimization process similar to any gradient-based method. This concept has been applied for geometrical inverse problems [7], in linear isotropic elasticity [8], and in the context of solving some design problems in steady-state heat conduction [9]. Another situation addressed in [10–12], consists in studying the influence of the insertion of a small inhomogeneity which is nonempty, but whose constitutive parameters are different from those of the background medium. This approaches was successfully investigate in the case of inhomogeneities conductor materials. For more details, we refer the reader to the recent works on shape reconstruction and stability analysis for some imaging problems with the topological derivative [13–15].

In this chapter, we extend the asymptotic analysis with respect to the insertion of local small inhomogeneity to the case of the insertion of local small coated inclusion. An asymptotic expansion of a given shape functional is derived with the help of a relevant adjoint method. The computed topological derivative allows us to solve numerically the minimization problem.

The chapter is organized as follows. In Section 2 we present the model problem and the shape optimization formulation. In Section 3 and 4 we perform the asymptotic expansions of the shape functional. The numerical results are presented in Section 5. The paper ends with some concluding remarks in Section 6.

2. The model problem

Let Ω be a bounded domain in $\mathbb{R}^2$ with Lipschitz boundary $\partial\Omega$ and let ω be an open subset of Ω composed of two different subdomains Ω_1 and Ω_2 where the subset Ω_2 is surrounded by the subset Ω_1. We denote by $\Gamma_2 := \partial\Omega_2$ and $\Gamma_1 \cup \Gamma_2 := \partial\Omega_1$ as depicted in **Figure 1** and we set $\Omega_0 := \Omega\backslash\overline{\Omega_1 \cup \Omega_2}$.

Throughout the chapter, we consider piecewise constant thermal conductivity

$$\sigma = \sigma_0\chi_{\Omega_0} + \sigma_1\chi_{\Omega_1} + \sigma_2\chi_{\Omega_2}, \tag{2}$$

where $\sigma_0, \sigma_1, \sigma_2 \in \mathbb{R}_+^*$ and χ_E denotes the indicator function of the set E. We assume further that there exists two constants c_0, c_1 such that

$$0 < c_0 \le \sigma_0, \sigma_1, \sigma_2 \le c_1.$$

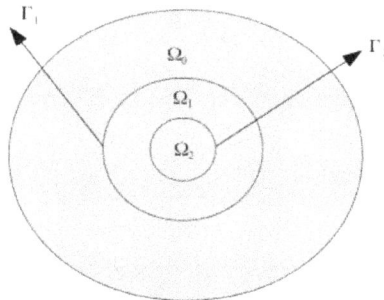

Figure 1.
The domain $\Omega = \Omega_0 \cup \Omega_1 \cup \Gamma_1 \cup \Gamma_2 \cup \Omega_2$.

For a given source term $f \in L^2(\Omega)$ and the Dirichlet data $g \in H^{1/2}(\partial\Omega)$, the temperature u_ω satisfies the following problem

$$
\begin{cases}
-\operatorname{div}(\sigma\nabla u_\omega) = f & \text{in } \Omega \\
[\![u_\omega]\!] = 0 & \text{on } \Gamma_i, i = 1, 2, \\
[\![\sigma\partial_n u_\omega]\!] = 0 & \text{on } \Gamma_i, i = 1, 2, \\
u_\omega = g & \text{on } \partial\Omega.
\end{cases}
\tag{3}
$$

For simplicity we take $g = 0$ by choosing a lifting function $G \in H^2(\Omega)$, $G = g$ in $\partial\Omega$ and modifying the left hand side that we still denote f. Then, the weak solution to problem (3) is defined by:

$$
\text{Find } u_\omega \in H_0^1(\Omega) \text{ such that } a(u_\omega, v) = l(v), \quad \forall v \in H_0^1(\Omega), \tag{4}
$$

where

$$
a(u_\omega, v) = \int_\Omega \sigma\nabla u_\omega \cdot \nabla v\, dx, \quad \text{and} \quad l(v) = \int_\Omega fv\, dx.
$$

The existence and uniqueness of the weak solution u_ω follows from the Lax-Milgram Lemma.

We consider the following shape minimization problem

Determine the position of $\omega \subset \Omega$ to obtain an appropriate layout temperature. (5)

To deal with numerical computation of problem (5), we consider two shape functionals. The first shape functional corresponds to the maximum temperature

$$
M(\omega, u_\omega) = \|u_\omega\|_{L^\infty(\Omega)}.
$$

Since the functional M is not differentiable, we can not use the topological derivative framework to perform the sensitivity analysis. Thus we use the shape functional J_p, for large $p \geq 2$ instead of the functional M:

$$
J_p(\omega, u_\omega) = \frac{1}{p}\int_\Omega |u_\omega|^p\, dx.
$$

The second shape functional may appear as a particular case of J_p but has its own physical and mathematical interests, corresponds to the minimization of the L^2 mean oscillations of the temperature:

$$
K(\omega, u_\omega) := \int_\Omega \left(u_\omega - \frac{1}{|\Omega|}\int_\Omega u_\omega dx \right)^2 dx.
$$

Then, the optimization problems read:

$$
\begin{cases}
\text{minimize } J_p(\omega, u_\omega) := \dfrac{1}{p}\int_\Omega |u_\omega|^p\, dx \\
\text{subject to } \omega \in \mathcal{O} \text{ and } u_\omega \text{ solves problem (3),}
\end{cases}
\tag{6}
$$

and

$$\begin{cases} \text{minimize } K(\omega, u_\omega) := \int_\Omega \left(u_\omega - \frac{1}{|\Omega|} \int_\Omega u_\omega \, dx \right)^2 dx \\ \text{subject to } \omega \in \mathcal{O} \text{ and } u_\omega \text{ solves problem (3).} \end{cases} \tag{7}$$

Where $\mathcal{O}$ is the admissible set:

$$\mathcal{O} = \{\omega \subset \Omega : P(\omega, \Omega) < \infty\},$$

and $P(\omega, \Omega)$ is the relative perimeter:

$$P(\omega, \Omega) := \sup \left\{ \int_\omega \nabla \cdot \varphi \, dx : \phi \in C_c^1(\Omega, \mathbb{R}^2), \|\varphi\|_\infty \le 1 \right\}.$$

3. Topological derivatives

Now, we assume that the domain $\Omega_2 := x_0 + \alpha \varepsilon B$ and Ω_1 is such that $\Omega_1 \cup \Omega_2 = x_0 + \varepsilon B$, where B is the unit ball in $\mathbb{R}^2$, $\varepsilon > 0$ and $0 < \alpha < 1$. We rewrite Ω_1^ε and Ω_2^ε instead of Ω_1 and Ω_2. This allows to perform an asymptotic expansion of the shape functional $J_p(\omega_\varepsilon)$ where $\omega_\varepsilon := \Omega_2^\varepsilon \cup \Omega_1^\varepsilon$. We also introduce $\Gamma_2^\varepsilon := \partial \Omega_2^\varepsilon$ and Γ_1^ε the outer boundary of Ω_1^ε.

In the perturbed domain, the state u_ε is solution to the following problem:

$$\begin{cases} -\text{div}(\sigma_\varepsilon \nabla u_\varepsilon) = f_\varepsilon \text{ in } \Omega \\ u_\varepsilon = 0 \text{ on } \partial\Omega \end{cases} \tag{8}$$

where

$$\sigma_\varepsilon = \sigma_2 \chi_{\Omega_2^\varepsilon} + \sigma_1 \chi_{\Omega_1^\varepsilon} + \sigma_0 \chi_{\Omega \setminus \overline{\Omega_1^\varepsilon \cup \Omega_2^\varepsilon}}, \text{ and } f_\varepsilon = f_2 \chi_{\Omega_2^\varepsilon} + f_1 \chi_{\Omega_1^\varepsilon} + f_0 \chi_{\Omega \setminus \overline{\Omega_1^\varepsilon \cup \Omega_2^\varepsilon}}.$$

The functions f_i are in L^2. The variational formulation associated with the problem (8) is defined by:

$$\begin{cases} \text{find } u_\varepsilon \in H_0^1(\Omega), \text{ such that} \\ a_\varepsilon(u_\varepsilon, v) = l_\varepsilon(v), \quad \forall v \in H_0^1(\Omega), \end{cases} \tag{9}$$

where

$$a_\varepsilon(u_\varepsilon, v) = \int_\Omega \sigma_\varepsilon \nabla u_\varepsilon \cdot \nabla v \, dx, \quad \text{and} \quad l_\varepsilon(v) = \int_\Omega f_\varepsilon v \, dx.$$

We denote by u_0 the background solution of the following problem:

$$-\text{div}(\sigma_0 \nabla u_0) = f_0, \text{ in } \Omega \quad u_0 = 0 \text{ on } \partial\Omega. \tag{10}$$

The following proposition describes in an abstract framework the adjoint method we will use to derive the asymptotic expansion of a given shape functional. For more details the reader is referred to [16] and the references therein.

Proposition 1 Let $\mathcal{H}$ be a Hilbert space. For all parameter $\varepsilon \in [0, \varepsilon_0[, \varepsilon_0 > 0$, consider a function $u_\varepsilon \in \mathcal{H}$ solving a variational problem of the form

$$a_\varepsilon(u_\varepsilon, v) = l_\varepsilon(v) \forall v \in \mathcal{H}, \tag{11}$$

where a_ε is a bilinear form and l_ε is a linear form on $\mathcal{H}$. For all $\varepsilon \in [0, \varepsilon_0[$, consider a functional $J_\varepsilon : \mathcal{H} \to \mathbb{R}$ that is Fréchet differentiable at u_0. Assume that the following hypotheses are satisfied.

H_1 There exist a scalar function $f(\varepsilon) \geq 0$ and two numbers $\delta a, \delta l$ such that

$$(a_\varepsilon - a_0)(u_0, v_\varepsilon) = f(\varepsilon)\delta a + o(\, f(\varepsilon)), \tag{12}$$

$$(l_\varepsilon - l_0)(v_\varepsilon) = f(\varepsilon)\delta l + o(\, f(\varepsilon)), \tag{13}$$

$$\lim_{\varepsilon \to 0} f(\varepsilon) = 0, \tag{14}$$

where $v_\varepsilon \in \mathcal{H}$ is an adjoint state satisfying

$$a_\varepsilon(\varphi, v_\varepsilon) = -DJ_\varepsilon(u_0)\varphi, \forall \varphi \in \mathcal{H}. \tag{15}$$

H_2 There exist two numbers δJ_1 and δJ_2 such that

$$J_\varepsilon(u_\varepsilon) = J_\varepsilon(u_0) + DJ_\varepsilon(u_0)(u_\varepsilon - u_0) + f(\varepsilon)\delta J_1 + o(\, f(\varepsilon)), \tag{16}$$

$$J_\varepsilon(u_0) = J_0(u_0) + f(\varepsilon)\delta J_2 + o(\, f(\varepsilon)). \tag{17}$$

Then the first variation of the cost function with respect to ε is given by

$$J_\varepsilon(u_\varepsilon) - J_0(u_0) = f(\varepsilon)(\delta a - \delta l + \delta J_1 + \delta J_2) + o(\, f(\varepsilon)). \tag{18}$$

3.1 Application to the model problem

In this subsection, we will give explicitly the variations $\delta a, \delta l, \delta J_1, \delta J_2$ and we perform the asymptotic expansion of the shape functional J_p. Analogously, we can derive in the same manner the asymptotic expansion of the shape functional K.

3.1.1 Variation of the bilinear form

In this subsection, we look on the asymptotic analysis of the variation

$$(a_\varepsilon - a_0)(u_0, v_\varepsilon) = \int_{\Omega_1^\varepsilon} (\sigma_1 - \sigma_0)\nabla u_0 \cdot \nabla v_\varepsilon dx + \int_{\Omega_2^\varepsilon} (\sigma_2 - \sigma_0)\nabla u_0 \cdot \nabla v_\varepsilon dx. \tag{19}$$

Let us first look at the behavior of the adjoint state v_ε solution of the following boundary value problem

$$\begin{cases} \operatorname{div}(\sigma_\varepsilon \nabla v_\varepsilon) = u_\omega |u_\omega|^{p-2} & \text{in } \Omega, \\ v_\varepsilon = 0 & \text{on } \partial\Omega. \end{cases} \tag{20}$$

Since u_ω is Hölder continuous, $u_\omega |u_\omega|^{p-2}$ is at least in $L^2(\Omega)$. Therefore problem (20) has a unique solution $v_\varepsilon \in H_0^1(\Omega)$.

We split in (19) v_ε into $v_0 + (v_\varepsilon - v_0)$ and introducing the" small" terms (which will be checked later)

$$\mathcal{E}_1(\varepsilon) := \int_{\Omega_1^\varepsilon} (\sigma_1 - \sigma_0)(\nabla u_0 \cdot \nabla v_0 - \nabla u_0(x_0) \cdot \nabla v_0(x_0))dx, \tag{21}$$

$$\mathcal{E}_2(\varepsilon) := \int_{\Omega_2^\varepsilon} (\sigma_2 - \sigma_0)(\nabla u_0 \cdot \nabla v_0 - \nabla u_0(x_0) \cdot \nabla v_0(x_0)) \, dx, \tag{22}$$

we obtain

$$(a_\varepsilon - a_0)(u_0, v_\varepsilon) = \pi \varepsilon^2 \left((1 - \alpha^2)(\sigma_1 - \sigma_0) + \alpha^2(\sigma_2 - \sigma_0) \right) \nabla u_0(x_0) \cdot \nabla v_0(x_0) \\ + \mathcal{F}_1(\varepsilon) + \mathcal{F}_2(\varepsilon) + \mathcal{E}_1(\varepsilon) + \mathcal{E}_2(\varepsilon), \tag{23}$$

where

$$\mathcal{F}_1(\varepsilon) := \int_{\Omega_1^\varepsilon} (\sigma_1 - \sigma_0) \nabla u_0 \cdot \nabla (v_\varepsilon - v_0) \, dx, \tag{24}$$

$$\mathcal{F}_2(\varepsilon) := \int_{\Omega_2^\varepsilon} (\sigma_2 - \sigma_0) \nabla u_0 \cdot \nabla (v_\varepsilon - v_0) \, dx. \tag{25}$$

We will now study the asymptotic of $\mathcal{F}_1$ and $\mathcal{F}_2$. Introducing the variation $\tilde{v}_\varepsilon := v_\varepsilon - v_0$, we obtain from (20) that $\tilde{v}_\varepsilon$ solves

$$\begin{cases} \Delta \tilde{v}_\varepsilon = 0 & \text{in } \Omega_1^\varepsilon \cup \Omega_2^\varepsilon \cup (\Omega \; \overline{\Omega_1^\varepsilon \cup \Omega_2^\varepsilon}), \\ [\sigma \partial_\nu \tilde{v}_\varepsilon] = -(\sigma_1 - \sigma_0) \nabla v_0 \cdot \nu & \text{on } \Gamma_1^\varepsilon, \\ [\sigma \partial_\nu \tilde{v}_\varepsilon] = -(\sigma_2 - \sigma_1) \nabla v_0 \cdot \nu & \text{on } \Gamma_2^\varepsilon, \\ \tilde{v}_\varepsilon = 0 & \text{on } \partial\Omega. \end{cases} \tag{26}$$

We set $V := \nabla v_0(x_0)$ and we approximate $\tilde{v}_\varepsilon$ by the solution h_ε^V of the auxiliary problem

$$\begin{cases} \Delta h_\varepsilon^V = 0 & \text{in } \Omega_1^\varepsilon \cup \Omega_2^\varepsilon \cup (\mathbb{R}^2 \; \overline{\Omega_1^\varepsilon \cup \Omega_2^\varepsilon}), \\ [\sigma \partial_\nu h_\varepsilon^V] = (\sigma_0 - \sigma_1) V \cdot \nu & \text{on } \Gamma_1^\varepsilon, \\ [\sigma \partial_\nu h_\varepsilon^V] = (\sigma_1 - \sigma_2) V \cdot \nu & \text{on } \Gamma_2^\varepsilon, \\ h_\varepsilon^V \to 0 & \text{at } \infty. \end{cases} \tag{27}$$

By shifting the coordinate system, we can assume for simplicity that $x_0 = 0$. For our case, we can compute explicitly the function h_ε^V using polar coordinates:

$$h_\varepsilon^V(x) = \begin{cases} (\beta + \gamma) V \cdot x & \text{if } x \in \Omega_2^\varepsilon, \\ \beta V \cdot x + \gamma(\alpha \varepsilon)^2 \dfrac{V \cdot x}{|x|^2} & \text{if } x \in \Omega_1^\varepsilon, \\ \left(\beta \varepsilon^2 + \gamma(\alpha \varepsilon)^2 \right) \dfrac{V \cdot x}{|x|^2} & \text{if } x \in \mathbb{R}^2 \backslash \overline{\Omega_1^\varepsilon \cup \Omega_2^\varepsilon}, \end{cases} \tag{28}$$

where

$$\beta := \frac{(\sigma_1 + \sigma_2)(\sigma_0 - \sigma_1) - \alpha^2(\sigma_1 - \sigma_2)(\sigma_0 - \sigma_1)}{(\sigma_1 + \sigma_2)(\sigma_1 + \sigma_0) + \alpha^2(\sigma_1 - \sigma_2)(\sigma_0 - \sigma_1)},$$

and

$$\gamma := \frac{2\sigma_0(\sigma_1 - \sigma_2)}{(\sigma_1 + \sigma_2)(\sigma_1 + \sigma_0) + \alpha^2(\sigma_1 - \sigma_2)(\sigma_0 - \sigma_1)}.$$

Its gradient is given by

$$
\nabla h_\varepsilon^V =
\begin{cases}
(\beta + \gamma)V & \text{if } x \in \Omega_2^\varepsilon, \\[2ex]
\beta V + \gamma(\alpha\varepsilon)^2 \left[\dfrac{V}{|x|^2} - 2V \cdot x\, \dfrac{x}{|x|^4} \right] & \text{if } x \in \Omega_1^\varepsilon, \\[2ex]
\left(\beta\varepsilon^2 + \gamma(\alpha\varepsilon)^2 \right) \left[\dfrac{V}{|x|^2} - 2V \cdot x\, \dfrac{x}{|x|^4} \right] & \text{if } x \in \mathbb{R}^2 \setminus \overline{\Omega_1^\varepsilon \cup \Omega_2^\varepsilon}.
\end{cases}
\tag{29}
$$

Denoting

$$
\mathcal{E}_3(\varepsilon) := \int_{\Omega_1^\varepsilon} (\sigma_1 - \sigma_0)\nabla u_0 \cdot \nabla\left(\tilde{v}_\varepsilon - h_\varepsilon^V\right) dx,
\tag{30}
$$

$$
\mathcal{E}_4(\varepsilon) := \int_{\Omega_1^\varepsilon} (\sigma_1 - \sigma_0)(\nabla u_0 - \nabla u_0(x_0)) \cdot \nabla h_\varepsilon^V\, dx,
\tag{31}
$$

$$
\mathcal{E}_5(\varepsilon) := \int_{\Omega_2^\varepsilon} (\sigma_2 - \sigma_0)\nabla u_0 \cdot \nabla\left(\tilde{v}_\varepsilon - h_{\varepsilon,}^V\right) dx,
\tag{32}
$$

and

$$
\mathcal{E}_6(\varepsilon) := \int_{\Omega_2^\varepsilon} (\sigma_2 - \sigma_0)(\nabla u_0 - \nabla u_0(x_0)) \cdot \nabla h_\varepsilon^V\, dx.
\tag{33}
$$

Then we obtain

$$
\mathcal{F}_1(\varepsilon) = \int_{\Omega_1^\varepsilon} (\sigma_1 - \sigma_0)\nabla u_0(x_0) \cdot \nabla h_\varepsilon^V\, dx + \mathcal{E}_3(\varepsilon) + \mathcal{E}_4(\varepsilon)
$$

$$
= (\sigma_1 - \sigma_0)\nabla u_0(x_0) \cdot \int_{\Omega_1^\varepsilon} \nabla h_\varepsilon^V\, dx + \mathcal{E}_3(\varepsilon) + \mathcal{E}_4(\varepsilon).
$$

Using polar coordinates and integrating by parts, yields

$$
\mathcal{F}_1(\varepsilon) = \pi\left(1 - \alpha^2\right)\varepsilon^2 \beta(\sigma_1 - \sigma_0)\nabla u_0(x_0) \cdot V + \mathcal{E}_3(\varepsilon) + \mathcal{E}_4(\varepsilon),
$$

$$
\mathcal{F}_2(\varepsilon) = \int_{\Omega_2^\varepsilon} (\sigma_2 - \sigma_0)\nabla u_0(x_0) \cdot \nabla h_\varepsilon^V\, dx + \mathcal{E}_5(\varepsilon) + \mathcal{E}_6(\varepsilon)
$$

$$
= \pi\alpha^2\varepsilon^2 (\beta + \gamma)(\sigma_2 - \sigma_0)\nabla u_0(x_0) \cdot V + \mathcal{E}_5(\varepsilon) + \mathcal{E}_6(\varepsilon).
$$

After rearrangement, we get

$$
(a_\varepsilon - a_0)(u_0, v_\varepsilon) = \pi\varepsilon^2 \Lambda \nabla u_0(x_0) \cdot \nabla v_0(x_0) + \sum_{i=1}^{6} \mathcal{E}_i,
$$

where

$$
\begin{aligned}
\Lambda &:= \left(1 - \alpha^2\right)(1 + \beta)(\sigma_1 - \sigma_0) + \alpha^2(\sigma_2 - \sigma_0)(1 + \beta + \gamma) \\
&= \frac{2(1 - \alpha^2)\sigma_0(\sigma_1 - \sigma_0)(\sigma_1 + \sigma_2) + 4\alpha^2\sigma_0\sigma_1(\sigma_2 - \sigma_0)}{(\sigma_1 + \sigma_2)(\sigma_1 + \sigma_0) - \alpha^2(\sigma_1 - \sigma_2)(\sigma_1 - \sigma_0)}.
\end{aligned}
\tag{34}
$$

3.1.2 Variation of the linear form

Let us now turn to the asymptotic analysis of the variation

$$(l_\varepsilon - l_0)(v_\varepsilon) = \int_{\Omega_1^\varepsilon} (f_1 - f_0) v_\varepsilon dx + \int_{\Omega_2^\varepsilon} (f_2 - f_0) v_\varepsilon dx. \tag{35}$$

We can rewrite (35) as

$$(l_\varepsilon - l_0)(v_\varepsilon) = \pi\varepsilon^2 \left[(1-\alpha^2)(f_1 - f_0) + \alpha^2(f_2 - f_0)\right] v_0(x_0) + \mathcal{E}_7(\varepsilon) + \mathcal{E}_8(\varepsilon),$$

where

$$\mathcal{E}_7(\varepsilon) = \int_{\Omega_1^\varepsilon} (f_1 - f_0)\tilde{v}_\varepsilon dx, \quad \mathcal{E}_8(\varepsilon) = \int_{\Omega_2^\varepsilon} (f_2 - f_0)\tilde{v}_\varepsilon dx.$$

Again, it will be proved that $\mathcal{E}_7$ and $\mathcal{E}_8$ are small terms. Consequently we set

$$\delta l = \pi\left[(1-\alpha^2)(f_1 - f_0) + \alpha^2(f_2 - f_0)\right] v_0(x_0).$$

3.1.3 Variation of the cost function

Expression of $J_{p,\varepsilon}(u_\varepsilon) - J_{p,\varepsilon}(u_0)$. For simplicity of the calculus, we assume that p is even, then we have

$$J_{p,\varepsilon}(u_\varepsilon) - J_{p,\varepsilon}(u_0) = \frac{1}{p}\int_\Omega |u_\varepsilon|^p dx - \frac{1}{p}\int_\Omega |u_0|^p dx$$

$$= \frac{1}{p}\int_\Omega |(u_\varepsilon - u_0) + u_0|^p - \frac{1}{p}\int_\Omega |u_0|^p dx$$

$$= \frac{1}{p}\sum_{k=0}^{p} \binom{p}{k} \int_\Omega u_0^{p-k}(u_\varepsilon - u_0)^k dx - \frac{1}{p}\int_\Omega |u_0|^p dx$$

$$= \frac{1}{p}\sum_{k=2}^{p} \binom{p}{k} \int_\Omega u_0^{p-k}(u_\varepsilon - u_0)^k dx + \int_\Omega u_0^{p-1}(u_\varepsilon - u_0)dx.$$

Therefore

$$J_{p,\varepsilon}(u_\varepsilon) - J_{p,\varepsilon}(u_0) - DJ_{p,\varepsilon}(u_0)(u_\varepsilon - u_0) = \mathcal{E}_9(\varepsilon),$$

where

$$\mathcal{E}_9(\varepsilon) := \frac{1}{p}\sum_{k=2}^{p} \binom{p}{k} \int_\Omega u_0^{p-k}(u_\varepsilon - u_0)^k dx.$$

We will prove in the next section that $\mathcal{E}_9(\varepsilon) = o(\varepsilon^2)$, and thus $\delta J_1 = 0$.
Expression of $J_{p,\varepsilon}(u_0) - J_{p,0}(u_0)$. We have

$$J_{p,\varepsilon}(u_0) - J_{p,0}(u_0) = \frac{1}{p}\int_\Omega |u_0|^p dx - \frac{1}{p}\int_\Omega |u_0|^p dx = 0.$$

Consequently $\delta J_2 = 0$. Now, we are ready to state the main result of this paper.

Theorem 1.1 The topological asymptotic expansion of the functional J with respect to the insertion of small coated inclusion ω_ε is given by

$$J_{p,\varepsilon}(u_\varepsilon) - J_{p,0}(u_0) = \varepsilon^2 G(x_0) + o(\varepsilon^2),$$

where

$$G(x_0) = \pi \Lambda \nabla u_0(x_0) \cdot \nabla v_0(x_0) + \pi[(1-\alpha^2)(f_1 - f_0) + \alpha^2(f_2 - f_0)]v_0(x_0),$$

(36)

and

$$\Lambda := \frac{2(1-\alpha^2)\sigma_0(\sigma_1 - \sigma_0)(\sigma_1 + \sigma_2) + 4\alpha^2\sigma_0\sigma_1(\sigma_2 - \sigma_0)}{(\sigma_1 + \sigma_2)(\sigma_1 + \sigma_0) - \alpha^2(\sigma_1 - \sigma_2)(\sigma_1 - \sigma_0)}.$$

Remark 1 When $\sigma_1 = \sigma_2$ and $\alpha = 0$, the topological derivative defined in (36) becomes

$$G(x_0) = 2\pi\rho\sigma_0\nabla u_0(x_0) \cdot \nabla v_0(x_0) + \pi(f_1 - f_0)v_0(x_0), \quad \rho = \frac{\sigma_1 - \sigma_0}{\sigma_1 + \sigma_0}.$$

(37)

Expression (37) is known in the literature when the inclusion ω is an homogenous disk; see for instance ([17], Thm 4.3).

4. Estimates of the remainders

In this section the estimation for the remainders on the topological asymptotic expansion are presented. The results are derived by using simple arguments from functional analysis.

4.1 Preliminary lemmas

Lemma 1

i. For any vector $V \in \mathbb{R}^2$, $x_0 \in \Omega$ and positive radius R, we have

$$\|h_\varepsilon^V\|_{L^2(\Omega)} = O\left(\varepsilon^{3/2}\right),$$

$$\|\nabla h_\varepsilon^V\|_{L^p(\Omega)} = O\left(\varepsilon^{2/p}\right) \forall p > 1,$$

$$\|h_\varepsilon^V\|_{L^p(\Omega \setminus B(x_0,R))} + \|\nabla h_\varepsilon^V\|_{L^p(\Omega \setminus B(x_0,R))} = O(\varepsilon^2) \quad \forall p \geq 1.$$

ii. Given a function $\psi : \Omega \to \mathbb{R}^2$ which is θ Hölder continuous $(0 < \theta < 1)$ in a neighborhood of x_0 and consider the solution w_ε of the system:

$$\begin{cases} -\text{div}(\sigma_\varepsilon \nabla w_\varepsilon) = 0 & \text{in } \Omega_1^\varepsilon \cup \Omega_2^\varepsilon \cup (\Omega \ \overline{\Omega_1^\varepsilon \cup \Omega_2^\varepsilon}), \\ [\sigma\partial_\nu w_\varepsilon,] = (\sigma_0 - \sigma_1)\psi \cdot \nu & \text{on } \Gamma_1^\varepsilon, \\ [\sigma\partial_\nu w_\varepsilon] = (\sigma_1 - \sigma_2)\psi \cdot \nu & \text{on } \Gamma_2^\varepsilon, \\ w_\varepsilon = 0 & \text{on } \partial\Omega. \end{cases}$$

(38)

Then, we have

$$\|w_\varepsilon - h_\varepsilon^{\psi(x_0)}\|_{H^1(\Omega)} = o(\varepsilon). \tag{39}$$

Proof. The estimates *i*) in Lemma 1 follow directly from (28) and (29). Now, we prove the second part. Let $\varphi \in H_0^1(\Omega)$ be an arbitrary test function, then from (38), we have

$$\int_\Omega \sigma_\varepsilon \nabla w_\varepsilon \cdot \nabla \varphi dx = (\sigma_0 - \sigma_1) \int_{\Gamma_1^\varepsilon} \psi \cdot \nu \varphi ds + (\sigma_1 - \sigma_2) \int_{\Gamma_2^\varepsilon} \psi \cdot \nu \varphi ds.$$

Using Green's formula together with (27), we obtain

$$\int_\Omega \sigma_\varepsilon \nabla h_\varepsilon^{\psi(x_0)} \cdot \nabla \varphi dx = (\sigma_0 - \sigma_1) \int_{\Gamma_1^\varepsilon} \psi(x_0) \cdot \nu \varphi ds + (\sigma_1 - \sigma_2) \int_{\Gamma_2^\varepsilon} \psi(x_0) \cdot \nu \varphi ds.$$

Denote $\Theta_\varepsilon := w_\varepsilon - h_\varepsilon^{\psi(x_0)}$. It follows that

$$\int_\Omega \sigma_\varepsilon \nabla \Theta_\varepsilon \cdot \nabla \varphi dx = (\sigma_0 - \sigma_1) \int_{\Gamma_1^\varepsilon} (\psi - \psi(x_0)) \cdot \nu \varphi ds + (\sigma_1 - \sigma_2) \int_{\Gamma_2^\varepsilon} (\psi - \psi(x_0)) \cdot \nu \varphi ds. \tag{40}$$

Using the change of variable, the θ-Hölder continuity of ψ in the vicinity of x_0 and the trace theorem, we get for ε small enough

$$\left| \int_{\Gamma_2^\varepsilon} (\psi - \psi(x_0)) \cdot \nu \varphi ds \right| = \varepsilon \left| \int_{\Gamma_2} (\psi(\varepsilon x) - \psi(x_0)) \cdot \nu \varphi(\varepsilon x) ds \right|$$

$$\leq c \varepsilon^{1+\theta} \|\varphi(\varepsilon x)\|_{H^{1/2}(\Gamma_2)}$$

$$\leq c \varepsilon^{1+\theta} \|\varphi(\varepsilon x)\|_{H^1(\Omega_2)}$$

$$\leq c \varepsilon^\theta \|\varphi\|_{L^2(\Omega_2^\varepsilon)} + c \varepsilon^{\theta+1} \|\nabla \varphi\|_{L^2(\Omega_2^\varepsilon)}.$$

From the Hölder inequality and the Sobolev imbedding theorem, we obtain

$$\|\varphi\|_{L^2(\Omega_2^\varepsilon)} \leq c \varepsilon^{\frac{1}{p}} \|\varphi\|_{\frac{2p}{L^{p-1}}(\Omega_2^\varepsilon)} \leq c \varepsilon^{1/p} \|\varphi\|_{H^1(\Omega)}, \quad \text{for any } p > 1.$$

Therefore

$$\left| \int_{\Gamma_2^\varepsilon} (\psi - \psi(x_0)) \cdot \nu \varphi ds \right| \leq c \left(\varepsilon^{\theta+1/p} + \varepsilon^{\theta+1} \right) \|\varphi\|_{H^1(\Omega)}.$$

Analogously, we can prove that

$$\left| \int_{\Gamma_1^\varepsilon} (\psi - \psi(x_0)) \cdot \nu \varphi ds \right| \leq c' \left(\varepsilon^{\theta+1/p} + \varepsilon^{\theta+1} \right) \|\varphi\|_{H^1(\Omega)}.$$

From (40) and the first part of Lemma 1, we deduce that

$$\left| \int_{\Omega} \sigma_{\varepsilon} \nabla \Theta_{\varepsilon} \cdot \nabla \varphi dx \right| \leq c'' \left(\varepsilon^{\theta+1/p} + \varepsilon^{\theta+1} + \varepsilon^2 \right) \|\varphi\|_{H^1(\Omega)}. \tag{41}$$

Choosing $\varphi = \Theta$ and $p \in (1, \frac{1}{1-\theta})$ in (41), yield

$$\|\Theta\|_{H^1(\Omega)} = o(\varepsilon),$$

and the proof is completed.

Lemma 2 We have the following estimates

$$\|u_{\varepsilon} - u_0\|_{H^1(\Omega)} = O(\varepsilon), \tag{42}$$

$$\|v_{\varepsilon} - v_0\|_{H^1(\Omega)} = O(\varepsilon), \tag{43}$$

$$\|u_{\varepsilon} - u_0\|_{L^2(\Omega)} = o(\varepsilon), \tag{44}$$

$$\|v_{\varepsilon} - v_0\|_{L^2(\Omega)} = o(\varepsilon). \tag{45}$$

Proof. From the Poincaré inequality, we deduce that

$$\int_{\Omega} |u_{\varepsilon} - u_0|^2 dx \leq C \left(\int_{\Omega} |\nabla(u_{\varepsilon} - u_0)|^2 dx \right), \tag{46}$$

for some constant C independent of ε. Then, it suffices to show that

$$\int_{\Omega} |\nabla(u_{\varepsilon} - u_0)|^2 dx \leq C\varepsilon^2.$$

From (9), we obtain immediately that

$$a_{\varepsilon}(u_{\varepsilon} - u_0, v) = -(a_{\varepsilon} - a_0)(u_0, v), \quad \forall v \in H_0^1(\Omega). \tag{47}$$

According to (19), we get

$$(a_{\varepsilon} - a_0)(u_0, v) = \int_{\Omega_1^{\varepsilon}} (\sigma_1 - \sigma_0) \nabla u_0 \cdot \nabla v dx + \int_{\Omega_2^{\varepsilon}} (\sigma_2 - \sigma_0) \nabla u_0 \cdot \nabla v dx.$$

Using the fact that ∇u_0 is uniformly bounded on Ω_1^{ε} and Ω_2^{ε}, we obtain

$$|(a_{\varepsilon} - a_0)(u_0, v)| \leq |\sigma_1 - \sigma_0| \sup_{\Omega_1^{\varepsilon}} |\nabla u_0| |\Omega_1^{\varepsilon}|^{1/2} \|\nabla v\|_{L^2(\Omega)}$$

$$+ |\sigma_2 - \sigma_0| \sup_{\Omega_2^{\varepsilon}} |\nabla u_0| |\Omega_2^{\varepsilon}|^{1/2} \|\nabla v\|_{L^2(\Omega)}$$

$$\leq C\varepsilon \|\nabla v\|_{L^2(\Omega)},$$

and from (47), we obtain

$$\int_{\Omega} |\nabla(u_{\varepsilon} - u_0)|^2 dx \leq C\varepsilon^2.$$

This proves the asymptotic formula (42). Analogously we derive the estimate (43). The proof of (44) and (45) follows straightforwardly from [4, Lemma 9.3].

4.2 Asymptotic behavior of the remainders

In this subsection, we shall prove that $\mathcal{E}_i(\varepsilon) = o(\varepsilon^2)$ for $i = 1 \ldots 9$. We have

$$\mathcal{E}_2(\varepsilon) = \int_{\Omega_2^\varepsilon} (\nabla u_0 \cdot \nabla v_0 - \nabla u_0(x_0) \cdot \nabla v_0(x_0)) dx.$$

Using the regularity of u_0 and v_0 near x_0 and Taylor-Lagrange expansion, we straightforwardly obtain $\mathcal{E}_2(\varepsilon) \leq c\varepsilon^3$, and thus $\mathcal{E}_2(\varepsilon) = o(\varepsilon^2)$. Similarly, we can prove that $\mathcal{E}_1(\varepsilon) = o(\varepsilon^2)$. Let's now prove that $\mathcal{E}_4(\varepsilon) = o(\varepsilon^2)$ and $\mathcal{E}_6(\varepsilon) = o(\varepsilon^2)$. We have

$$\mathcal{E}_4(\varepsilon) := \int_{\Omega_1^\varepsilon} (\sigma_1 - \sigma_0)(\nabla u_0 - \nabla u_0(x_0)) \cdot \nabla h_\varepsilon^V dx,$$

and

$$\mathcal{E}_6(\varepsilon) := \int_{\Omega_2^\varepsilon} (\sigma_2 - \sigma_0)(\nabla u_0 - \nabla u_0(x_0)) \cdot \nabla h_\varepsilon^V dx.$$

Using Cauchy-Schwarz inequality, yields

$$|\mathcal{E}_6(\varepsilon)| \leq |\sigma_2 - \sigma_0| \left(\int_{\Omega_2^\varepsilon} |(\nabla u_0 - \nabla u_0(x_0))|^2 dx \right)^{1/2} \left(\int_{\Omega_2^\varepsilon} |\nabla h_\varepsilon^V|^2 dx \right)^{1/2}$$

$$\leq |(\beta + \gamma)(\sigma_2 - \sigma_0)| \|V\| \alpha \sqrt{\pi} \varepsilon \left(\int_{\Omega_2^\varepsilon} |(\nabla u_0 - \nabla u_0(x_0))|^2 dx \right)^{1/2}.$$

From the regularity of u_0 near x_0 and Taylor-Lagrange expansion, we obtain the bound $|\mathcal{E}_6(\varepsilon)| \leq c\varepsilon^{5/2}$. Analogously, we can show that $\mathcal{E}_4(\varepsilon) = o(\varepsilon^2)$. Let's now focus on $\mathcal{E}_3$ and $\mathcal{E}_5(\varepsilon)$. Using Hölder inequality, we obtain

$$|\mathcal{E}_3(\varepsilon)| = \left| \int_{\Omega_1^\varepsilon} (\sigma_0 - \sigma_1) \nabla u_0 \cdot \nabla (\tilde{v}_\varepsilon - h_\varepsilon^V) dx \right|$$

$$\leq |\sigma_0 - \sigma_1| \sup_{x \in \Omega} |\nabla u_0(x)| |\Omega_1^\varepsilon|^{1/2} \|\nabla (\tilde{v}_\varepsilon - h_\varepsilon^V)\|_{L^2(\Omega_1^\varepsilon)}$$

$$\leq C\varepsilon \|\nabla (\tilde{v}_\varepsilon - h_\varepsilon^V)\|_{L^2(\Omega_1^\varepsilon)}.$$

From Lemma 1, we deduce that

$$\|\nabla (\tilde{v}_\varepsilon - h_\varepsilon^V)\|_{L^2(\Omega_1^\varepsilon)} \leq \|\nabla (\tilde{v}_\varepsilon - h_\varepsilon^V)\|_{L^2(\Omega)} = o(\varepsilon).$$

Therefore, we conclude that $\mathcal{E}_3(\varepsilon) = o(\varepsilon^2)$. By the same techniques, we prove that

$$\mathcal{E}_5(\varepsilon) = o(\varepsilon^2).$$

Let's now check that $\mathcal{E}_7(\varepsilon) = o(\varepsilon^2)$ and $\mathcal{E}_8(\varepsilon) = o(\varepsilon^2)$. We have

$$|\mathcal{E}_7(\varepsilon)| \leq c \int_{\Omega_1^\varepsilon} |v_\varepsilon - v_0| dx.$$

From the Hölder inequality, we obtain

$$|\mathcal{E}_7(\varepsilon)| \leq c\varepsilon^{2/r}\|v_\varepsilon - v_0\|_{L^s(\Omega_1^\varepsilon)} \leq c\varepsilon^{2/r}\|v_\varepsilon - v_0\|_{L^s(\Omega)},$$

for all $r, s \in [1, +\infty]$ satisfying $\frac{1}{r} + \frac{1}{s} = 1$. Due to Lemma 2, we conclude that $\mathcal{E}_7(\varepsilon) = o(\varepsilon^2)$. Similarly, we obtain

$$\mathcal{E}_8(\varepsilon) = o(\varepsilon^2).$$

Let's now focus on

$$\mathcal{E}_9(\varepsilon) := \frac{1}{p}\sum_{k=2}^{p}\binom{p}{k}\int_\Omega u_0^{p-k}(u_\varepsilon - u_0)^k \, dx.$$

The Hölder inequality combined with the estimates in Lemma 2, yield

$$\mathcal{E}_9(\varepsilon) = o(\varepsilon^2).$$

5. Numerical experiments

For the numerical computation of the state and the adjoint, we use the finite element method. The computational domain Ω is the unit disk centered at the origin. The term source and the conductivities are given by

$$f_0 = 0, \quad f_1 = 0, \quad f_2 = 20 \text{ and } \sigma_0 = 1, \quad \sigma_1 = 0.5, \quad \sigma_2 = 30.$$

All the numerical computations are done under Matlab R2018a. To solve the optimization problem, we apply a fast non-iterative algorithm, based on the following steps:

Non-iterative algorithm.

---

a. Compute the solution of the problem (8) and the solution of the adjoint problem (20) in the unperturbed domain.

b. Compute the topological gradient G in (36).

c. Find the point x_0 where the topological derivative is the most negative.

d. Locate the inclusion ω at the point x_0.

5.1 Example 1

In this example, the Dirichlet boundary condition is given by $g = \sin(\theta), \theta \in [0, 2\pi]$. We present some numerical experiments using the functional J_p for different values of p ($p = 2, 10, 50$) and the functional K. The coated inclusion ω to be located in order to minimize the objective function, is composed of two concentric disks Ω_1 and Ω_2 with radius $r_1 = 0.2$ and $r_2 = 0.1$. We compute the position x_0 of ω using the proposed algorithm.

In **Figures 2–5**, the x_1-coordinate of the center of the inclusion is fixed to zero and the x_2-coordinate ≤ 0. We observe that the shape functional is decreasing with respect to the variation of the second coordinate x_2. **Figures 6–8** show the image of

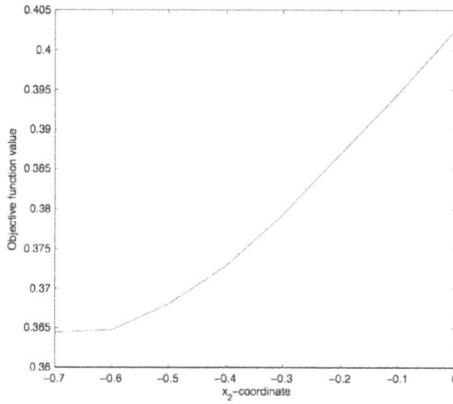

Figure 2.
Values of the objective function J_2 for variation of the x_2-coordinate of the center of the inclusion.

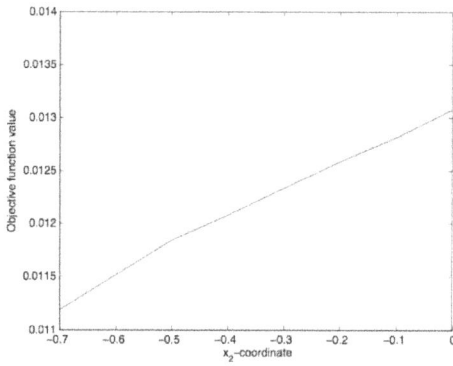

Figure 3.
Values of the objective function J_{10} for variation of the x_2-coordinate of the center of the inclusion.

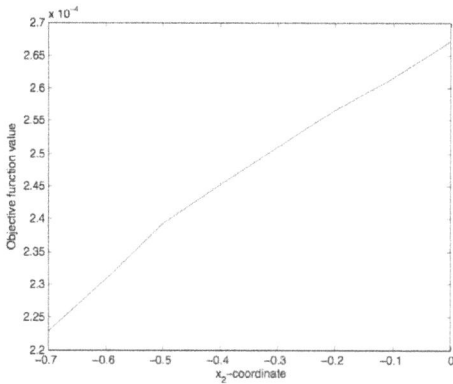

Figure 4.
Values of the objective function J_{50} for variation of the x_2-coordinate of the center of the inclusion.

topological gradient and the image of temperature distribution after the minimization process. **Figure 9** shows the image of the topological gradient corresponding to the shape functional J_{50}. For $p \geq 50$, we observe that the shape functional tends to

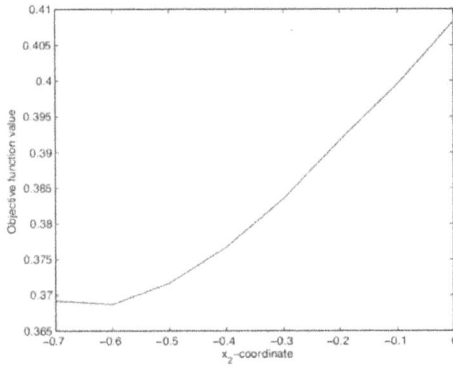

Figure 5.
Values of the objective function K for variation of the x_2-coordinate of the center of the inclusion.

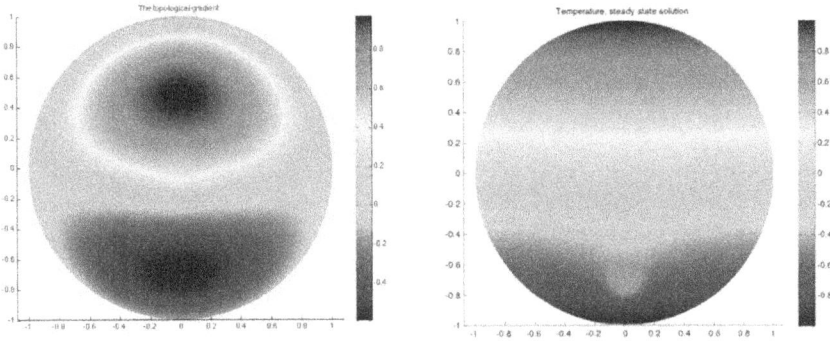

Figure 6.
On the left the topological derivative of the functional J_2 and on the right the temperature distribution relative to the position $x_0 = (-0.0115, -0.6915)$ of the coated inclusion given by Algorithm.

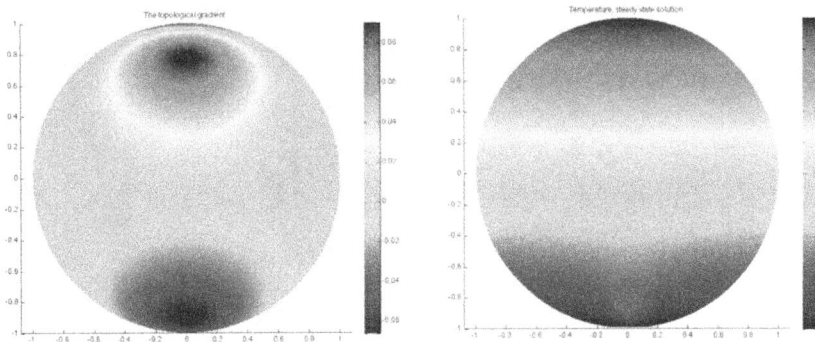

Figure 7.
On the left the topological derivative of the functional J_{10} and on the right the temperature distribution. $x_0 = (-0.0035, -0.9109)$ is the position given by Algorithm. In order to locate the inclusion far from the boundary $\partial\Omega$, we have taken an approximation of x_0, that is $x_a = (0, -0.75)$ for the optimization process.

115

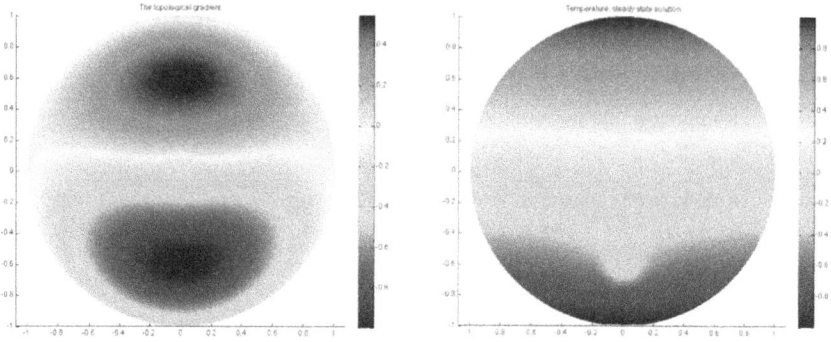

Figure 8.
On the left the topological derivative of the functional K and on the right the temperature distribution with respect the position $x_0 = (-0.0046, -0.5850)$ given by Algorithm.

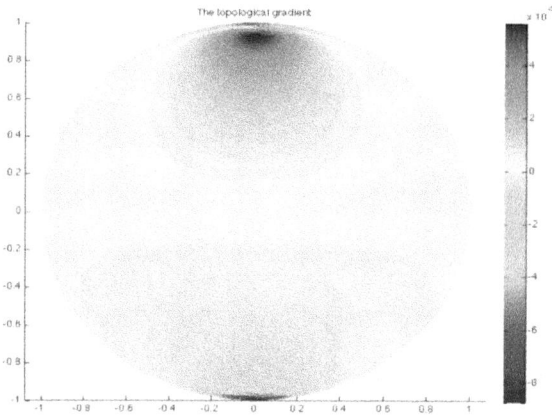

Figure 9.
The topological derivative of the functional J_{50}.

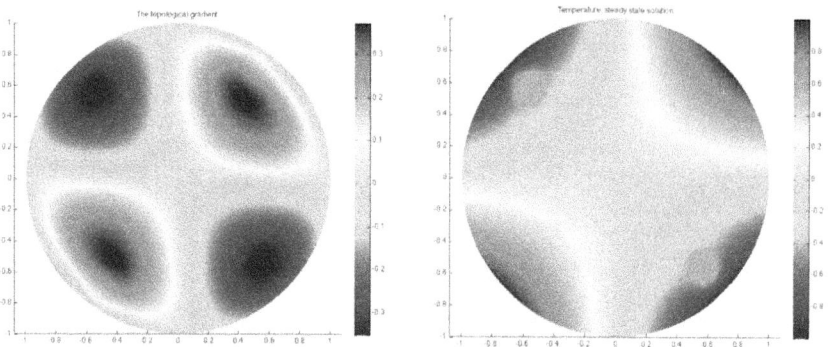

Figure 10.
On the left the topological derivative of the functional K and on the right the temperature distribution when the positions of the coated inclusions are given by Algorith.

zero and the image of the topological gradient is most negative on the boundary $\partial\Omega$. A suitable boundary condition that takes into account the effect of radiation and convection could be relevant in this case to solve properly the minimization problem.

5.2 Example 2

In this example, the Dirichlet boundary condition is given by $g = \sin(2\theta), \theta \in [0, 2\pi]$. We show some numerical experiments in the case of two coated inclusions using the proposed algorithm corresponding to the shape functional K.

Figure 10 depicts the image of the topological gradient of the shape functional K and the image of temperature distribution after the optimization process.

6. Conclusion

In this chapter, we have considered an optimization problem for a heat distribution in order to get a suitable thermal environment. We have performed the asymptotic expansion of the proposed shape functions with respect to the insertion of small coated inclusions. We have used a one shot algorithm based on the topological derivative to solve numerically the optimization problem. Numerical results are presented showing the efficiency of method. The proposed method can be extend to solve practical problems, like the thermal optimization of electrical multicables.

Author details

Zakaria Belhachmi[1†], Amel Ben Abda[2†], Belhassen Meftahi[2†]
and Houcine Meftahi[2,3*†]

1 Université de Haute Alsace, Mulhous, France

2 ENIT of Tunis, Tunisia

3 Institute of Computer Science Isikef/University of Jendouba, Tunisia

*Address all correspondence to: houcine.meftahi@enit.utm.tn

† These authors contributed equally.

IntechOpen

References

[1] G. Allaire. *Shape optimization by the homogenization method*, volume 146. Springer Science & Business Media, 2012.

[2] D. Bucur and G. Butazzo. Variational methods in shape optimization problems, 2006.

[3] A. Henrot and M. Pierre. *Variation et optimisation de formes*, volume 48 of *Mathématiques & Applications (Berlin) [Mathematics & Applications]*. Springer, Berlin, 2005. Une analyse géométrique. [A geometric analysis].

[4] F. Murat and L. Tartar. Calculus of variations and homogenization. In *Topics in the mathematical modelling of composite materials*, pages 139–173. Springer, 1997.

[5] O. Pironneau. *Optimal shape design for elliptic systems*. Springer Science & Business Media, 2012.

[6] J. Sokolowski and J.-P. Zolesio. Introduction to shape optimization. In *Introduction to Shape Optimization*, pages 5–12. Springer, 1992.

[7] S. Chaabane, M. Masmoudi, and H. Meftahi. Topological and shape gradient strategy for solving geometrical inverse problems. *Journal of Mathematical Analysis and applications*, 400(2):724–742, 2013.

[8] S. Garreau, P. Guillaume, and M. Masmoudi. The topological sensitivity for linear isotropic elasticity. *ECCMa 99*, 1999.

[9] A. A. Novotny, R. A. Feijóo, E. Taroco, and C. Padra. Topological sensitivity analysis. *Computer Methods in Applied Mechanics and Engineering*, 192(7):803–829, 2003.

[10] C. Alves and H. Ammari. Boundary integral formulae for the reconstruction of imperfections of small diameter in an elastic medium. *SIAM Journal on Applied Mathematics*, 62(1):94–106, 2001.

[11] D. J. Cedio-Fengya, S. Moskow, and M. S. Vogelius. Identification of conductivity imperfections of small diameter by boundary measurements. continuous dependence and computational reconstruction. *Inverse Problems*, 14(3):553–595, 1998.

[12] M. S. Vogelius and D. Volkov. Asymptotic formulas for perturbations in the electromagnetic fields due to the presence of inhomogeneities of small diameter. *ESAIM: Mathematical Modelling and Numerical Analysis*, 34(04):723–748, 2000.

[13] H. Ammari, H. Kang, M. Lim, and H. Zribi. Conductivity interface problems. part i: small perturbations of an interface. *Transactions of the American Mathematical Society*, 362(5): 2435–2449, 2010.

[14] D. Auroux. From restoration by topological gradient to medical image segmentation via an asymptotic expansion. *Mathematical and Computer Modelling*, 49(11):2191–2205, 2009.

[15] P. Gangl, U. Langer, A. Laurain, H. Meftahi, and K. Sturm. Shape optimization of an electric motor subject to nonlinear magnetostatics. *SIAM J. Sci. Comput*, 37(6):132–125, 2015.

[16] S. Amstutz. Sensitivity analysis with respect to a local perturbation of the material property. *Asymptotic analysis*, 49(1):87–108, 2006.

[17] S. Amstutz. A penalty method for topology optimization subject to a pointwise state constraint. *ESAIM: Control, Optimisation and Calculus of Variations*, 16(03):523–544, 2010.

Chapter 7

A Hybrid Approach for Solving Constrained Multi-Objective Mixed-Discrete Nonlinear Programming Engineering Problems

Satadru Roy, William A. Crossley and Samarth Jain

Abstract

Several complex engineering design problems have multiple, conflicting objectives and constraints that are nonlinear, along with mixed discrete and continuous design variables; these problems are inherently difficult to solve. This chapter presents a novel hybrid approach to find solutions to a constrained multi-objective mixed-discrete nonlinear programming problem that combines a two-branch genetic algorithm as a global search tool with a gradient-based approach for the local search. Hybridizing two algorithms can provide a search approach that outperforms the individual algorithms; however, hybridizing the two algorithms, in the traditional way, often does not offer advantages other than the computational efficiency of the gradient-based algorithms and global exploring capability of the evolutionary-based algorithms. The approach here presents a hybridization approach combining genetic algorithm and a gradient-based approach with improved information sharing between the two algorithms. The hybrid approach is implemented to solve three engineering design problems of different complexities to demonstrate the effectiveness of the approach in solving constrained multi-objective mixed-discrete nonlinear programming problems.

Keywords: multi-objective, mixed-discrete, constrained, nonlinear, genetic algorithm, gradient-based optimization

1. Introduction

Many engineering design problems require simultaneous optimization of multiple, often competing, objectives. Unlike in single-objective optimization, a multi-objective problem with competing objectives has no single solution. An optimum solution with respect to only one objective may not be acceptable when measured with respect to the other objectives. Multi-objective problems have a number of solutions called the Pareto-optimal set, named after Vilfred Pareto [1], that represent the range of best possible compromises amongst the objectives. Traditional gradient-based optimization algorithms are capable of addressing the multi-objective problems by converting the problem into a single-objective formulation.

On the other hand, evolutionary algorithms (EAs)[1] are well suited for the multi-objective problems as they can evolve to a set of designs that represent the Pareto frontier in a single run of the algorithm [2, 3]. As a result, EAs often find application to address multi-objective problems. Despite the popularity of these algorithms to solve a wide range of problems, they, like all non-gradient meta-heuristic searches, have issues with computational cost and rate of convergence to the Pareto frontier. After some number of generations, the candidate solutions may begin to exhibit little or no improvement. Modified versions of these algorithms exist which improve the convergence rate [4]. However, hybridizing EAs with an efficient gradient-based algorithm may significantly improve the convergence rate and has demonstrated the ability to solve multi-objective problems more efficiently than the EA alone [3]. Hybridization of an EA with a gradient-based local search algorithm has started to gain popularity owing to its promising capabilities to address the demerits of many optimization algorithms when used independently.

The genetic algorithm (GA) [5] is a class of EA and is a well-known population-based global search algorithm. Apart from its ability to explore the design space, GA is also capable of handling both discrete and continuous type design variables. This makes the GA an ideal choice to address problems that combine both discrete and continuous variables. However, the GA, like other EAs, does not provide any proof of convergence, and the GA cannot directly enforce constraints. Commonly, constraint handling for a GA search uses a penalty approach such that the fitness function reflects the objective function value and accounts for violated constraints. This generally requires the use of penalty multipliers to adjust the "strength" with which the penalty impacts the fitness function and selecting suitable penalty multipliers is often difficult. Further, for multi-objective problems, the different scaling or magnitude of the objectives can complicate selecting appropriate penalty multipliers.

On the other hand, Sequential Quadratic Programming (SQP) [6], is a well-known gradient-based search algorithm that directly handles constraints and provides proof of convergence to local optima using Karush-Kuhn-Tucker (KKT) optimality criteria [7]. Because SQP uses gradient information, it is a computationally efficient search algorithm. However, SQP cannot handle discrete design variables or discontinuous functions and has difficulty with multi-modal functions. Therefore, both of these (GA and SQP) well-known optimization algorithms have their own pros and cons that limit their individual applicability to fully address constrained multi-objective problems that combine both continuous and discrete type design variables. Combining the GA with SQP creates a hybrid approach that improves the overall optimization process for constrained mixed-discrete nonlinear programming problems (MDNLP).

The chapter presents a combination of the two-branch tournament GA for multi-objective problems with an SQP-based local search implementation of the goal attainment problem formulation allowing an improved information sharing between the two algorithms. To the best of the authors' knowledge, there exists no work that emphasizes the process of hybridization combining an N-branch tournament selection GA with the goal attainment formulation as the local search in a compatible manner and then demonstrates application of the approach to solve a hard-to-solve constrained multi-objective, mixed-discrete nonlinear optimization problem. Later in the chapter, the hybrid approach is applied to solve a three-bar truss problem, a ten-bar truss problem, and a greener aircraft design optimization

---

[1] Here, the term "evolutionary algorithm" encompasses all population-based search algorithms that use features inspired by biological evolution.

problem – all representatives of constrained multi-objective, mixed-discrete nonlinear programming problem. The truss problems have basis in test problems for structural optimization, and the motivation to select a greener aircraft design optimization problem arises from the increased concern about the environmental impact of the growing air transportation system.

2. Literature review

The ability of the EAs to evolve to a Pareto-frontier as the generation progresses makes them an ideal choice for several multi-objective optimization problems. Vector Evaluated GA (VEGA), proposed by Schaffer [8] back in 1985, is one of the earlier versions of multi-objective GA. Several multi-objective EAs are developed since then including Multi-Objective Genetic Algorithm (MOGA) [9], Strength Pareto Evolutionary Algorithm [10], Non-dominated Sorted Genetic Algorithm (NSGA) [11] to mention a few popular ones.

Coello [2, 12] has conducted comprehensive literature surveys of various evolutionary multi-objective techniques. Konak et al. [13] compared various multi-objective optimization algorithms and provides a set of guidelines to follow while developing a multi-objective algorithm. Their effort primarily lies in guiding researchers with very little background in MOGA and making them familiar with the ideas and approaches of multi-objective optimization.

One such multi-objective algorithm named Non-dominated Sorting Genetic Algorithm (NSGA), developed by Srinivas and Dev [11] – arguably one of the most widely used multi-objective EAs – uses the concept of non-dominated sets originally proposed by Goldberg in his book on Genetic Algorithm and Machine Learning [14]. The NSGA approach maintains sets of non-dominated individuals, with the first set of individuals not dominated by any other individuals in the population. The second set finds the new set of non-dominated individuals after excluding the individuals from the first set. This step continues until all the individuals in the population are categorized inside the non-dominated sets.

A majority of these multi-objective algorithms, in some form, require an assignment of a scalar measure of a fitness value to the individuals in the population. As an example, MOGA [9] and NSGA [11] assign a fitness value based on a ranking scheme depending on the individual's levels of domination. The two-branch tournament selection genetic algorithm presented by Crossley et al. [15] uses a tournament selection scheme that chooses parents considering both the objectives directly in the fitness functions. The individuals are evaluated based on their fitness across both the objectives. The overall process remains the same as that of a traditional GA. However, the only difference appears in the tournament selection operator. During the tournament selection step, the algorithm selects 50% of the parents based on the fitness value associated with the first objective, that is, the individuals are evaluated solely with respect to the first objective without consideration of the other objective. These selected parents are by nature strong in objective 1, or Φ_1-strong. Similarly, the tournament selects the remaining 50% of the parents based on the fitness value associated with the second objective. This second 50% are Φ_2-strong parents. With this parent selection approach, randomly choosing the selected parents to pair off for crossover, ideally would result in the following distribution of matches: 25% $\Phi_1 - \Phi_1$ type parents, 25% $\Phi_2 - \Phi_2$ type parents, and 50% are mixed i.e., $\Phi_1 - \Phi_2$ type parents.

The hybrid approach, presented in this chapter, uses this two-branch tournament selection GA as the global search optimizer and combines with a gradient-based approach to refine the search using a novel information sharing concept in the process of hybridization. The unique tournament selection strategy of the two-branch tournament GA allows to understand the underlying trait of the parents, i.e., if they are Φ_1 or Φ_2 strong, and this information is later leveraged during the crossover step to obtain children with certain desired traits.

Another challenge with multi-objective EAs is their ability to enforce constraints. Unlike gradient-based methods, which use constraint gradient information to guide the search in the feasible direction, no such constraint gradient is available for EAs. There have been several efforts to handle the constraints for EAs; however, not all of these methods strictly or directly enforce the problem constraints. The penalty function approach is arguably the most widely known of the various approaches to handle constraints in EAs. Assuming a minimization problem, this approach adds a penalty to the objective function when constraints are violated [14].

Another simple approach includes ignoring any infeasible design solution; because this does not differentiate between constraints that are close to the constraint boundaries and those that are far apart, this constraint handling method is inefficient.

Binh and Korn [16] suggested a method to assign fitness to individuals based on combining both the objective function vector as well as the degree to which the individual violates the constraint. Infeasible individuals are categorized into different classes based on how close or how far they are to the constraints boundaries.

Fonseca and Fleming [17] proposed a priority-based constraint handling strategy where search is first driven for feasibility followed by optimality by assigning high priority to constraints and low priority to objective functions. Although there are various techniques to "handle" constraints in EAs, "enforcing" them in a robust way is still an open issue. This is another motivation to pursue the hybrid approach that leverages the efficacy of gradient-based search to enforce the problem constraints.

Further, these population-based searches have issues with computational cost and rate of convergence to the Pareto frontier. After some number of generations, the candidate solutions may begin to exhibit little or no improvement. Modified versions of the algorithms work to improve the convergence rate [4, 18]; however, hybridizing EAs or GAs with an efficient gradient-based algorithm can significantly improve the convergence rate, thereby reducing the computational cost. Hybridization of an EA or GA with a gradient-based local search algorithm is not new. There are numerous references demonstrating how hybridization may improve the quality of the search for both single objective and multi-objective problem formulations; these include, but are not limited to, those appearing in [3, 19–32]. The local search can be considered as the local learning that takes place in an individual throughout its lifespan. Some of the approaches apply the local search to the final non-dominated set, while some techniques apply local search to all or many individuals of the population as the generation progresses.

The effort here extends the previous effort by Lehner and Crossley [27] to include a multi-objective formulation and combine the advantage of the hybrid approach with an novel information sharing technique between the global and the local search. The two-branch tournament selection GA algorithm globally explores the design space handling both discrete and continuous type variables, while the gradient-based approach sees only the continuous variables in a goal attainment formulation and seeks to efficiently refine the population based on the information passed on by the top-level GA while enforcing all the problem constraints.

3. Methodology and approach

The hybrid approach presented in this chapter combines the two-branch tournament GA (see **Figure 1**) for the global search [15] and the goal attainment SQP

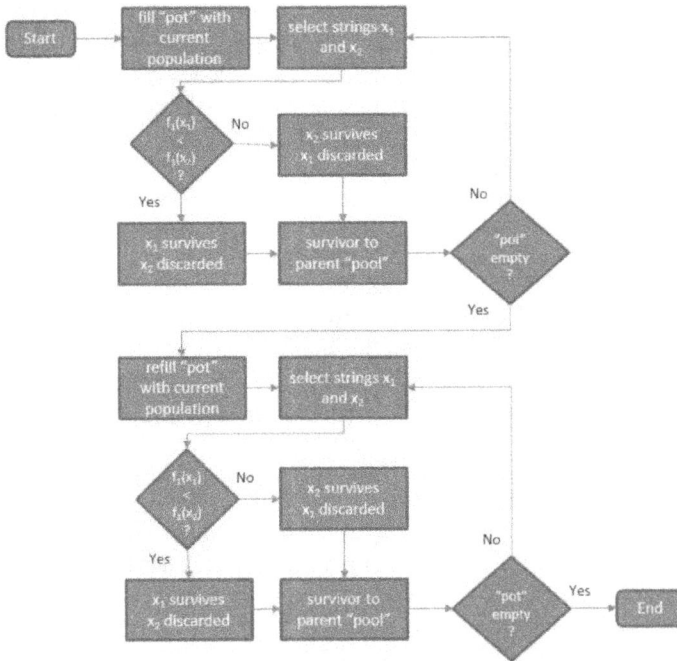

Figure 1.
Original two branch tournament selection GA for two objective problems. Adapted from reference [15].

algorithm provided in the function *fgoalattain* available from the MATLAB Optimization Toolbox [33] as the local search. For solution via hybrid approach, the problem statement contains two levels, as appears below.

3.1 Level I: Two-branch tournament genetic algorithm

The top level of the problem, which the GA sees as its optimization problem, is a bound constrained (i.e., only side constraints on the continuous design variables) multi-objective minimization problem that uses the two-branch tournament selection technique with some modification to include the local search. This level includes both discrete and continuous design variables of the original problem. The continuous variables in this level, x_c^0, are the initial values (starting point) for the local search problem. This way, the GA acts like a guide for a sequential multi-start approach as it searches the combined discrete and continuous design space. The top level formulation appears below:

$$\underset{x_d, x_c^0}{\text{Maximize}} : \left\{ \begin{array}{c} f_1(x_d, x_c^0) \\ f_2(x_d, x_c^0) \end{array} \right\}$$

Subject to : $\qquad\qquad\qquad\qquad\qquad\qquad\qquad\qquad$ (1)

$$(x_c)_i^L \le (x_c)_i \le (x_c)_i^U$$

$$(x_d)_i \in A, B, C, D, \dots \text{(discrete variables)}$$

In the original two-branch tournament selection GA, the tournament step selects 50% of the parents based on the fitness value associated with the first objective.

These parents are by nature strong in objective 1 or Φ_1-strong. Similarly, the tournament selects the remaining 50% of the parents based on the fitness value associated with the second objective. This second 50% are Φ_2-strong parents. With this parent selection approach, randomly choosing the selected parents to pair off for crossover would result, on average, in the following distribution of matches: 25% Φ_1-Φ_1 type parents, 25% Φ_2-Φ_2 type parents, and 50% are mixed i.e., Φ_1-Φ_2 type parents. This has the effect of generating many compromise solutions near the middle of the Pareto frontier, potentially limiting the spread and quality of the Pareto-front. The approach described in this chapter improves the spread and quality of the Pareto front by pairing off the parents in a more prescribed manner. A flowchart depicting how the modified two-branch tournament GA interacts with the gradient-based (SQP) for local search appears in **Figure 2**.

With a given goal f_i^G, a starting point for the continuous variables x_c^0 and a set of discrete values x_d, the goal attainment problem formulation, for each individual in the GA-level population, seeks to find the optimal design x_c^*. The goal attainment problem formulation also assigns the fitness value to the individuals, thereby waiving off the need of fitness evaluation at the GA-level (level I). Using the fitness information of these populations, a new set of goals are generated for the next iteration following the tournament selection, crossover and mutation steps of the two-branch tournament GA algorithm. The resulting design x_c^* need not conform to the binary Gray coding scheme implemented to represent the chromosome of each individual in the population. The effort here employs a Lamarckian strategy [20], that updates the chromosomes of the individual to conform to the gray-coding scheme of the GA.

Figure 3 demonstrates the parent selection process of the new two-branch tournament selection GA and the goal assignment technique with a simple example. The approach starts with a population size of $8n$, where n is any positive integer (**Figure 3** assumes a population size of 8; i.e., $n = 1$). After the two-branch tournament selection process, $4n$ parents are Φ_1-strong and the other $4n$ parents are Φ_2-strong. These parent groups remain in two separate parent pools. An additional step after the two branch selection process further categorizes these parents into

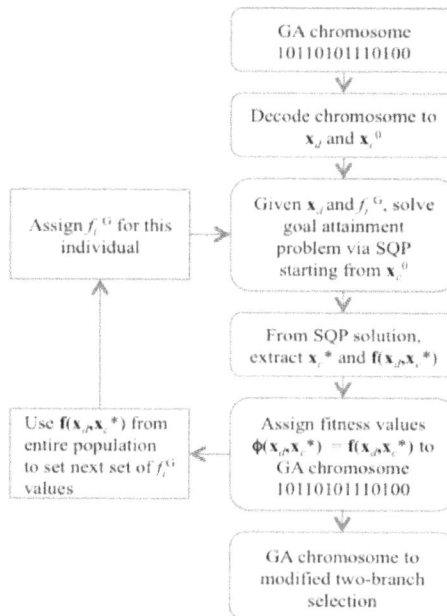

Figure 2.
Modified two branch tournament selection GA and SQP interaction.

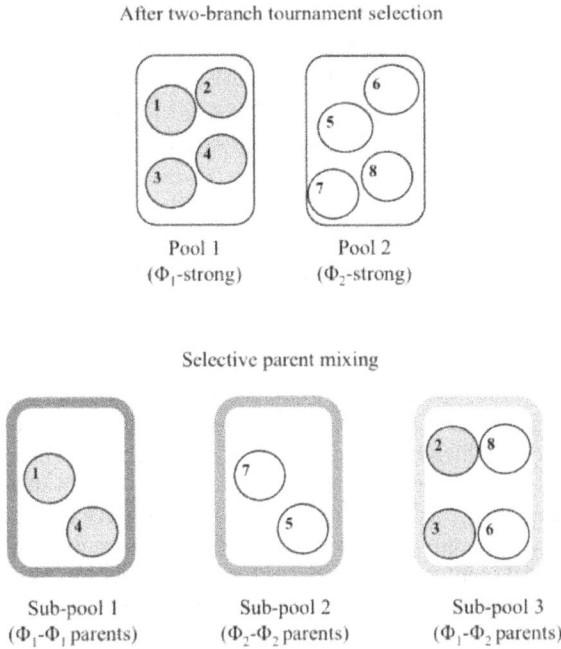

Figure 3.
Selective parent mixing strategy.

sub-pools to ensure a prescribed mix of Φ_1-strong and Φ_2- strong parents for crossover. To begin, half of the parents from pool 1 (which contains Φ_1-strong parents) are randomly moved to sub-pool 1. This sub-pool contains Φ_1-Φ_1 type parents paired for crossover and leads to offspring that will likely be Φ_1-strong. Similarly, half of the parents from pool 2 are randomly moved to sub-pool 2 to form Φ_2-Φ_2 type paired parents. Sub-pool 3 pairs parents so that a Φ_1-strong parent and a Φ_2-strong parent form children via the crossover operation. These would create children that have features from both Φ_1 and Φ_2 strong parents. This modification to the original two-branch tournament selection approach leads to a more pre-scribed, yet diversified, set of parents in each pool for the crossover, somewhat analogous to the idea of breeding for plant hybridization.

3.2 Level II: sequential quadratic programming

The lower-level problem presented to the SQP algorithm refines the population of the GA by searching the continuous variable space and helps the hybrid algorithm converge to the Pareto frontier at a faster rate. The *fgoalattain* algorithm, available in MATLAB, converts the multi-objective algorithm into a single-objective optimiza-tion problem by converting all the objectives into a set of inequality constraints and minimizes a slack variable γ (also called the attainment factor) as the objective.

$$
\text{Given} : x_c^0, x_d, f_i^G
$$
$$
\text{Minimize} : \gamma
$$
$$
\text{Subject to} : f_i(x_c) - \alpha_i \gamma \le f_i^G
$$
$$
g_j(x_c) \le 0
$$
$$
h_k(x_c) = 0
$$
$$
(x_c)_i^L \le (x_c)_i \le (x_c)_i^U
$$

(2)

This goal attainment formulation seeks to attain values for the objectives close to a set of predefined goal values, f_i^G, without violating any of the problem constraints $g_i(x) \leq 0$ and $h_k(x) = 0$. The weight values, α_i, are set as the absolute of the corresponding goal values, f_i^G, based on the guidance in [34]. This prevents scaling issues with objectives of various dimensions and magnitudes. The solution to this problem describes the set of continuous variables x_c^* that minimizes γ and satisfies all constraints; the values of $f_i(x_c^*)$–the fitness value of the individual in the population– are returned to the GA-level for the use in the two-branch tournament selection.

The *fgoalattain* formulation needs a defined goal point in the objective space, and the algorithm tries to find a design as close as possible to these goal values. **Figure 4** illustrates the goal point assignment task for each newly created individual in the sub-pools following the example presented in **Figure 3**. In **Figure 4**, the points indicate the child "designs" from a set of parents; e.g., C1$_{(1-4)}$ is the first child from the crossover of parent 1 and parent 4. The color of the symbol indicates the parent sub-pool from which the child designs were generated. Therefore, in this example, there are two children generated from parents 1 and 4 in sub-pool 1, which are indicated with the light blue color to match **Figure 3**. There are four children generated from sub-pool 2 and two children from sub-pool 3.

To assign the goal point values, the hybrid approach first identifies the local ideal point in each generation. This ideal point is the combination of the lowest f_1 and f_2 values in the current population. For this effort, the utopia point (which includes some tolerance to give the utopia smaller–or better–f_1 and f_2 values than the ideal point with the intent of encouraging under achievement in the goal attainment problem) is set as 0.95 times the local ideal point. In subsequent generations, any new objective value smaller than the corresponding value in the current utopia point replaces that current value in the utopia point. This makes the utopia point dynamic with each generation. For two-objective problems, two perpendicular lines originate from the utopia point and extend infinitely into the objective space. These straight lines appear as dashed lines in **Figure 4**.

To assign a goal point to an individual, the approach defines a vector that originates from an individual and ends to where the vector intersects with either of the dotted lines. The point of intersection becomes the goal point for that individual. Children of parents from sub-pool 1, the Φ_1-strong sub-pool, receive a goal vector with slope of zero in the objective space. These are the horizontal arrows in **Figure 4**.

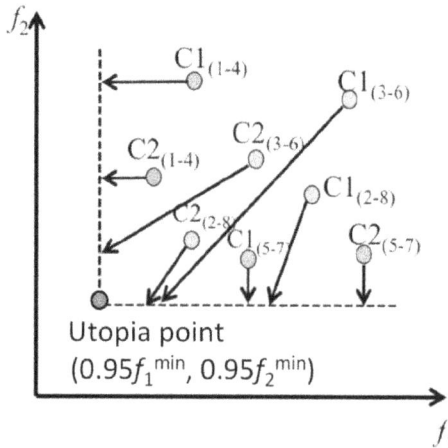

Figure 4.
Goal assignment technique.

These goal points would seek the most improvement along the direction of objective 1. Similarly, children of parents from sub-pool 2 receive a goal vector with 90 degree slope in the objective space. This ensures improvement along the direction of objective 2. Lastly, the children from sub-pool 3 parents receive goal vectors relative to their spatial location in the objective space. An individual closer towards objective 1 will have a vector inclined more towards improvement in objective 1 and vice versa.

Referring back to **Figure 3**, parents 1 and 4 from sub-pool 1 create children $C1_{(1-4)}$ and $C2_{(1-4)}$. This indicates the first child of parent 1 and 4 and the second child of parent 1 and 4 respectively. During the SQP search, these children have goal points that will minimize along the direction of f_1 without increasing their current values of f_2. $C1_{(5-7)}$ and $C2_{(5-7)}$ result from sub-pool 2, and the local search will seek to improve f_2 without increasing f_1. $C1_{(3-6)}$, $C2_{(3-6)}$, $C1_{(2-8)}$ and $C2_{(2-8)}$ all result from sub-pool 3, and they will have different goal points for their local searches to improve both f_1 and f_2. This modified parent selection and goal assignment strategy, via the hybrid formulation, seeks to exploit the tournament selection process of the two-branch tournament GA and tailor the local search for children, depending on traits of their parents.

Although the approach seems robust in enforcing constraints via goal attainment formulation, there may be instances when no feasible solution exists to the goal attainment formulation for a given set of discrete variables. In such cases, the local search will not be able to return a feasible solution and the fitness function receives a severe penalty in the GA-level in an effort to discard such discrete design choices from the population. This severe penalty has some resemblance to the approach of ignoring infeasible designs that was criticised above; however, because the situation where no locally-feasible design exists results from a specific combination of discrete variables, there is no analog to having a "nearly feasible" design with a slightly violated constraint. Severely penalizing such infeasible designs for certain combinations of discrete variable choices, in this context, is appropriate.

4. Application to engineering design problems

To demonstrate the efficacy of the hybrid approach in solving constrained multi-objective MDNLP problems, we solve three different engineering test problems with varying difficulties - a three-bar truss, a ten-bar truss, and greener aircraft design problem.

4.1 Three-bar truss problem

For the three-bar truss problem (see **Figure 5**), the problem formulation includes the objectives of minimizing the weight of the truss and minimizing the deflection of the free node. The deflection of a node is calculated as the resultant of the deflections in both the x and y directions. The problem consists of six design variables, of which three are continuous and three are discrete. The continuous variables describe the cross-sectional area of the three bars while the discrete variables describe the material selection properties of these bars. The details of the continuous design variables and their design bounds appear in **Table 1**. For this problem, four discrete material selection choices are available for each element and include aluminum, titanium, steel, and nickel options. The yield stress for every bar acts as a constraint for the problem (total three constraints), not allowing the stress in the bar to go beyond that upper limit. References [35, 36] provide more details about the three-bar truss problem. For the hybrid approach, the GA population is limited to 8 individuals while setting the upper limit for the number of generations

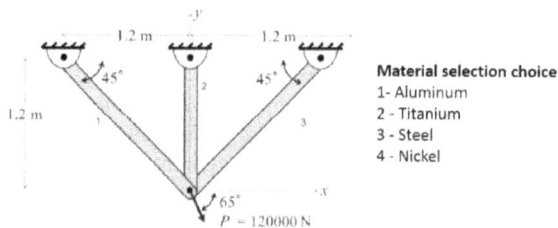

Figure 5.
Three bar truss problem.

Design variables	Lower bound	Upper bound
Cross-sectional area of bar 1 [cm^2]	0	5
Cross-sectional area of bar 2 [cm^2]	0	5
Cross-sectional area of bar 3 [cm^2]	0	5

Table 1.
Continuous variables for three-bar truss problem.

to 50. The probability of crossover is set to 0.5 and the mutation rate is fixed at 0.005. The continuous and discrete variables uses 8 and 2 bits respectively in the Gray-coded binary scheme.

The resulting Pareto frontier for the three-bar truss problem appears in **Figure 6(a)**. The plot shows the Pareto frontier has a good spread, leading to a total of 248 non-dominated points as solutions to the optimization problem. The visible trend in the non-dominated design set indicates that as the weight of the three bar truss system increases, they are accompanied by similar increases in the cross-sectional area of the bars with the material selection choice gradually shifting to steel for all the three bars. Aluminum or nickel never appeared as the material selection choice in the first two bars. The designs visible in the top left corner of the Pareto front in **Figure 6(a)** correspond to high displacement and low weight designs. The separated cluster of points (six designs) visible at the bottom right corner of the Pareto frontier corresponds to low displacement and high weight designs, with the maximum weight design having a material combination of all steel bars.

For the three-bar truss problem, only 64 possible combinations of discrete design variables exist. Hence, it is possible to perform a complete enumeration of the discrete design space and get a sense of the shape of the true Pareto front and help assess the performance of the hybrid approach. This led the authors to compare the hybrid approach (and the original two-branch tournament selection GA[2]) with a gradient-based weighted sum approach for this three-bar truss problem. The weighted sum approach converts the multi-objective problem formulation into a single objective problem by assigning weights to both the objectives and solves the single objective problem with the gradient-based approach using MATLAB's *fmincon* solver [33].

First, the objectives are normalized using the utopia point. Next, objective 1 is assigned a weight w that varies from 0 to 1 in a step increment of 0.05. The weight for the second objective is set to $1 - w$. For each possible combination of discrete

---

[2] The original two-branch tournament selection GA was proposed for unconstrained problems. In this example, the problem constraints in the original two-branch tournament selection GA (used for comparison) are handled using an exterior penalty approach.

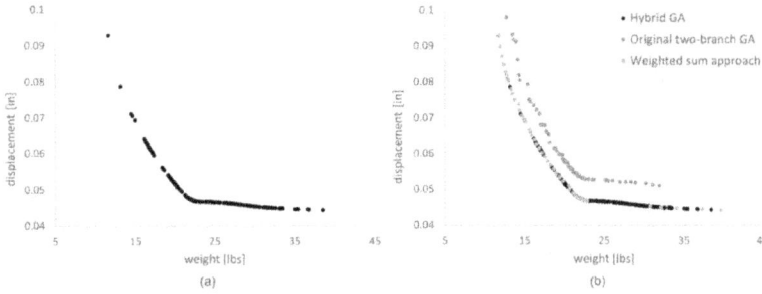

Figure 6.
Pareto front for the three-bar truss problem and its comparison with the other approaches. (a) Pareto front for the three-bar truss problem using the hybrid approach. (b) Comparison of Pareto frontier obtained using the hybrid approach, a weighted sum approach and the original two-branch tournament GA approach.

variable choice and a given weight pair, the approach leads to a single point in the objective space. The weighted sum approach then conducts gradient-based search for all 21 different weight pairs corresponding to each of the 64 possible discrete combination choices. The resulting Pareto frontier using the weighted sum approach is compared with the hybrid approach and the original two-branch tournament GA approach in **Figure 6(b)**. The original two-branch tournament GA finds an inferior set of solutions, possibly due to the lack of local search feature, and the set of solutions also has a reduced spread across the Pareto frontier. On the other hand, the weighted sum approach with complete enumeration on the material selection choices has a slightly better spread compared to the hybrid approach but with fewer non-dominated points.

Figure 7 compares how the Pareto frontier evolved with generations using the original two-branch tournament GA and the proposed hybrid approach. As expected, without the local search feature, the original two-branch tournament selection GA shows distinct improvement in both the quality and the spread of the Pareto front as the generation progresses. That is, the black diamonds (non-dominated set after second generation) are replaced with better non-dominated designs as the generation progresses. However, in the hybrid case, we start to see the shape of the final Pareto front immediately after the second generation. As the generation progresses further, more points get added to the list of non-dominated designs. This is due to the multi-start approach where the top-level GA populates various possible combinations of the discrete material selection choices and the local gradient-based search then improves these designs by varying the continuous design variables. The hybrid approach is able to rapidly get to the final Pareto front

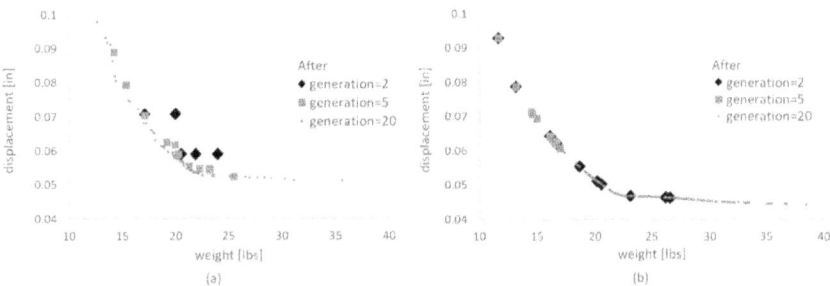

Figure 7.
Evolution of the non-dominated sets as the generation progresses. (a) Original two-branch tournament selection GA. (b) Proposed hybrid approach.

at the expense of increased number of function evaluations needed by the gradient-based local search.

4.2 Ten-bar truss problem

Next, the hybrid approach solves a more difficult and challenging version of the three-bar truss problem – a ten-bar truss. Similar to the three-bar truss problem, the ten-bar truss has the competing objectives that include minimizing the weight of the ten-bar truss system and minimizing the resultant displacement of any of the free nodes. The displacement is taken as the absolute of the maximum calculated displacement among all the bar elements. This problem consists of twenty design variables – ten continuous type and ten discrete type. The continuous variables describe the cross-sectional diameters of the ten bars, ranging from 0.1 cm^2 to 40 cm^2, while the discrete variables specify the material selection properties of these bars. Like the three-bar problem, the four discrete material choices available for each bar include aluminum, titanium, steel, and nickel. However, this problem has over one million possible combinations of the discrete choices ($4^{10} = 1,048,576$) making complete enumeration of the discrete design space computationally prohibitive, unlike the three-bar truss. References [35, 36] provide more details about the ten-bar truss problem considered in this study.

Figure 8(a) compares the Pareto front obtained using the hybrid approach after 20 GA generations with the Pareto frontier obtained using the two-branch tournament selection GA after 100 generations. The figure shows both the approaches performed well for this problem with the two-branch tournament selection GA resulting a better spread in the low weight/high displacement region of the objective space, whereas the hybrid GA has a better spread in the low displacement/high weight region. **Figure 8(b)** shows how the non-dominated set evolved as the generation progresses using the hybrid approach. We see a similar trend as that of the three-bar truss problem. That is, there is not much significant change in the final shape of the Pareto front other than the increase in the number of non-dominated designs as the generation progresses. However, this time there is slight improvement in the quality of the Pareto front (the red non-dominated set obtained after generation 20 is slightly better than the blue or the black non-dominated designs obtained at generation 5 and 2 respectively).

For the three-bar example, a majority of the improvements across the objective space are due to the gradient-based local search's ability to obtain designs with better cross-sectional area. With only 64 possible material selection combinations,

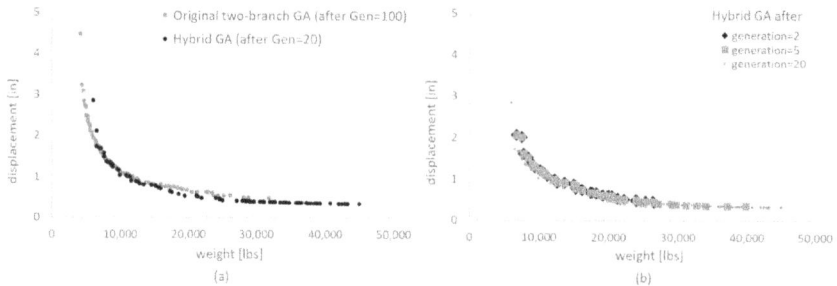

Figure 8.
Ten bar truss problem results. (a) Comparison between the original two-branch tournament GA (after 100 GA generation) and the hybrid approach (after 20 GA generations) for the 10 bar truss problem. (b) Evolution of non-dominated set as the generation progresses for the 10 bar truss problem using the hybrid approach.

there are not many discrete material selection options to explore. On the other hand, for the ten-bar truss problem, a vast majority of the improvement is due to the ability of the GA to find a better material selection combination rather than fine-tuning the cross sectional variables. It is not possible to seek further improvement in the Pareto front just by varying the continuous variables, so the local search saturates as appear in the case of black (diamonds) and blue (squares) non-dominated designs. After few more GA iterations, the algorithm is able to find better combinations of material selection that lead to further improvement in the Pareto front (red dots).

4.3 Greener aircraft design problem

The third application problem solved using the hybrid approach is the greener aircraft design problem. Here, a "greener" aircraft design problem provides an example to demonstrate the efficacy of the hybrid algorithm and its ability to solve such MDNLP problems. The intent is to find aircraft designs that represent the best possible trade-offs among performance, economics, and environmental metrics which essentially makes this a multi-objective problem. Further, with the inclusion of discrete technologies, the problem becomes MDNLP in nature.

The aircraft design optimization problem employs the NASA sizing code FLOPS [37] to evaluate discrete design configurations and perform the sizing and performance calculations of the candidate aircraft designs. The sizing code accepts both continuous and discrete design variables as input and returns the aircraft gross weight along with environmental metrics (fuel weight, which corresponds to CO_2 emissions, and NO_X emissions) and total operating cost. Simple models simulating the potential "greener technologies" are modeled in MATLAB [33] and then integrated with FLOPS for the performance calculations. The goal of the aircraft sizing problem is to develop an aircraft with 2940 nmi design range with a seat capacity of 162 seats in two classes. A brief description of the greener aircraft design optimization problem appears below. For more details about the aircraft design problem, we encourage the readers to see Ref. [38].

4.3.1 Description of the continuous variables

The problem includes ten continuous variables that define the wing and the engine parameters of the aircraft. The details of these continuous design variables and their design bounds appear in **Table 2**.

4.3.2 Simulating the discrete technologies

This aircraft design optimization study models three types of discrete technologies. **Table 3** lists the set of discrete technologies considered in this study. To model composite material selection choice on various aircraft components, the approach here uses a binary variable for each of the aircraft components that includes wing, fuselage, tail, and nacelle. A value of one represents composites being present while a value of zero represents no composite materials in that structure. The second discrete variable includes the eight possible combinations of the location and the number of engines. Lastly, eight combinations of laminar flow technologies are included for this problem, depending on whether it is natural laminar flow (NLF) or hybrid laminar flow control (HLFC) technology and the number of components on which it is applied (as listed in **Table 3**). References [38–40] describe the various discrete technologies used in this study in further detail.

Design variables	Lower bound	Upper bound
Aspect Ratio	8	12
Taper Ratio	0.3	0.5
Thickness to Chord Ratio	0.09	0.17
Wing Area [ft^2]	1,000	1,500
Wing Sweep at 25 percent [deg]	0	40
Thrust per engine [lbs]	20,000	30,000
By-Pass Ratio	5	10
Turbine Inlet Temperature [R]	3010	3510
Overall Pressure Ratio	35	55
Fan Pressure Ratio	1.6	1.7

Table 2.
Continuous variables for aircraft design problem.

Laminar Flow Technologies	Engine Position	Composite Material Choices			
		Wing	Fuselage	Nacelle	Tail
NLF-Wing	2 wing	Yes	Yes	Yes	Yes
HLFC-Wing	2 fuselage	No	No	No	No
HLFC-Wing + Nacelle	2 wing +1 fuselage				
HLFC-Wing + Tail	3 fuselage				
HLFC-Wing + Tail + Nacelle	4 wing				
NLF-Wing + HLFC-Tail	2 wing +2 fuselage				
NLF-Wing + HLFC-Nacelle	1 fuselage				
NLF-Wing + HLFC-Tail + HLFC-Nacelle	4 wing +1 fuselage				

Table 3.
Discrete technologies for aircraft design problem.

The problem also has four constraints that appear in **Table 4**. The constraints ensure that the design solution meets the desired field length criteria, has sufficient ground clearance, and sets a maximum limit on the amount of allowable fuel carrying space in the fuselage.

The aircraft design optimization problem considers two different pairs of competing objectives. The first pair involves simultaneous minimization of the aircraft fuel weight (index of CO_2 emissions) and the total operating cost of the aircraft, and the second pair involves minimizing the NO_x emissions and the total operating cost of the aircraft. The GA population has been limited to 48 individuals while setting the upper limit for the number of generations to 50 as before. The maximum

Take-off field length [ft]	$\leq 8,000$
Landing field length [ft]	$\leq 7,500$
Landing gear length [in]	≤ 150
Fuselage fuel capacity [lbs]	$\leq 28,800$

Table 4.
Problem constraints.

number of function evaluations for the SQP minimization (using MATLAB's fmincon) have been limited to the default value of 100 times the total number of continuous variables for this study. For certain combinations of discrete technology selection choices, the gradient-based approach may not find a feasible solution. In such cases, as mentioned in the methodology section, those designs are assigned high penalty for elimination in the subsequent generations.

4.3.3 Results for aircraft design problem

4.3.3.1 Objective pair - fuel weight vs. total operating cost.

Figure 9 shows the set of 24 non-dominated designs for the competing objective pair – aircraft fuel weight and total operating cost. The aircraft fuel weight, analogous to fuel burn, is directly related to the amount of CO_2 produced during the trip. The Pareto frontier consists of designs employing combinations of composite structures, eight different engine position(s), and eight different laminar flow technologies, modeled as a part of the greener technology approaches described in the previous section.

The design point ND1 (for Non-Dominated design number 1) in **Figure 9** corresponds to highest total operating cost (also lowest fuel weight) and makes use of NLF technology on the wing and HLFC technology on the nacelles and tail, along with two wing-mounted engines. This design also features composite wings, fuselage, and nacelles. The use of composite structures leads to a decrease in the fuel consumption (due to the reduction in aircraft empty weight) at the expense of increased total operating cost (due to increase in the manufacturing and maintenance costs associated with composite materials). The design with the lowest total operating cost (ND24) makes use of NLF technology on the wing and HLFC technology on the nacelles and tail, along with two wing-mounted engines as well. But, this design has no composite components and, hence, has the lowest total operating cost according to the models used in this study.

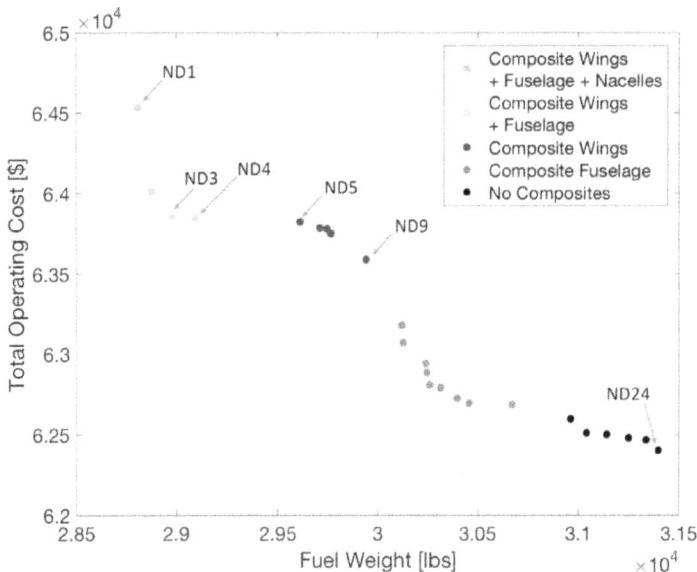

Figure 9.
The non-dominated set for objective pair – aircraft fuel weight (index for CO₂ production) and the total operating cost.

All the non-dominated designs employ NLF technology on the wings and HLFC technology on both the nacelle and tail, along with two wing-mounted engine configuration. The laminar flow technologies tend to reduce the skin friction drag of the aircraft, making the design more aerodynamically efficient, and reducing its fuel consumption for a given mission range. All the non-dominated designs employ these technologies in various forms (NLF or HLFC) to reduce fuel burn, depicting the importance of employing these technologies in near future "greener" aircraft design.

An interesting region in the Pareto frontier from an airline's standpoint would be near the points ND1 and ND3 (or ND2), where a substantial decrease in total operating cost is possible for a marginal increase in the aircraft fuel weight (index of CO_2 production per trip). Considering non-dominated designs ND1 and ND3, a nearly 1% reduction in total operating cost is possible to achieve for only a 0.6% increase in the total fuel weight needed for the mission, as one move from ND1 to ND3. Similar trends for the objective pair in consideration are also observed for designs ND9, ND10, and ND11.

4.3.3.2 Objective pair - NO_X emissions and total operating cost

The Pareto front corresponding to the NO_X emissions and the total operating cost objective pair appears in **Figure 10** and has 24 non-dominated designs. The non-dominated designs have different geometric design variable values that best match the different discrete greener aircraft technologies to arrive at the trade-off between the NO_X emissions and the total operating cost.

The design with minimum NO_X emissions and maximum total operating cost (ND1) employs a three-engine configuration with one fuselage-mounted and two wing-mounted engines, along with a composite wing. The laminar flow technologies on this design include NLF technology on wings and HLFC technology on the nacelles and tail. The maximum NO_X emitting design with minimum total operating cost (ND24) employs a two-engine configuration with wing-mounted engines, along with NLF technology on the wings and HLFC technology on the tail, and a composite nacelle. All the non-dominated designs, except the one with maximum NO_X emissions, employ NLF technology on the wings and HLFC technology on both the nacelle and tail. As we move from left to right along the Pareto frontier in

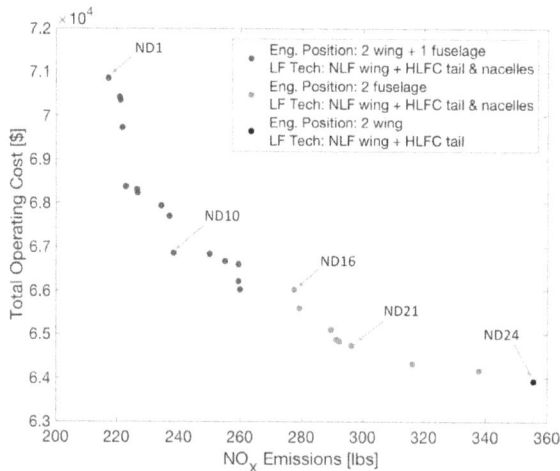

Figure 10.
The non-dominated set for objective pair – NOX emissions versus total operating cost.

Figure 10, the aircraft engines change from a three-engine configuration (two wing-mounted and one fuselage-mounted) to a two-engine (wing-mounted) configuration, thereby reducing the total operating cost.

An interesting region from the airline's point of view is the near the points ND2, ND3, ND4 and ND5, where a nearly vertical portion is visible in the top left portion of the Pareto frontier (refer to **Figure 10**). Moving from left to right in this region, a substantial decrease in total operating cost is possible for a marginal increase in the NO_x emissions of the aircraft. A plausible design from an airline's perspective– among the obtained non-dominated designs–would be the ND10 design. The reason for this observation is that a substantial increase in total operating cost will be incurred if further reduction in NO_x emissions are desired, while any effort to further reduce the total operating cost will lead to very high NO_x emissions, which is not desired from an environment standpoint.

Given there is some degree of randomness associated with the genetic operations in the GA, subsequent runs of the hybrid GA for the two objective pairs find a slightly different number of non-dominated designs points. However, the basic trait of the Pareto frontier, in terms of the discrete choices, did not alter; only the density of points in the Pareto frontier varied with different runs.

5. Conclusions

This chapter describes a hybrid multi-objective algorithm that makes use of an efficient gradient-based SQP algorithm for fitness evaluation inside a GA in a learning approach. The combination allows the GA to evolve a population of designs in the direction of the Pareto frontier while the SQP algorithm enforces constraints, eliminating the need for penalty multipliers or other special constraint handling methods and refines the values of the continuous design variables. The selective parent mixing and unique sets of goal point assignment to the individual lead to a distinct improvement in convergence and the quality of the Pareto frontier from a previous variation of this approach. When applied to various constrained MDNLP engineering design problems, the hybrid algorithm shows the ability to identify promising designs.

Although the ability of the hybrid approach to solve difficult constrained MDNLP problems is demonstrated in this chapter, the methodology relies heavily on the constraint enforcing ability and efficient searching of the continuous design space via the local gradient-based SQP algorithm that requires some estimates (either numerically or analytically) of the gradients of the objectives and the constraints with respect to the continuous design variables. A major advantage of a gradient-based approach besides being able to enforce the problem constraints (hence, the motivation to hybridize) is that the computational cost needed to compute the gradients is nearly independent of the number of design variables [41] when using adjoint-based methods to estimate the derivatives. This allows the gradient-based approach to efficiently solve problems with a very large number of design variables. However, if the objectives are encapsulated in a black-box function and are computationally very expensive to evaluate, then it may not be possible to directly implement a gradient-based search and may require a surrogate-based design optimization approach [40, 42, 43].

Nomenclature

α_i	Weight vector for the relative under/over-attainment of objective
$f_i(x)$	Value of the objective

f_i^G	Goal value for objective
$g_i(\mathbf{x})$	Nonlinear inequality constraints
γ	Attainment factor
$h_i(\mathbf{x})$	Nonlinear equality constraints
n	Population size
x_c	Continuous design variable
x_d	Discrete design variable
x^L	Design variable lower bound
x^U	Design variable upper bound

Abbreviations

EA	Evolutionary algorithm
GA	Generic algorithm
HLFC	Hybrid laminar flow control
MDNLP	Mixed-discrete nonlinear programming
ND	Non-dominated design
NLF	Natural laminar flow
NSGA	Non-dominated Sorted Genetic Algorithm
SPEA	Strength Pareto Evolutionary Algorithm
SQP	Sequential Quadratic Programming

Author details

Satadru Roy†, William A. Crossley*† and Samarth Jain†
Purdue University, West Lafayette, USA

*Address all correspondence to: crossley@purdue.edu

† These authors contributed equally.

IntechOpen

References

[1] Cirillo R. The Economics of Vilfredo Pareto. Routledge; 1979. ISBN: 978-1-136-27816-7

[2] Coello Coello C A. A Comprehensive Survey of Evolutionary-Based Multiobjective Optimization Techniques. Knowledge and Information Systems 1, 1999;269–308. DOI: 10.1007/BF03325101

[3] Deb K, Goel T. A Hybrid Multi-objective Evolutionary Approach to Engineering Shape Design. In: Zitzler E, Thiele L, Deb K, Coello Coello C A, and Corne D, editors. Evolutionary Multi-Criterion Optimization. Lecture Notes in Computer Science, vol 1993. Springer; 2001. p. 385-399. DOI: 10.1007/3-540-44719-9_27

[4] Deb K, Goel T. Controlled Elitist Non-dominated Sorting Genetic Algorithms for Better Convergence. In: Zitzler E, Thiele L, Deb K, Coello Coello C A, and Corne D, editors. Evolutionary Multi-Criterion Optimization. Lecture Notes in Computer Science, vol 1993. Springer;2001. p. 67-81. DOI: 10.1007/3-540-44719-9_5

[5] Holland J H. Adaptation in Natural and Artificial Systems: An Introductory Analysis with Applications to Biology, Control and Artificial Intelligence. MIT Press;1992. 232 p. ISBN: 978-0-262-08213-6

[6] Sequential Quadratic Programming. In: Nocedal J, Wright S J, editors. Numerical Optimization. Springer Series in Operations Research and Financial Engineering. Springer;2006. p. 529-562. DOI: 10.1007/978-0-387-40065-5_18

[7] Fundamentals of Unconstrained Optimization. In: Nocedal J, Wright S J, editors. Numerical Optimization. Springer Series in Operations Research and Financial Engineering. Springer;

2006. p. 10-29. DOI: 10.1007/978-0-387-40065-5_2

[8] Schaffer J. Some Experiments in Machine Learning Using Vector Evaluated Genetic Algorithms [thesis]. 1985

[9] Fonseca C M, Fleming P J. Genetic Algorithms for Multiobjective Optimization: FormulationDiscussion and Generalization. In: Proceedings of the 5th International Conference on Genetic Algorithms; 1993. San Francisco: Morgan Kaufmann Publishers Inc.;1993. p. 416-423

[10] Zitzler E, Thiele L. An Evolutionary Algorithm for Multiobjective Optimization: The Strength Pareto Approach. 1998

[11] Srinivas N, Deb K. Muiltiobjective optimization using nondominated sorting in genetic algorithms. Evolutionary Computation. 1994; 221-248. DOI: 10.1162/evco.1994.2.3.221

[12] Coello C A. An updated survey of GA-based multiobjective optimization techniques. ACM Computing Surveys. 2000;109-143. DOI: 10.1145/358923.358929

[13] Konak A, Coit D W, Smith A E. Multi-objective optimization using genetic algorithms: A tutorial. Reliability Engineering & System Safety. Special Issue - Genetic Algorithms and Reliability. 2006;992-1007. DOI: 10.1016/j.ress.2005.11.018

[14] Goldberg D E. Genetic Algorithms in Search, Optimization, and Machine Learning. Addison-Wesley Publishing Company;1989. 372 p. ISBN: 978-0-201-15767-3

[15] Crossley, W A, Cook A M, Fanjoy D W, Venkayya V B. Using the Two-Branch Tournament Genetic Algorithm

for Multiobjective Design. AIAA Journal. 1999;37:261-267. DOI: 10.2514/2.699

[16] To T B, Korn U. MOBES: A Multiobjective Evolution Strategy for Constrained Optimization Problems. In: Proceedings of the third international conference on genetic algorithms;1997. p. 176-182

[17] Fonseca, C M, Fleming, P J. Multiobjective Optimization and Multiple Constraint Handling with Evolutionary Algorithms-Part I: A Unified Formulation. IEEE Transactions on Systems, Man, and Cybernetics, Part A: Systems and Humans. 1998;28:26-37. DOI: 10.1109/3468.650319

[18] Deb K, Agrawal S, Pratap A, Meyarivan, T. A Fast Elitist Non-dominated Sorting Genetic Algorithm for Multi-objective Optimization: NSGA-II. In: Schoenauer M, Deb K, Rudolph G, Yao X, Lutton E, Merelo J J, Schwefel H, editors. Parallel Problem Solving from Nature PPSN VI. Springer; 2000. p. 849-858. DOI: 10.1007/3-540-45356-3_83

[19] Whitley D, Gordon V S, Mathias Keith. Lamarckian evolution, the Baldwin effect and function optimization. In: Davidor Y, Schwefel H, Männer R, editors. Parallel Problem Solving from Nature — PPSN III. Springer;1994. p. 5-15. DOI: 10.1007/3-540-58484-6_245

[20] Houck C R, Joines J A, Kay M G. Utilizing Lamarckian Evolution and the Baldwin Effect in Hybrid Genetic Algorithms. 1996

[21] Xiao X, Dow E R, Eberhart R, Miled Z B, Oppelt R J. A hybrid self-organizing maps and particle swarm optimization approach. Concurrency and Computation: Practice and Experience. 2004;16:895-915. DOI: 10.1002/cpe.812

[22] Kumar A, Sharma D, Deb K. A hybrid multi-objective optimization

procedure using PCX based NSGA-II and sequential quadratic programming 2007 IEEE Congress on Evolutionary Computation. 2007;ISSN: 1941-0026. p. 3011–3018. DOI:10.1109/CEC.2007.4424855

[23] Isaacs A, Ray T, Smith W. A Hybrid Evolutionary Algorithm With Simplex Local Search 2007 IEEE Congress on Evolutionary Computation. 2007; ISSN: 1941-0026. p. 1701-1708. DOI:10.1109/CEC.2007.4424678

[24] Satoru H, Tomoyuki H, Mitsunori M. Hybrid optimization using DIRECT, GA, and SQP for global exploration 2007 IEEE Congress on Evolutionary Computation. 2007; ISSN: 1941-002. p. 1709-1716 DOI:10.1109/CEC.2007.4424679

[25] Gil C, Márquez A, Baños R, Montoya M G, Gómez J. A hybrid method for solving multi-objective global optimization problems. Journal of Global Optimization. 2007;38:265-281. DOI: 10.1007/s10898-006-9105-1

[26] Majig M, Hedar A, Fukushima M. Hybrid evolutionary algorithm for solving general variational inequality problems. Journal of Global Optimization. 2007;38:637-651. DOI: 10.1007/s10898-006-9102-4

[27] Lehner S, Crossley W A. Hybrid Optimization for a Combinatorial Aircraft Design Problem In: 9th AIAA Aviation Technology, Integration, and Operations Conference (ATIO); 2009. DOI: 10.2514/6.2009-7116

[28] Zhong X, Fan W, Lin J, Zhao Z. Hybrid Non-dominated Sorting Differential Evolutionary Algorithm with Nelder-Mead. Second WRI Global Congress on Intelligent Systems. 2010;1: 306-311. DOI: 10.1109/GCIS.2010.198

[29] Wang X, Tang L. A PSO-Based Hybrid Multi-Objective Algorithm for Multi-Objective Optimization Problems.

In: Tan Y, Shi Y, Chai Y, Wang G, editors. Advances in Swarm Intelligence. ICSI 2011. Lecture Notes in Computer Science, vol 6729, Berlin, Heidelberg: Springer; 2011. DOI: 10.1007/978-3-642-21524-7_4

[30] Li X, Du G. BSTBGA: A hybrid genetic algorithm for constrained multi-objective optimization problems. Computers & Operations Research. 2013;40:282-302. DOI: 10.1016/j.cor.2012.07.014

[31] Žilinskas A, Žilinskas J. A hybrid global optimization algorithm for non-linear least squares regression. Journal of Global Optimization. 2013;56:265-277. DOI: 10.1007/s10898-011-9840-9

[32] Lohpetch D, Jaengchuea S. A Hybrid Multi-objective Genetic Algorithm with a New Local Search Approach for Solving the Post Enrolment Based Course Timetabling Problem Advances in Intelligent Systems and Computing. Springer International Publishing. 2016; 195-206 DOI: 10.1007/978-3-319-40415-8_19

[33] The Mathworks Inc., MATLAB version 2017a

[34] MathWorks Solve multiobjective goal attainment problems - MATLAB fgoalattain https://www.mathworks.com/help/optim/ug/fgoalattain.html [Accessed 4 February 2020]

[35] Roy S. Multi-objective Optimization using a Hybrid Approach for Constrained Mixed Discrete Non-linear Programming Problems - Applied to the Search for Greener Aircraft [thesis]. Purdue University; 2012.

[36] Jain S. A Multi-Fidelity Approach to Address Multi-Objective Constrained Mixed-Discrete Nonlinear Programming Problems With Application to Greener Aircraft Design [thesis]. Purdue University; 2018.

[37] McCullers L A, FLOPS, Software Package, Ver. 8.12 NASA Langley Research Center. Hampton, VA 2010

[38] Lehner S, Crossley W. A. Combinatorial Optimization to Include Greener Technologies in a Short-to-Medium Range Commercial Aircraft In: The 26th Congress of ICAS and 8th AIAA ATIO DOI: 10.2514/6.2008-8963

[39] Roy S, Crossley W A. Hybrid Multi-Objective Combinatorial Optimization Technique with Improved Compatibility between GA and Gradient-Based Local Search. In: Proceedings of 12th AIAA Aviation Technology, Integration, and Operations (ATIO) Conference and 14th AIAA/ISSMO Multidisciplinary Analysis and Optimization Conference; September 2012. AIAA; 2012

[40] Jain S, Crossley W A, Roy S. A Multi-Fidelity Approach to Address Multi-Objective Mixed-Discrete Nonlinear Programming Problems. In: 2018 Multidisciplinary Analysis and Optimization Conference; June 2018. AIAA; 2018

[41] Gray J S, Hwang J T, Martins J R R A, Moore K T, Naylor B A. OpenMDAO: An open-source framework for multidisciplinary design, analysis, and optimization. Structural and Multidisciplinary Optimization. 2019; 59:1075-1104. DOI: 10.1007/s00158-019-02211-z

[42] Jones D R, Schonlau M, Welch W J. Efficient Global Optimization of Expensive Black-Box Functions. Journal of Global Optimization. 1998;13:455-492. DOI: 10.1023/A:1008306431147

[43] Roy S, Crossley W A, Moore K T, Gray J S, Martins J R R A. Monolithic Approach for Next-Generation Aircraft Design Considering Airline Operations and Economics. Journal of Aircraft. 2019;56:1565-1576. DOI: 10.2514/1.C035312

Optimization Multicriteria Scheduling Criteria through Analytical Hierarchy Process and Lexicographic Goal Programming Modeling

Azzabi Lotfi, Azzabi Dorra and Abdessamad Kobi

Abstract

The rapidity of technological development and multi-criteria decision-making (MCDM) has enabled a diversity of models and multi-criteria decision support methods. Then, in Multi-criteria Decision Making problems dealing with qualitative criteria and uncertain information is suitable for the experts in order to express their judgments. It is common that the group of experts involved in such problems have different degrees of knowledge about the criteria. MCDM problems have been solved in the literature by using different methods, In this chapter we propose the multicriteria methodology to solve problem scheduling criteria based of application the Analytical hierarchy process methods and lexicographic goal programming.

Keywords: Multicriteria decision making, Analytical Hierarchy Process, Lexicographic Goal Programming, problem schedeling criteria

1. Introduction

Production scheduling is the technique of production control, the purpose of which is to enable the production program to be carried out on time, at minimum cost.

The objective is to find a schedule, an optimal program from the M possible scheduling sequences, where j is the number of jobs and M the number of machines.

The scheduling problem becomes even more complex when it occurs in an open and dynamic environment, where changes in the number of jobs or machines can occur at any time.

The scheduling problem has been addressed, mainly due to its combinatorial aspects, dynamic nature, and applicability in manufacturing systems [1] and many scheduling methods have been developed, based on different techniques such as heuristics, linear programming, constraint satisfaction techniques, La grangienne relaxation, neighborhood search techniques (eg annealing by simulation or taboo search) and genetic algorithms [2]. The aim of this chapter is the application of the optimization methodology to the problem of scheduling criteria based on a multicriteria approach.

2. General overview

The spill of products requested by customers was produced in a production workshop, however due to low productivity and high production costs, production workshops are not generally suitable for high volume production [3]. There is a need for production systems capable of producing a wide variety of products which can cost as little as mass production [4].

Scheduling systems can be considered as a special case of business information systems. Planning is defined as the allocation of resources to jobs over time.

It is a decision-making process which makes it possible to optimize one or more objectives [5].

The objectives can be the minimization of the production time, the average flow time, the delay of the works, the manufacturing costs ... , planning has an important role in many manufacturing and production systems.

The problems of planning seeking to optimize the time of realization of the project while respecting a certain number of constraints, are for the most part NP-difficult.

Research has mainly focused on finding optimal (or near optimal) solutions for static models with respect to various measures.

These approaches mostly have used the implicit assumption of static environments without any kind of failures. Extensive literature reviews on static deterministic scheduling can be found in [6–9].

Predictive planning has now become the planning of production systems [10, 11]. For example, machine breakdowns, arrival of urgent work, change of due date, etc. [12] addressed the nature of the gap between scheduling theory and the practice of scheduling [13], in their research on intelligent time control real in manufacturing systems, said the comparison of scheduling theory and practice showed very little correspondence between the two. Cowling and Johansson [14] found a large gap between scheduling theory and practice, and stated that scheduling models and algorithms are incapable of using real-time information. Until very recently, the problem of programming in the presence of real-time events, called dynamic programming. In this chapter, we focus on a number of issues that have arisen in recent years with dynamic planning in manufacturing systems.

We are mainly concerned with the question of knowing how to manage the occurrence of events in real time during the execution of a given schedule in the workshop?

In order to close this gap between scheduling models and procedures, and their implementation in a real manufacturing setting, the former should be translated into a system supporting scheduling decisions in a company.

Among the various activities that must be present in the development of a production planning system is the design of the system architecture. Despite the importance of the architecture of production planning systems, planning research has often overlooked this topic because the related literature is scarce and does not provide researchers with a complete view of the planning system [15].

3. Multicriteria decision analysis methods

However, the MCDA approach has certain drawbacks linked to the subjective nature of the preferences granted by decision-makers. Specially, the criteria weighting profile can be considered as the uncertain part as it relies heavily on subjective measurements and has a great influence on the final result.

Linkov offers traditional risk analysis and Monte Carlo simulation taking into account the uncertainties underlying the point estimates, to which all considerations and calculations are reduced [16].

Historically Multiple Criteria Evaluation methods were developed to select the best alternative from a set of competing options [17].

3.1 Analytical hierarchy process methods

The Analytical Hierarchy Process (AHP) is a method that was invented by Professor Thomas Saaty. It provides a decision-making structure that takes into account factors weighed in as a group [18].

The analytical hierarchy process for decision making is a relative measurement theory based on pairwise comparisons. Pairwise comparison matrices are formed either by providing judgments to estimate dominance using absolute numbers from the 1 to 9 fundamental scales of the AHP, or by directly constructing pairwise dominance ratios using real measurements.

The process of synthesis of weighting and addition applied in the hierarchical structure of the AHP combines multidimensional measurement scales into a single "one-dimensional" priority scale [19].

The steps of the AHP method are as follows:

- Construction of the hierarchy, it is an abstraction of the structure of the problem used to study; the interaction with the components of the problem and their effect on the final solution, it allows the problem to be broken down into a hierarchy of interconnected data. At the top of the hierarchy, we find the objective, and at the lower level, the elements contributing to achieve this objective, the last level is that of actions (**Figure 1**).

- To proceed to comparisons by elemental pairs of each hierarchic relative level to an element of the hierarchic superior level. This step permits to build matrix of comparisons. Values of those matrix are obtained by the judgments transformation in numerical values according to the ladder of Saaty (Ladder of binary comparisons), everything respecting the principle of reciprocity (**Table 1**)

$$(Ea, Eb) = \frac{1}{Pc(Eb, Ea)} \tag{1}$$

- To determine the relative elemental importance calculating primary vectors to correspond of the maximal values of comparisons matrix.

- To verify the judgments coherence. One to calculate at first, the indicator coherence IC:

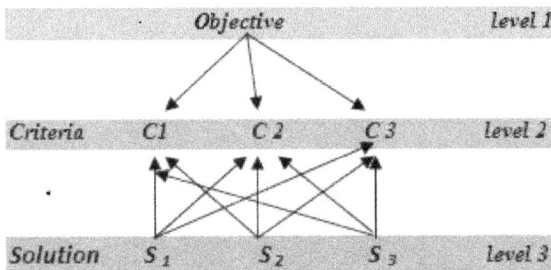

Figure 1.
Construction of the hierarchy.

$$IC = \frac{\lambda max - n}{n - 1} \tag{2}$$

Where: λ max is the primary maximal value running in matrix of comparisons by pairs; and n: is a large number of comparative elements. Then; the ratio of coherence (RC) defines by:

$$RC = 100 . \frac{IC}{ACI} \tag{3}$$

Where: ACI is the means coherence indicator of obtained generating aleatory matrix of judgment equalizes height. The means of indicator coherence is identified in the following **Table 2**.

A value of RC inferior to 10% is generally acceptable; otherwise, comparisons by pairs must be examined again to reduce the incoherence.

- To settle the relative performance of each action:

$$P_k\left(e_1^k\right) = \sum_{J=1}^{nk-1} P_{K-1}\ \left(e_i^{k-1}\right) P_k\frac{e_i^k}{e_i^{k-1}} \tag{4}$$

With: $P_k\left(e_1^k\right) = 1$, and: nk $-$ 1 are a large number of elements of the hierarchic level k-1, Pk $\left(e_i^k\right)$ is the terms priority to the element e^i to the hierarchic level k [20].

Importance grade	Define
1	Importance equalizes of both elements.
3	Importance weak person of a relative element to another.
5	Importance strong or determinant of a relative element to other.
7	Importance attested of a relative element to another.
9	Importance absolved of a relative element to another
2,4,6,8	Intermediate values with two values neighbor.

Table 1.
Saaty scales.

Matrix dimension	Aleatory coherence ACI
1	0.00
2	0.00
3	0.58
4	0.90
5	1.12
6	1.24
7	1.32
8	1.41
9	1.45
10	1.49

Table 2.
Means coherence indicator.

3.2 The lexicographic goal programming approach (LGP)

The LGP is an extension of linear programming (LP), was originally introduced by [21] and further presented by [22], and others. This technique was developed to handle multi-criteria situations within the general framework of LP. In the variant of the lexicographic Goal Programming, the objectives are ranked in order of priority, as the relative importance given to them by the decision maker. The mathematical formulation corresponding to this variant consists of a vector the deviations ordered on different objectives, which implies a minimization in the order of the different priority levels q [23]. The mathematical program is written as follows:

$$lex. \operatorname*{Min.L}_{\underline{x} \in A} = [l1(\delta^-, \delta^+), \, ... \, , lq(\delta^-, \delta^+)] \tag{5}$$

Subject to:

$$a_{1j}x_j + \delta^-_1 + \delta^+_1 \le g_1;$$
$$a_{2j}x_j + \delta^-_2 + \delta^+_2 \le g_2; \tag{6}$$
$$...$$
$$a_{nj}\,x_j + \delta^-_n + \delta^+_n \le g_n;$$
$$\delta^-_i, \delta^+_i \ge 0 \text{ pour } (i = 1, 2, \, ... \, , n)$$
$$x_j \ge 0 \text{ pour}(j = 1, 2, \, ... \, , n) \tag{7}$$

4. Application: process of cutting the electric cables

It is difficult to imagine modern cars without electronics. Computer board, control lamp, reversing sensors ... were designed to increase the comfort level and safety of the motorist. Electrical motor, the intensities being involved cover a range of about 0.5 A, a dashboard light bulb and up to several hundred amperes for a starter. But, any short circuit involves very high currents and can easily set fire to the vehicle and cause severe burns on contact elements short circuit.

In this paper, we propose a methodology for multicriteria optimization criteria considered essential to establish the correct ordering of the machines to avoid the loss of time and minimize the rate of waste and waste (**Figure 2**).

4.1 The analytical hierarchy process to ranking criteria scheduling

In the process of cutting the electric cables, we have identified six criteria:

- L1: qualification of workers,

- L2: safety culture,

- L3: equipment performance,

- L4: scheduling equipment,

- L5: production technology,

- L6: cost of upgrading waste.

Step 1: structure combination of scheduling criteria (**Figures 3** and **4**)
Step 2: Pairwise comparisons of the elements of each hierarchical level with respect to an element of higher level

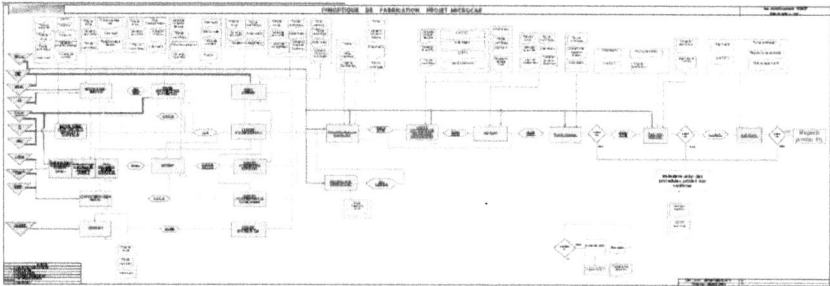

Figure 2.
Cut marking process flow chart setting.

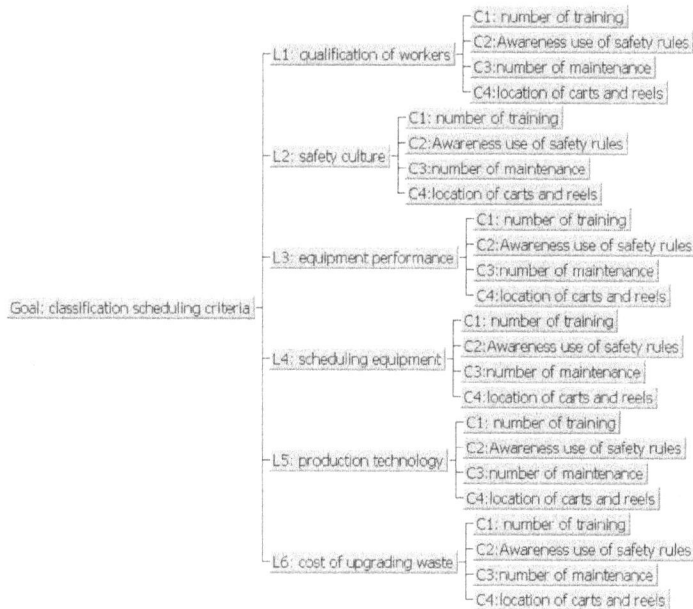

Figure 3.
Hierarchical structure of scheduling criteria.

Figure 4.
Pairwise comparison of criteria.

The pairwise comparison of criteria by the matrix above has identified the choice of the decision maker on the degree of importance of each criterion. For the decision maker, the most important criterion is "cost of upgrading waste", and the test that has no great fatality of the decision of the prioritization of scheduling criteria machines is "production technology".

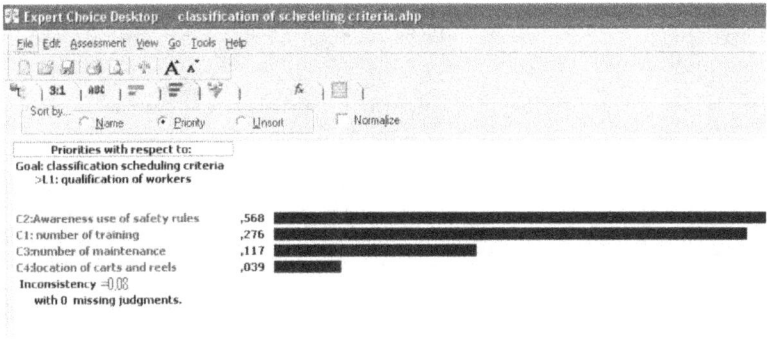

Figure 5.
Eigenvector of the criterion L1. The alternative is the most important criterion L1" qualification of workers", is C2 "Awareness use of safety rules" how long has the value of the eigenvector is greater.

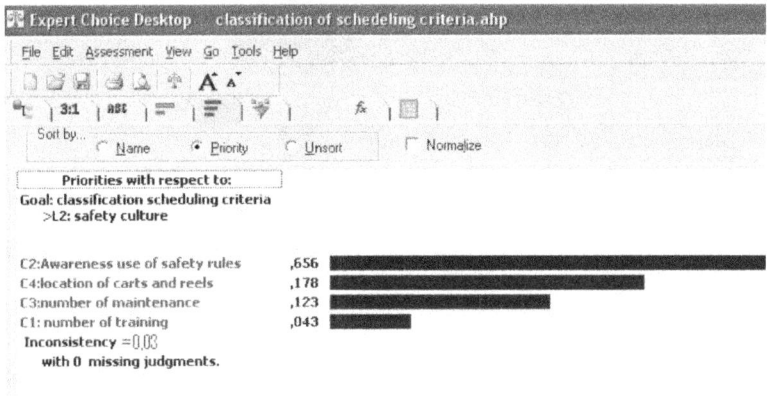

Figure 6.
Eigenvector of the criterion L2. The alternative is the most important criterion L2" qualification of workers", is then C2 "Awareness use of safety rules" how long has the value of the eigenvector is greater.

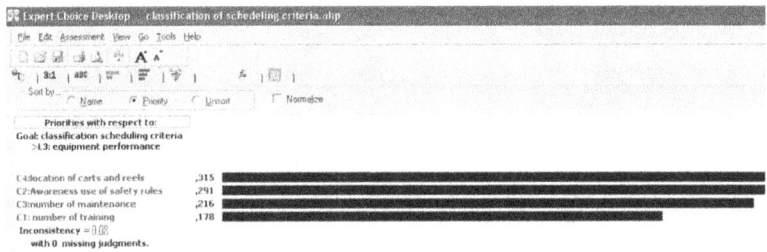

Figure 7.
Eigenvector of the criterion L3. For the eigenvector criterion L3 "equipment performance" alternative C4 "location of carts and reels", is the largest, and the criterion C2 is also essential according to the criterion L2.

Step 3: Determination of the relative importance of elements by calculating the eigenvectors corresponding to the maximum eigenvalue of comparison matrix (**Figures 5–10**):

Step 4: Check the consistency of judgments. All judgments are consistent (**Table 3**)

Step 5: Calculation of performance

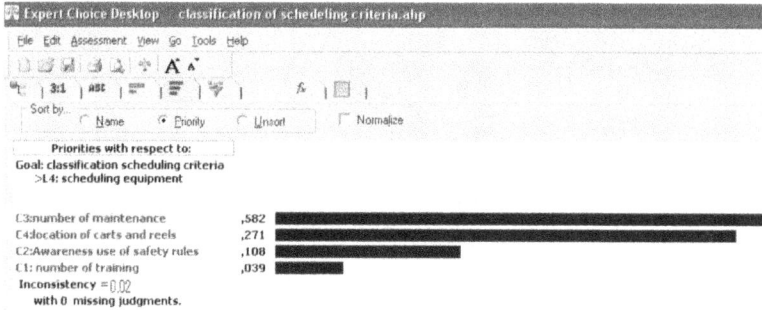

Figure 8.
Eigenvector of the criterion L4. For the eigenvector criterion L4 "scheduling equipment", alternative C3'number of maintenance" is the essential.

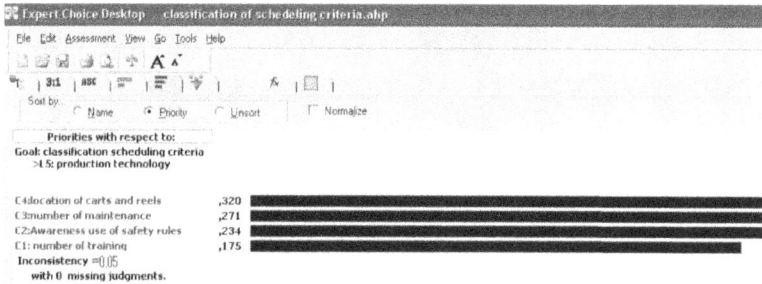

Figure 9.
Eigenvector of the criterion L5. The alternative of C4" location of carts and reels", is considered essential and priority according to the criterion L5.

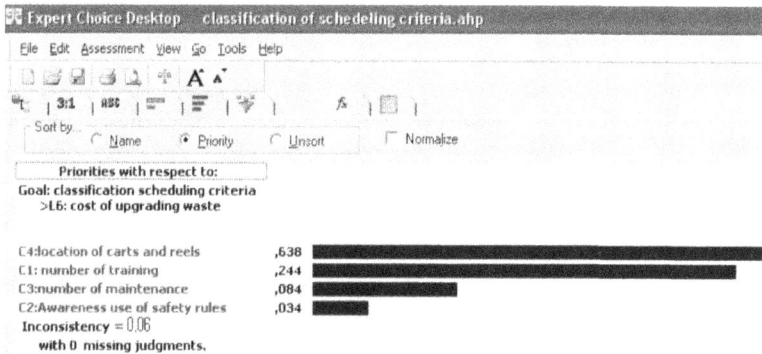

Figure 10.
Eigenvector of the criterion L6. For the eigenvector criterion L6" cost of upgrading waste ", then, the alternative of C4" location of carts and reels", is considered essential and priority.

Eigenvector	Ratio of coherence	Décision
Eigenvector L1	0.08	Consistency accepted
Eigenvector L2	0.03	Consistency accepted
Eigenvector L3	0.08	Consistency accepted
Eigenvector L4	0.02	Consistency accepted
Eigenvector L5	0.05	Consistency accepted
Eigenvector L6	0.06	Consistency accepted

Table 3.
Judgment consistency ratios.

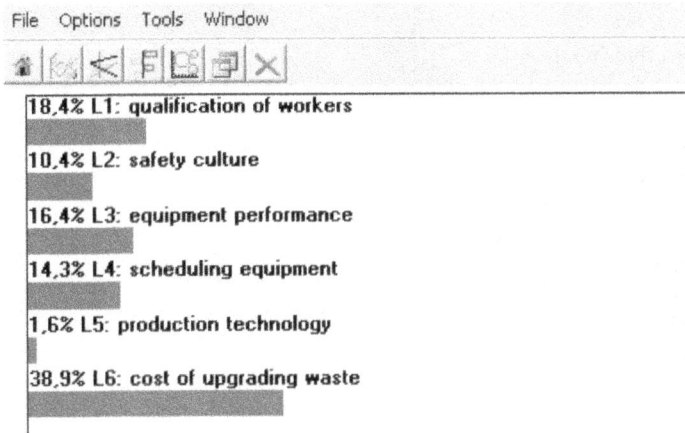

Figure 11.
Prioritization of scheduling criteria ranking.

Prioritization method AHP has established the classification scheduling criteria as follows (**Figure 11**).

The decision maker considers that the scheduling criteria L6 is a priority to set up an optimal schedule.

4.2 Scheduling optimization criteria according to the prioritization by the method of lexicographic goal programming

In this section we will make multicriteria optimization of six criteria identifies previously, in order of priority Realize by AHP. The problem is to find an optimal solution of scheduling in a cutting process by acting on all the criteria considered indispensable. We propose:

a1: cost per worker qualification;
a2: cost of training per worker safety culture;
a3: maintenance cost per machine;
a4: cost of extending the area of machine scheduling;
a5: cost of purchasing software programmable machine;
a6: cost of recycling machine.

Then, the goal to which the decision maker for each criterion to minimize the deviation are as follows (**Table 4**):

Criterion	Permissible limit
L1*: qualification of workers	1600£
L2*: safety culture	2000£
L3*: equipment performance	5000£
L4*: scheduling equipment	1600£
L5*: production technology	35000£
L6*: cost of upgrading waste	500£

Table 4.
Permissible limit of criterion.

In order of priority for the AHP method the objective functions of each criterion is as follows:

The objective function L6 is:
Max L = δ-6
Subject to:
$800x_1 + δ-1 + δ + 1 \leq 1600$;
$400x_2 + δ-2 + δ + 2 \leq 2000$;
$50x_3 + δ-3 + δ + 3 \leq 5000$;
$200x_3 + δ-4 + δ + 4 \leq 1600$;
$50x_4 + δ-5 + δ + 5 \leq 35000$;
$20x_5 + δ-6 + δ + 6 \leq 500$;
δ-i, δ + i ≥ 0 pour (i = 1,2, ... ,6)
$x_j ≥ 0$ pour (j = 1,2, ... ,6)

The objective function L1 is:
Max L = δ-1
Subject to
$800x_1 + δ-1 + δ + 1 \leq 1600$;
$400x_2 + δ-2 + δ + 2 \leq 2000$;
$50x_3 + δ-3 + δ + 3 \leq 5000$;
$200x_3 + δ-4 + δ + 4 \leq 1600$;
$50x_4 + δ-5 + δ + 5 \leq 35000$;
$20x_5 + δ-6 + δ + 6 \leq 500$;
δ-i, δ + i ≥ 0 pour (i = 1,2, ... ,6)
$x_j ≥ 0$ pour (j = 1,2, ... ,6)

The objective function L3 is:
Max L = δ-3
Subject to
$800x_1 + δ-1 + δ + 1 \leq 1600$;
$400x_2 + δ-2 + δ + 2 \leq 2000$;
$50x_3 + δ-3 + δ + 3 \leq 5000$;
$200x_3 + δ-4 + δ + 4 \leq 1600$;
$50x_4 + δ-5 + δ + 5 \leq 35000$;
$20x_5 + δ-6 + δ + 6 \leq 500$;
δ-i, δ + i ≥ 0 pour (i = 1,2, ... ,6)
$x_j ≥ 0$ pour (j = 1,2, ... ,6)

The objective function L4 is:
Max L = δ-4

Subject to
$800x1 + \delta-1 + \delta + 1 \leq 1600;$
$400x2 + \delta-2 + \delta + 2 \leq 2000;$
$50x3 + \delta-3 + \delta + 3 \leq 5000;$
$200x3 + \delta-4 + \delta + 4 \leq 1600;$
$50x4 + \delta-5 + \delta + 5 \leq 35000;$
$20x5 + \delta-6 + \delta + 6 \leq 500;$
$\delta-i, \delta + i \geq 0$ pour $(i = 1,2, \ldots ,6)$
$xj \geq 0$ pour $(j = 1,2, \ldots ,6)$

The objective function L2 is:
Max L $= \delta-2$
Subject to
$800x1 + \delta-1 + \delta + 1 \leq 1600;$
$400x2 + \delta-2 + \delta + 2 \leq 2000;$
$50x3 + \delta-3 + \delta + 3 \leq 5000;$
$200x3 + \delta-4 + \delta + 4 \leq 1600;$
$50x4 + \delta-5 + \delta + 5 \leq 35000;$
$20x5 + \delta-6 + \delta + 6 \leq 500;$
$\delta-i, \delta + i \geq 0$ pour $(i = 1,2, \ldots ,6)$
$xj \geq 0$ pour $(j = 1,2, \ldots ,6)$

The objective function L5 is:
Max L $= \delta-5$
Subject to
$800x1 + \delta-1 + \delta + 1 \leq 1600;$
$400x2 + \delta-2 + \delta + 2 \leq 2000;$
$50x3 + \delta-3 + \delta + 3 \leq 5000$
$200x3 + \delta-4 + \delta + 4 \leq 1600;$
$50x4 + \delta-5 + \delta + 5 \leq 35000;$
$20x5 + \delta-6 + \delta + 6 \leq 500;$
$\delta-i, \delta + i \geq 0$ pour $(i = 1,2, \ldots ,6)$
$xj \geq 0$ pour $(j = 1,2, \ldots ,6)$

By using the software LINDO, the ideal solution obtained for optimize the scheduling problem in a cutting process is (**Table 5**).

The solution is satisfactory with the support of the decision maker's preferences. It is obvious that the implementation of the prioritization criterion that the lexicographic goal programming, the solution is much improved. The level of satisfaction

Criterion	Result
L6: cost of upgrading waste	1600ε
L1: qualification of workers	2000ε
L3: equipment performance	5000ε
L4: scheduling equipment	1600ε
L2: safety culture	35000ε
L5: production technology	500ε
Z*	45700ε

Table 5.
Optimal solution.

achieved for the six objectives is 100%. Achieved this level of satisfaction implies that all specifications are met.

5. Conclusions

In this paper, we presented a multicriteria methodology to optimize the problem of scheduling in cutting process of electrical cables.

Then, this multicriteria approach has debited by the classification criteria with the AHP methods for identify the prioritization, and in the second parts we have optimized this criteria with the lexicographic goal programming. This methodology has given optimal results as long as it is based on the preferences and the intervention of any decision-maker during the optimization process.

Author details

Azzabi Lotfi[1]*, Azzabi Dorra[2] and Abdessamad Kobi[3]

1 ESSCA School of Management, Angers, France

2 High School Sacre Cœur La Salle of Angers, France

3 Polytech Angers, University of Angers, France

*Address all correspondence to: lotfi_azzabi@yahoo.fr

IntechOpen

References

[1] Shen, W. (2002): Distributed manufacturing scheduling using intelligent agents. IEEE Intell Syst,;17 (1):88–94.

[2] Leitao, P. Restivo, F. (2008): A holonic approach to dynamic manufacturing scheduling robotics and computer-integrated manufacturing. 24, 625–634.

[3] Sun, H. Huang, H.-C. Jaruphongsa, W. (2009): A Genetic Algorithm for the Economic Lot Scheduling Problem Under the Extended Basic Period and Power-of-Two Policy. Journal of Manufacturing Science and Technology, 2: 29–34.

[4] Parviz, F. Neda T. Amir, J. Fariborz, J.(2010): A hybrid algorithm to solve the problem of re-entrant manufacturing system scheduling. Journal of Manufacturing Science and Technology 3 268–278.

[5] Pinedo, M.L., (2007): Planning and Scheduling in Manufacturing and Services, 3ed. Springer, Berlin.

[6] Jain, A. S. Meeran, S. (1999): Deterministic job-shop scheduling: Past, present and future, European Journal of Operational Research, 113 (2), 390–434.

[7] Pinedo, M. (1995): Scheduling theory, algorithms and systems. First edition, Prentice Hall.

[8] Pinedo, M. (2002): Scheduling theory, algorithms and systems. Second edition, Prentice Hall.

[9] Weirs, V. C. S. (1997): A review of the applicability of OR and AI scheduling techniques in practice. Omega International Journal of Management Science, 25 (2), 145–153.

[10] Shafaei, R. Brunn, P. (1999) : Workshop scheduling using practical

(inaccurate) data Part 2: An investigation of the robustness of scheduling rules in a dynamic and stochastic environment, International Journal of Production Research, 37 (18), 4105–4117.

[11] Vieira G. E., Hermann J. W., Lin E. (2000): Predicting the performance of rescheduling strategies for parallel machine systems. Journal of Manufacturing Systems, 19 (4), 256–266.

[12] MacCarthy, B. L. Liu, J. (1993): Addressing the gap in scheduling research: a review of optimization and heuristic methods in production scheduling. International Journal of Production Research, 31 (1), 59–79.

[13] Shukla, C. S. Chen, F. F. (1996): The state of the art in intelligent real-time FMS control: a comprehensive survey. Journal of Intelligent Manufacturing, 7, 441–455.

[14] Cowling, P. I. Johansson, M. (2002): Using real-time information for effective dynamic scheduling, European Journal of Operational Research, 139 (2), 230–244.

[15] Sha, D.Y. Hsing-Hung, L. (2010): A multi-objective PSO for job-shop scheduling problems. Expert Systems with Applications 37, 1065–1070

[16] Linkov, I. Satterstrom, F. K. (2006): From comparative risk assessment to multi criteria decision analysis and adaptive management: Recent developments and applications. Environment International 32: 1072–1093.

[17] Sharifi M.A., Boerboom L., Shamsudin K. B., Veeramuthu L. (2006): Spatial multiple criteria decision analysis in integrated planning for

public transport and land use
development study in klang valley,
Malaysia. ISPRS Technical Commission
II Symposium, Vienna, 12–14 July.

[18] Saaty, T.L (2001): The Analytic
Network Process: Decision Making with
Dependence and Feedback. RWS
Publications.

[19] Saaty T.L. (2002):Fundamentals of
Decision Making with the Analytic
Hierarchy Process, paperback, RWS
Publications, 4922 Ellsworth Avenue,
Pittsburgh, PA 15213–12807.

[20] Saaty. T.L, (1989): Group Decision
Making and the AHP in the Analytic
Hierarchy Process: Application and
Studies. Springer-Verlag.

[21] Charnes, A. and Cooper, W.W.
(1977): Goal programming and multiple
objective optimization, European
Journal of Operational Research, Vol.1,
No.1, pp.39–45.

[22] Ijiri Y. (1965):Management Goals
and Accounting for Control,
Amsterdam: North-Holland.

[23] Romero, C. (1991): Handbook of
critical issues in goal programming»,
Pergamon Press, Oxford.

Chapter 9

Versatility of Simulated Annealing with Crystallization Heuristic: Its Application to a Great Assortment of Problems

Tiago G. Goto, Hossein R. Najafabadi, Guilherme C. Duran,
Edson K. Ueda, André K. Sato, Thiago C. Martins,
Rogério Y. Takimoto, Hossein Gohari, Ahmad Barari
and Marcos S.G. Tsuzuki

Abstract

This chapter is related to several aspects of optimization problems in engineering. Engineers usually mathematically model a problem and create a function that must be minimized, like cost, required time, wasted material, etc. Eventually, the function must be maximized. This function has different names in the literature: objective function, cost function, etc. We will refer to it in the chapter as objective function. There is a wide range of possibilities for the problems and they can be classified in different ways. At first, the values of the parameters can be continuous, discrete (integers), cyclic (angles), intervals, and combinatorial. The result of the objective function can be continuous, discrete (integers) or intervals. One very difficult class of problems have continuous parameters and discrete objective function, this type of objective function has very weak sensibility. This chapter shows the versatility of the simulated annealing showing that it can have different possibilities of parameters and objective functions.

Keywords: Simulated Annealing, Optimization Problems, Cutting Packing, Electrical Impedance Tomography, Topological Optimization, Curve Interpolation

1. Introduction

Optimization problems are common in engineering, and the problem must be optimized regarding one function (usually called as cost function). The cost function has several parameters used to model the optimization problem. Several methods have been proposed in the literature to optimize functions, and the method is selected according the cost function and parameters characteristics.

Deterministic methods usually have an initial point, called as seed. Using the gradient of the cost function, the optimization method starts on the initial point and converges to the optimal parameter value. The parameters are usually continuous [1].

IntechOpen

However, engineering problems have problems which cannot be solved by this approach.

Some cost functions have several minimum, and the convergence strongly depends on the seed (or initial point). To overcome this issue, several meta-heuristics were proposed, like Genetic Algorithms (GA) [2, 3], Particle Swarm Optimization (PSO) [4], Simulated Annealing (SA) [5], and others. Optimization methods based on meta-heuristics do not need the gradient information. For some problems, it is very hard to determine the gradient. This is particularly true for problems with discrete cost function. The existing meta-heuristics, usually, are focused on some specific type of parameters. For example, PSO can be applied to engineering problems with continuous parameters. GA can be applied to engineering problems with integer or fixed precision parameters.

SA was first proposed to solve combinatorial problems [5]. Later on, Corana et al. [6] extended the SA to incorporate continuous parameters. It was shown that the proposal made by Corana et al. [6] only converged to the global optimum in specific problems. Ingber [7] improved the SA to overcome this issue, however he used the cost function gradient for some calculations. Martins and Tsuzuki [8] proposed the SA with crystallization heuristic which does not use any gradient information.

This chapter shows the versatility of the SA with crystallization heuristic. Initially, the SA with crystallization heuristic is explained in Section 2. It was used to solve different problems related to Cutting and Packing, particularly the case with discrete cost function and cyclic continuous parameters [8–10] will be explained in Section 3. A third application is Topological Optimization (TO), it has a large number of parameters, like the EIT problem. This application is explained in Section 4. The curve interpolation problem, explained in Section 5, has two different types of parameters: continuous and integer. It is relevant to point that the SA with crystallization heuristic has been applied with success to interval based objective function evaluation [11–13]. The conclusions are in Section 6.

2. Simulated annealing with crystallization heuristic

The SA is an optimization algorithm that uses a random local search and is able to find the global optimum solution. It was proposed by [5, 14], and it was inspired by the metal annealing process, which starts with a high temperature and reduces the temperature in small increments until reaching the minimum temperature. In each step, the new solution is accepted if it improves the current solution. Otherwise, it is only accepted if a probability factor $P(T)$ is more than a random number [10, 14, 15]. The $P(T)$ can be determined by

$$P(T) = \exp^{-\frac{\Delta E}{T}} \tag{1}$$

where T is the current temperature, and ΔE is the the energy state variation which is the difference between the objective function $F(x)$ of the new candidate a and the current solution b, as presented in $\Delta E = F(a) - F(b)$.

Algorithm 1 describes a possible implementation of a conventional SA. It initializes with a random solution and an initial temperature, which is an important parameter. There are two loops. The external loop controls the temperature. The temperature decreases according to a geometric cooling schedule with factor α, and it stops when the frozen state is reached. The cooling schedule, in this case the value of α, must be carefully chosen, as the convergence to the optimal distribution strongly depends on it. In the majority of the applications explained here, we used $\alpha = 0.98$. The internal loop, performs the random local search, until the thermal equilibrium is reached. The random search consists of modifying a single parameter and a new candidate x^* is obtained. The new objective function value is evaluated and compared with the cost of the current solution. Then, it is either accepted or rejected according to (1). The convergence conditions are usually controlled by algorithm parameters. They influence the quality of the final result as well as the speed of the algorithm.

Algorithm 1: Conventional SA

1: $i \leftarrow 0$
2: $x \leftarrow$ < random initial solution >
3: $T_0 \leftarrow$ < initial temperature >
4: **while** < Global condition not satisfied > **do**
5:     **while** < Local condition not satisfied > **do**
6:         $x^* \leftarrow$ < modify a parameter >
7:         $\Delta E = F(x^*) - F(x)$
8:         **if** $\Delta E < 0$ or $random(0,1) < e^{-\Delta E/T_i}$ **then**
9:             $x \leftarrow x^*$
10:         **end if**
11:     **end while**
12:     $i \leftarrow i + 1$
13:     $T_i \leftarrow T_{i-1} * \alpha$
14: **end while**

The random search in the SA algorithm can use different strategies to improve finding a new solution. One of them is the SA with crystallization heuristic proposed by [9, 16]. The SA has two phases when the search is performed: exploration and refinement. **Figure 1** shows the two phases and their connection with the crystallization factor. The exploration of the domain space, usually happens in higher temperatures. The refinement of the solution happens in lower temperatures. There is no clear interface between the two phases.

For problems with combinatorial and integer parameters, the generation of the new candidate can be easily performed. In the case of combinatorial parameters, the algorithm exchanges the position of two parameters. In the case of integer parameters, the algorithm increments, decrements or keeps the current value. There are

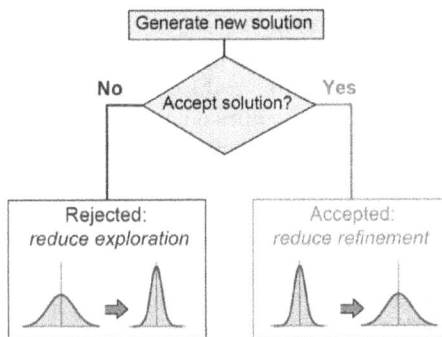

Figure 1.
The crystallization heuristic represented by SA controlling each parameter crystallization factor.

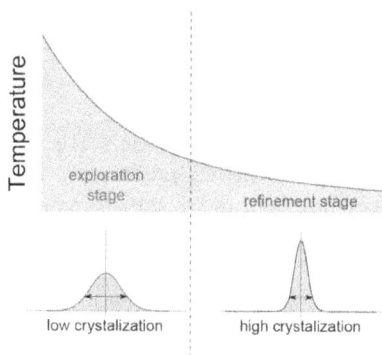

Figure 2.
During the optimization process, the SA has phases: exploration and refinement. In the exploration phase, the parameters have low crystallization and can perform larger jumps. The inverse happens in the refinement phase where the parameters have high crystallization.

possibilities, considering that integer parameters have lower and upper bounds, it is possible to generate a new value performing a random selection in the range.

The control of continuous parameters is challenging. The crystallization heuristic proposed by [9, 16] considers a fixed maximum step, and the SA controls the probability density. **Figure 1** shows that wider probability density allows larger jumps with higher probability. While, thinner probability density allows smaller jumps with higher probability. However, a larger jump still possible to happen with very small probability. This feature allows the SA to escape from local minima.

Each continuous parameter has a crystallization factor, which is represented by c_j. Considering the crystallization heuristic, the new candidate is determined by summing c_j times a random number, this procedure determines a Bates distribution. It is represented by

$$x^* = x + \frac{1}{c_j} \sum_{k=1}^{c_j} \text{random}\left(-1/2, 1/2\right) \cdot \Delta r_j \cdot e_j. \tag{2}$$

where x is the current solution, x^* is the new candidate, they are represented by vectors. Δr_j is the fixed step size associated with continuous parameter j. e_j represents the selected parameter, e_j is a vector with all elements equal to zero except one, the position associated with the j-th parameter. A common value assigned to Δr_j depends on the search interval, as $\Delta r_j = \left(max_j - min_j\right)/4$. This value will provide enough exploration for the algorithm.

It remains to explain the procedure in which the SA controls the crystallization factor for each parameter. The SA modifies only one parameter at a time, and according the decision of accepting or rejecting the new candidate, an action is performed. The procedure is represented in **Figure 2**. If the new candidate is rejected, it is assumed that the SA is performing exploration and to increase accepted solutions the crystallization factor associated with this parameter is increased. The increase in the crystallization factor will reduce the probability of larger jumps for this parameter. On the other hand, if the new candidate is accepted, it is assumed that the SA is performing refinement and to increase exploration the crystallization factor associated with this parameter is reduced. The reduction of the crystallization factor will enlarge the probability of larger jumps for this parameter.

Algorithm 2: SA with Crystallization Heuristic.

1: $i \leftarrow 0; c \leftarrow (1, 1, \ldots, 1)$
2: $x \leftarrow$ < random initial solution >
3: $T_0 \leftarrow$ < initial temperature >
4: $\Delta r \leftarrow$ < range of the parameters >
5: **while** < Global condition not satisfied > **do**
6:    **while** < Local condition not satisfied > **do**
7:       $j \leftarrow$ < select parameter to modify >
8:       $x^* \leftarrow x + \frac{1}{c_j} \sum_1^{c_j} random\left(-\frac{1}{2}, \frac{1}{2}\right) \cdot \Delta r_j \cdot e_j$
9:       $\Delta E = F(x^*) - F(x)$
10:      **if** $\Delta E < 0$ **then**
11:         $x \leftarrow x^*$
12:         **if** $c_j > 1$ **then**
13:           $c_j \leftarrow c_j - 1$ ▷ Positive Feedback
14:         **end if**
15:      **else**
16:         **if** $random(0, 1) < e^{-\Delta E / T_i}$ **then**
17:           $x \leftarrow x^*$
18:           **if** $c_j > 1$ **then**
19:             $c_j \leftarrow c_j - 1$ ▷ Positive Feedback
20:           **end if**
21:         **else**
22:           $c_j \leftarrow c_j + 1$ ▷ Negative Feedback
23:         **end if**
24:      **end if**
25:    **end while**
26:    $i \leftarrow i + 1$
27:    $T_i \leftarrow T_{i-1} * \alpha$
28: **end while**

Algorithm 2 describes the implementation of SA with crystallization heuristic. In this algorithm, the crystallization heuristic adjusts the modification to increase the acceptance of new solutions. The crystallization factor c is initialized with 1 for all continuous parameters, and the fixed step size Δr_i is defined as 25% of the search range. For example, if the search happens between -100 and $+100$, in this case $\Delta r_i = 50$.

3. Application in cutting and packing

The field of operations research concentrates many real world applications of optimization techniques in engineering, as it essentially e aims to increase the efficiency of industry operations [8, 9, 16–19]. Interest in this area has grown recently due to advances in optimization algorithms and the importance of the waste and pollution reduction, which is usually an effect of an improved solution.

Among operations research subjects, cutting and packing (C&P) problems can be highlighted due to its importance and singular combination of geometry and optimization. These characterises are even more prominent in irregular packing problems, which involves simple polygonal items. Essentially, C&P problems consists of assigning a set of small items to a set of set of large containers, maintaining some geometric restrictions, while minimizing an objective function.

In this section, a SA based solution for the irregular packing problem is proposed. It adopts a discrete objective function, as it is related to the number of items in the container, while using continuous parameter for items rotations.

3.1 Irregular bin packing problem with free rotations

In the 2D irregular single bin packing problem, given a collection of items, one must place a subset of these inside a rectangular container with the aim of minimizing the unused space inside the bin. Each item is represented by a simple polygon and may be rotated by any angle. The main geometric restrictions dictates that no two items may overlap and there should be no protrusion from the container.

Consider a set of items $\mathcal{P} = \{P_1, P_2, \dots, P_n\}$ and a rectangular container $\mathcal{C}$. The layout can be represented by the translation vector $\{t_1, t_2, \dots, t_n\}$, a rotation vector $\{r_1, r_2, \dots, r_i\}$ and an assignment set $\mathcal{T}$, which contains the subset items placed inside the container. The irregular bin packing problem can be described as the minimization of the difference between the area of the container $A(\mathcal{C})$ and the assigned set of items $A(\mathcal{T})$, as

$$
\begin{aligned}
\text{minimize} \quad & A(\mathcal{C}) - A(\mathcal{T}) \\
\text{subject to} \quad & i(P_i(r_i) \oplus t_i) \cap (P_j(r_j) \oplus t_j) = \emptyset \ i,j \in \mathcal{T} \text{ and } i < j \\
& (P_i(r_i) \oplus t_i) \subseteq \mathcal{C} \ 1 \le i \in \mathcal{T} \\
& 0 \le r_i < 2\pi \ 1 \le i \in \mathcal{T} \\
& t_i \in \mathfrak{R}^2 \ 1 \le i \in \mathcal{T}
\end{aligned}
\tag{3}
$$

where $P(r)$ represents an item rotate by r, operator $i(P)$ expresses only the interior of P and $P \oplus t$ indicates a translation t applied to P.

3.2 Solution using SA

One of the main challenges in irregular packing is the complexity of managing the geometric constraint, which results in fewer proposed solutions in the literature [20]. In order to optimize the layout without compromising the geometric feasibility, two main strategies are often employed. The first is to define a constructive heuristic, which places one item at a time, usually at the bottom-left position. In this approach, the optimization algorithm only controls the placement order; a popular solution is to employ genetic algorithms [2, 21]. The alternative is to allow the items to move freely inside the container while applying penalization on the item overlaps [22–24].

In [9], however, a different approach was adopted: a SA based algorithm was employed to directly control the placement order, as well as the position and rotation of each item. Items were placed sequentially, using the parameters given by the SA, until no more items fitted the container. Then, the objective function, which was the unoccupied area of the container, which is directly related to the subset of placed items, was evaluated.

The main difficulty was the definition of the item position parameter, which should always correspond to a valid placement, i.e., without overlap or protrusion from the container. It was given by the collision free region (CFR), a polygon describing the allowed placement region, as shown in **Figure 3**. The CFR can be obtained using modified Boolean operations on polygons, described in [25]. A continuous parameter evaluated to a position along the perimeter of the CFR, and the closest vertex was chosen as the placement position of the item. The rotation

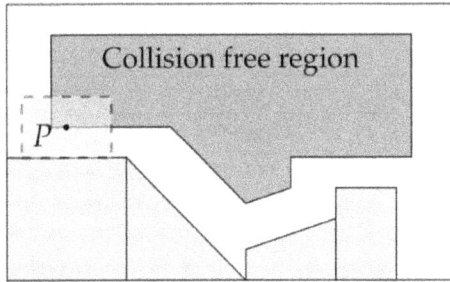

Figure 3.
Example of collision free region for item P.

was defined by a second controlled variable, and it had to be applied prior to the determination of the CFR.

Therefore, the solution optimization basically consisted of a SA controlling the placement and rotation parameter of each item in the layout. At each iteration, one parameter for a single item was changed, or two items were swapped in the placement order. Then, the cost was evaluated and the new solution was accepted according to (1). The crystallization factor described in Section 2 was applied to the rotation parameters.

3.3 Results and discussion

Six broken glass puzzle instances were created to evaluate the performance of the algorithm. These instances have a known optimal solution, which are displayed in **Figure 4**. Therefore, the effectiveness of the algorithm can be measure by its success rate, which relates to the number of executions converging to the optimal solution.

The tests were executed on a Phenom 9550 2.21GHz and the convergence condition for the SA was that (1) the cost variation for the final temperature was zero, and (2) the final solution has the lowest cost found. The initial temperature was adjusted by admitting initial 50% acceptance and a geometric cooling schedule was adopted with $\alpha = 0.98$ (as shown in Algorithm 1).

Table 1 shows the results data obtained for 30 executions of each instance. Results show that only very simple problems were solved optimally in every executions. This indicates that the SA algorithm should be complemented with heuristics approaches to increase the convergence rate. Nevertheless, given the complexity of the problem with free rotations, the fact that it found the optimal solutions for all instances is important, as some could not be achieved by enforcing simple heuristics such as bottom-left or larger first.

One important characteristic of the irregular bin packing problem with free rotations is that the objective function is discrete and some of the parameters are continuous. This is illustrated in **Figure 5**, which shows discrete cost values for different values for the leftmost item rotation. The SA solution was not affected by

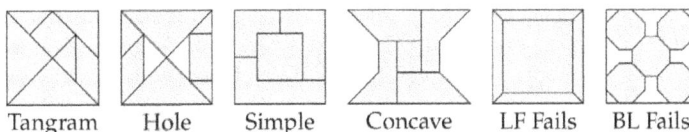

Figure 4.
Instances for the irregular bin packing problem.

Instance	N_{items}	N_{conv}	$T_{conv}(s)$	P_{conv}
Tangram	7	130,169	722.83	58.8%
Hole	7	12,847	28.89	93.5%
Simple	4	24,604	49.58	100.0%
Concave	5	17,912	43.96	86.7%
LF Fails	5	4,426	10.61	63.6%
BL Fails	5	117	0.16	100.0%

Table 1.
The results for the SA based irregular bin packing solution. N_{items}: total number of items. N_{conv}: average number of iterations. T_{conv}: average time untill convergence (in seconds). P_{conv}: success rate.

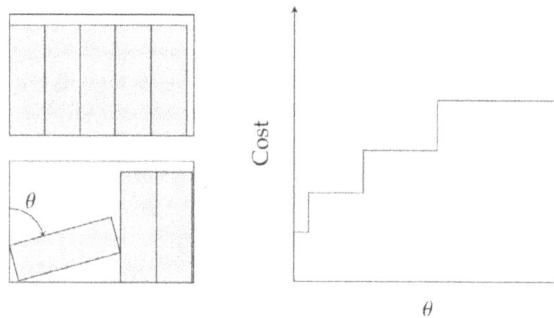

Figure 5.
Discrete cost behaviour with single continuous parameter variation.

this issue, as the results show, it can handle both continuous and discrete parameters and objective function. Moreover, the continuous rotation convergence for each item was improved by adopting the crystallization factor.

4. Application in TO

TO is a mathematical approach to determine distribution of material in a design domain such that the performance becomes maximized. The definition of performance could be different in each application. The physic of problem and desired application determines the objective function and the constraints. Depending on the problem, different methods have been developed in the literature [26, 27].

For the TO problems with well defined objective functions and constraints, the gradient based algorithms have been widely used [28–30]. The sensitivity information converges the final results to the optimized topology. This method shows fast convergence and low computational costs. The gradient based TO methods use Optimality Criteria (OC), Method of Moving Asymptotes (MMA), Sequential Linear Programming (SLP), etc. to optimize the objective functions [31, 32]. In the cases that the objective function or its derivatives are not mathematically modeled or hard to calculate, using the non-gradient based algorithms are more advantages [33, 34]. In the non-gradient based TO methods, GA and SA are the two most popular algorithms of optimization [3, 35–37]. Even these methods have high computational costs, but can optimize the topology without need to calculation of the derivatives and sensitivity information.

Using the SA method in the non-gradient based TO is beneficial because of reaching global minimum as well as provide the information of convergence.

The information of convergence such as the number of accepted and rejected solutions could be used in the evaluation of TO results. The available SA for TO uses random search to generate new solutions, while by using SA with crystallization heuristic [10], the search for new accepted solutions has been improved. After finishing the optimization process, a density filter can reduce gray area and discontinued regions.

4.1 SA for non-gradient based TO

The structural TO to minimize compliance in beams is a classic problem. This problem has been solved with gradient based methods in the literature [38] and the results have been used to verify TO with crystallization heuristic SA (as described in Section 2). The problem of minimizing compliance can be represented as the problem of minimizing strain energy, modeled as

$$\text{minimize } S = U^T K U, \text{ subjected to}: \ KU = \mathcal{F} \qquad (4)$$

where $\mathcal{F}$, U, and K are, respectively, external force, elastic deformation, and stiffness. The constraint is volume fraction that represents the final optimized topology should have less or equal volume of the design domain. By discretizing the design domain to N square elements, the total strain energy can be calculated as

$$S = \sum_{e=1}^{N} (x_e)^p u_e^T k_e u_e \qquad (5)$$

where x_e is the density of each element varying from a minimum value (to avoid singularity in matrix calculation) to 1. p is the penallization parameter, u_e is elastic deformation for element e and k_e is the stiffness for element e. The penalization factor p penalizes the intermediate gray area in the Solid Isotropic Material with Penalization (SIMP) method to reduce gray area. In this case, the TO has continuous parameters and it is solved using SA with crystallization heuristic, as described in Section 2.

A new heuristic is included, after reaching the thermal equilibrium, the domain is regularized by filtering. The new density of each element after filtering gets some effects from the adjacent elements, as

$$x_{\text{filter}} = \frac{\sum_{e=1}^{N} w_e x_e}{\sum_{e=1}^{N} w_e} \qquad (6)$$

where w_e is the weighting function which is the filter radius minus distance of each adjacent element. It should be noted that the weighting function is zero outside of the filter radius and the density changes just inside the filer domain. The design domain and loading conditions for cantilever and half-MBB beam problems are shown in **Figure 6**. A comparison of obtained compliance from proposed method and some results from the literature is shown in **Table 2**.

As shown in **Table 2**, the results from this non-gradient based TO method are very close to the gradient based results. The main advantage of the proposed method is that there is no need to calculate derivatives of the objective function.

4.2 SA for multi-objective TO

The TO objective function can be complex and combine two or more objective functions. In such situations, the solution is not necessarily unique and comes with a

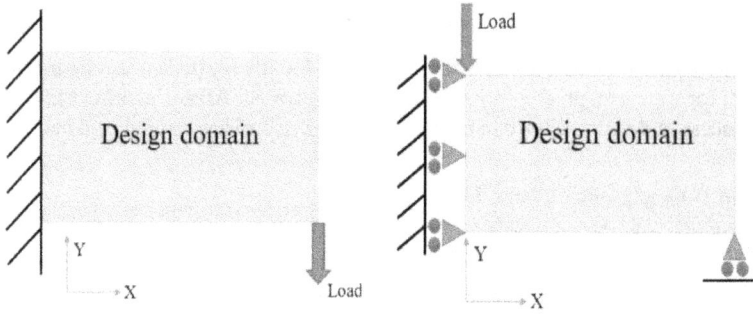

Figure 6.
The design domain and loading for cantilever beam (left) and half-MBB beam (right).

Volume fraction	Cantilever in [38]	Cantilever in SA	Half-MBB in [38]	Half-MBB in SA
0.5	75.05	79.87	82.37	89.65
0.6	62.89	65.70	69.19	73.36
0.7	55.27	57.10	60.61	63.39
0.8	49.89	51.29	54.85	56.63
0.9	46.29	47.36	50.57	51.79

Table 2.
The results for compliance of cantilever and half-MBB beam for different volume fractions.

set of optimum solutions called Pareto Front. The Pareto Front curve shows the solutions where none of the objective functions can be improved without degrading the other objective functions. The Pareto Front curve can be used for trade-off the suitable solution within this set instead of considering the full range of every parameter. The traditional TO usually optimizes one objective function while considering the other objective functions fixed or as a constraint [39]. The SA showed also this ability to incorporate multiple objective functions, like the one presented by the CoAnnealing [40] and AMOSA [41].

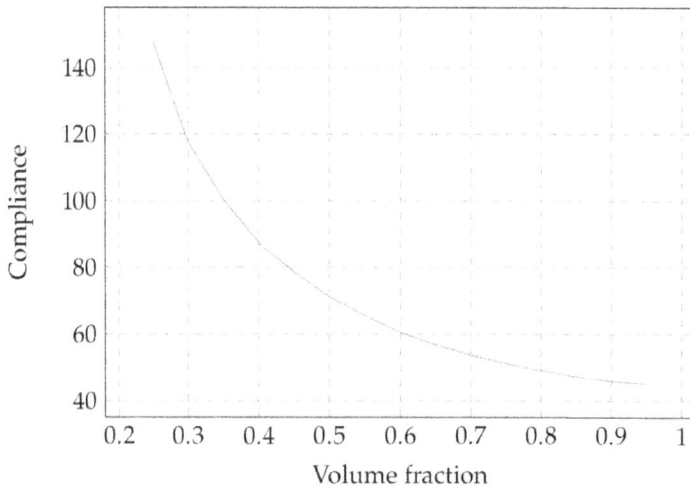

Figure 7.
The Pareto Front curve for minimization of compliance and weight in cantilever problem.

The CoAnnealing was used to solve TO problems and the Pareto Front was obtained; particularly, compliance and volume fraction as considered as cost functions in a cantilever beam, as shown in **Figure 7**. The points on the curve showed a good agreement with the results for that volume fraction and compliance [42]. The results shows that Coannealing can be used for the multiple objective TO problems.

5. Application to curve fitting

The problem of curve fitting is an essential Computer Aided Design (CAD) problem with applications in various engineering fields including but not limited to digital metrology, robotics, path planning, and data modeling. The problem consists of constructing a curve from a set of discrete points, as shown in **Figure 8**. The two main types of curve fitting can be defined, as an approximating curve fitting, when the constructed curve approximates the location of the points in the dataset, and an interpolating curve fitting, when the curve passes exactly through the set of points. The curve approximation has application in many engineering problems when a certain level of uncertainty exists in the dataset. This always can be expected in data points collected by an experimental process, when sources of uncertainties including the equipment errors, environmental effects, human errors, measurement resolution, etc., represent a combined level of uncertainty [43]. Due to the non-systematic nature of the uncertainties in the data, the approximating curve fitting process aims to recognise the true pattern in the data points instead of exactly passing through them [44]. The three major computation tasks to complete the approximating curve fitting include Point Measurement Planning (PMP), Substitute Geometry Estimation (SGE), and Deviation Zone Evaluation (DZE). Reducing or controlling the level of uncertainty in the constructed curve have been studied comprehensively at PMP by proper selection of the datapoints [45]. It is also addressed in SGE by improving the curve or surface fitting algorithms, using an enhanced optimization processes to avoid trapping in local minima, and by using iterative fitting approaches with monitoring some indicators for the level of uncertainty [46]. Various approaches have been also presented to measure and monitor the level of uncertainty typically at DZE stage by modeling the pattern and nature of the resulting deviations of the processed data points from the approximated curves or surfaces [47].

On the contrary, the curve interpolation is applicable when the datapoints are known to be fairly accurate and are conducted with no accommodation for any level

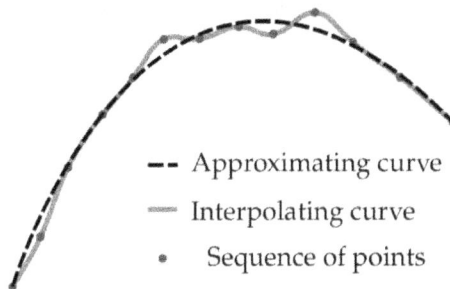

Figure 8.
Approximating curve (dashed) versus the interpolating curve (red).

of uncertainty in the datapoints. **Figure 8** presents a set of datapoints with its corresponding approximating and interpolating curves. The set of datapoints is shown by blue dots, the interpolating curve is presented by solid red line, and the approximating curve is presented by black dashed line.

Since a certain level of uncertainty typically exists in the most of the engineering problems, developing methodologies to control or reduce the level of uncertainty in the finally constructed curve is highly important. In this section, a SA approach is proposed to determine an approximation curve from a sequence of points. In this approach, the control points are continuous parameters and index corresponding points of the sequence are discrete parameters. All these parameters are adjusted by SA. This study was started by Ueda et al. [48]. The developed methodology employs piece-wise Bézier curve structure to solve the problem, as explained in the followings.

5.1 Piece-wise Bézier curve

There are several curve structures that can be used for the purpose of solving a curve fitting problem. However, each one has it own advantages and drawbacks. In a Bézier curve, the control points influence the entire curve globally including the regions that the curve is already fitted. On the other hand, the control points in a B-spline have only a local influence, i.e., by changing a group of control points only a certain region of the curve is modified. One feature of the Bézier curve structure that is beneficial in the presented fitting approach here is that the resulting Bézier curve always interpolates the first and last control point, while in the B-spline curve structure, these points usually are not interpolated. It is possible to interpolates these points in a B-spline. However, a higher number of optimization parameters is needed to achieve such feature, compared to the Bézier curve.

A piece-wise Bézier curve overcome the problem of the global influence of the control point. This curve is a sequence of cubic Bézier curves as shown in **Figure 9**, in which the last control point of curve $\mathbf{p}_3$ is the first control point of the following curve. The determination of the second control point of the second curve $\mathbf{p}_4$ is given by

$$\mathbf{P}_4 = \mathbf{P}_3 - \beta \cdot (\mathbf{P}_2 - \mathbf{P}_3), \tag{7}$$

with β being a proportional factor that ensures the weak-G1 continuity between the curves, i.e., the tangent vector of the end of the first curve has the same direction but not necessary the same intensity of the tangent of the start of the second curve. Ueda et al. [49] proposed an algorithm for automatic evaluate the number of piece-wise Bézier curves necessary to interpolate a sequence of points.

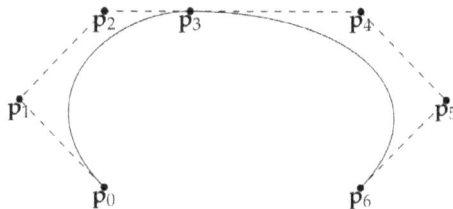

Figure 9.
Piece-wise cubic Bézier curve with 2 curve segments, points $\mathbf{p}_0, \mathbf{p}_1, \mathbf{p}_2$ and $\mathbf{p}_3$ defines the first curve and points $\mathbf{p}_3, \mathbf{p}_4, \mathbf{p}_5$ and $\mathbf{p}_6$ defines the second one.

5.2 SA for the approximation curve determination

Using a piece-wise Bézier curve has a advantage in two aspects. First, the curve is divided in several pieces, in which each curve segment approximates only part of the sequence. Second, it is known that a Bézier curve always interpolates its end points. The used cost function is

$$F(\{\mathbf{p}_i\}) = W \cdot \sum_{k=1}^{m-1} \|\mathbf{d}_k - \mathbf{P}_v(\mathbf{d}_k)\|^2 + (1 - W) \cdot L(\mathbf{P}(u)). \tag{8}$$

with $\mathbf{p}_i$ are the control points defining the curve, $\mathbf{d}$ is the sequence of points, $\mathbf{P}_v(\mathbf{d}_k)$ is the projection of point $\mathbf{d}_k$ in the curve $\mathbf{P}$, $L(\mathbf{P}(u))$ is the length of the curve, $\mathbf{P}$ is a piece-wise Bézier curve and W is a weight factor. This object function consists of two parts. $\sum_{k=1}^{m-1} \|\mathbf{d}_k - \mathbf{P}_v(\mathbf{d}_k)\|^2$ is the first part and it represents the discrepancy between the piece-wise curve and the sequence of points. The first and last points in the sequence are not considered as they already are on the determined piece-wise Bézier curves. For every point in the sequence $\mathbf{d}_k$ it is determined its projection in the piece-wise Bézier curve $\mathbf{P}_v(\mathbf{d}_k)$ [50]. $L(\mathbf{P}(u))$ is the second part, representing a regularization. It is used to avoid the over-fitting problem, the algorithm will search for short and smooth curves (and avoid long and sinuous curves).

SA needs to adjust the parameters defining a piece-wise Bézier curve that minimizes (8). Each piece-wise Bézier curve has two internal control points (as shown in **Figure 9**). The coordinates for each of these control points are continuous parameters and they are controlled using the crystallization heuristic. The connection point between two piece-wise Bézier curves (as point $\mathbf{p}_3$ shown in **Figure 9**) is represented as an integer parameter. It is an index representing a point from the given sequence.

Consider the example from **Figure 10**, the continuous parameters are the coordinate of control points $\mathbf{p}_1$, $\mathbf{p}_2$ and $\mathbf{p}_5$. Control point $\mathbf{p}_4$ is determined by modifying β from (7). Control points $\mathbf{p}_0$ and $\mathbf{p}_6$ are the first and last point of the sequence of points; and control point $\mathbf{p}_3$ is a connecting point. The definition of this point is the discrete parameter adjusted by the SA.

The continuous parameters are controlled by the crystallization heuristic. The discrete parameters have specific operator, a random value is picked from a

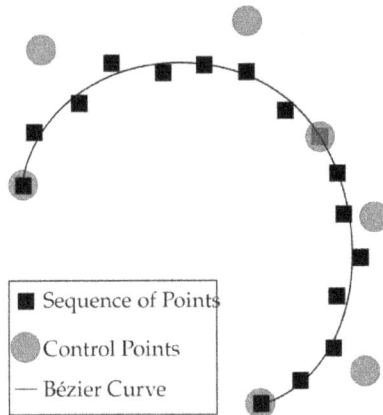

Figure 10.
Each Bézier curve segment approximates just one part of the sequence of points.

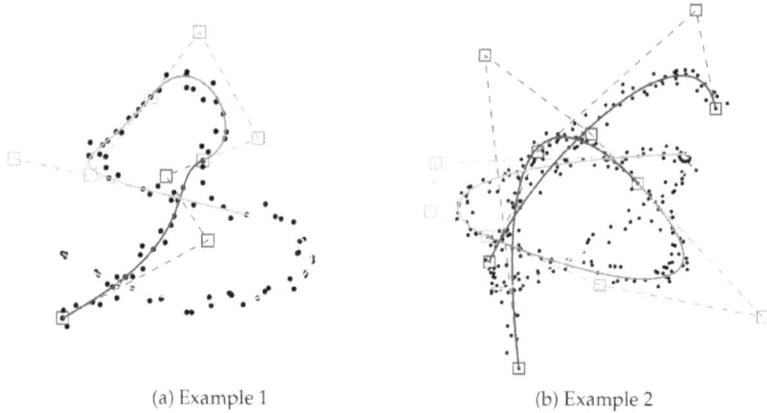

(a) Example 1   (b) Example 2

Figure 11.
Two artificial created examples tested by the proposed algorithm. Black points are the noisy sampled points, each curve segment is in a different color with its respective control points represented as squares.

Example	N_{pts}	N_{segm}	N_{iter}	Cost
1	101	4	508,404	10.37
2	300	5	473,680	134.97

Table 3.
The results for the SA based curve interpolation solution. N_{pts}: total number of points. N_{Segm}: number of curve segments. N_{iter}: number of iterations. Cost: objective function.

uniform distribution between $[0, 1]$, if this value is greater than 0.5 the discrete parameter is increased by 1, otherwise it is decreased by 1.

5.3 Results and discussion

A curve is defined and sampled, and a random noise is added to each sampled point creating an artificial example to be tested by the proposed algorithm. Two example were tested, shown in **Figure 11**. **Figure 11(a)** is sampled with 101 points and **Figure 11(b)** is sampled with 300 points. The initial temperature is set as 100 and for each temperature there are 1000 iteration or 500 accepted solutions, and an adaptive cooling schedule [50] was adopted with a variable α. A minimum of 5 accepted solutions is used as stop criteria.

Table 3 show the result of these two test examples. The objective function is higher in the second example, once N_{pts} is higher as well the curve is longer. N_{iter} is close in both example, there are less than 7% of difference between both tests.

The proposed interpolation algorithm was applied in different applications showing its versatility and robustness. Ueda et al. [51] proposed an algorithm to determine the curves which interpolates the boundary of a point cloud in space. Ueda et al. [52] proposed an algorithm to determine the defect zone from a 3D scanned turbine blade. Ueda et al. [53] also proposed an algorithm to determine open boundaries in point clouds with symmetry (like injuries in a skull).

6. Conclusions

This chapter describes the SA with crystallization heuristic and three different applications: cutting and packing, topology optimization and curve interpolation.

The cutting and packing problem determines the layout for a set of items to be placed inside a container with fixed dimensions. The cost function is the unused area inside the container. The items can be translated and rotated in the process of determining the layout. This cost function is discrete and the parameters are continuous. As there is no need of any gradient information, it can optimize discrete cost functions.

The topology optimization problem has continuous parameters. The number of parameters is much larger when compared with the cutting and packing problem. Usually, the cost function is one of the two possibilities: volume fraction and compliance. As there are two possibilities for cost function, it is considered the application of the CoAnnealing, which is a multi-objective version of the SA with crystallization heuristic. The last application is the curve interpolation. It has two different types of parameters: continuous and integer. The objective function is composed of two parts: the error between the the given points and the interpolating curve, and the length of the curve. The length of the curve is a regularization to force the curve to be shorter and smoother. The curve interpolation algorithm was used in different applications.

These examples show that the SA with crystallization heuristic is very generic, versatile and easy to implement.

Acknowledgements

G. C. Duran is supported by CNPq (process 140.299/2020-3). H. R. Najafabadi is supported by FAPESP (grant 2019/03453-2). M. S. G Tsuzuki is partially supported by CNPq (process 311.195/2019-9). A. K. Sato is supported by FUSP/Petrobras.

Nomenclature

α	Temperature Geometric Cooling Parameter
$A()$	Function Area
β	Proportional Factor to Ensure Weak-G1 continuity
C	Rectangular Conteiner
C_j	Crystallization Factor
ΔE	Energy Variation
Δr_j	Step Variation for Variable j
e_j	Position Vector for Variable j (all elements are zero, except one)
F_e	External Force
$F(x)$	Objective Function
k_e	Stiffness for Element e
$i(P_k)$	Interior of the Item
$L(\mathbf{P}(u))$	Length of Curve $\mathbf{P}(u)$
K	Stiffness
N	Number of Square Elements
p	Penalization Parameter
$\mathbf{P}_k$	Control Point
$\mathcal{P}$	Set of Items
P_k	Item
$P_k(r)$	Item Rotated by r
$P_k \oplus t$	Item Translated by t
$\mathbf{P}(u)$	Piece-wise Bézier Curve

$\mathbf{P}_v(\mathbf{d}_k)$	Projection of Point $\mathbf{d}_k$ in Curve $\mathbf{P}(u)$
$\mathfrak{R}$	Space of Continuous Numbers
r	Rotation
S	Strain Energy
t	Translation
T	Assigned Set of Items
T_i	Temperature at iteration i
U	Elastic Deformation
u_e	Elastic Deformation for Element e
x	Current Solution
x^*	Solution Candidate
x_e	Density of Element e
w_e	Weighting Function
W	Weight Used in the Approximation Curve

Abbreviations

AMOSA	Archived Multiobjective Simulated Annealing
CAD	Computer-Aided Design
CFR	Collision Free Region
C&P	Cutting & Packing
EIT	Electrical Impedance Tomography
FEM	Finite Element Method
GA	Genetic Algorithm
MAP	Maximum a Posteriori
MMA	Method of Moving Asymptotes
OC	Optimality Criteria
PDF	Probability Density Function
PSO	Particle Swarm Optimization
RANSAC	Random Sample Consensus
SA	Simulated Annealing
SIMP	Solid Isotropic Material with Penalization
SLP	Sequential Linear Programming
TO	Topology Optimization

Author details

Tiago G. Goto[1,2†], Hossein R. Najafabadi[2†], Guilherme C. Duran[2†], Edson K. Ueda[2†], André K. Sato[2†], Thiago C. Martins[2†], Rogério Y. Takimoto[2†], Hossein Gohari[3†], Ahmad Barari[3†] and Marcos S.G. Tsuzuki[2*†]

1 Universidade Federal de Rondonópolis, Rondonópolis, Brazil

2 Computational Geometry Laboratory, Escola Politénica da USP, São Paulo, Brazil

3 Ontario Tech University, Oshawa, Canada

*Address all correspondence to: mtsuzuki@usp.br

† These authors contributed equally.

IntechOpen

References

[1] J. V. Burke and M. C. Ferris, "A Gauss-Newton method for convex composite optimization," *Math Program*, vol. 71, no. 2, pp. 179–194, 1995.

[2] P. R. Pinheiro, B. Amaro Júnior, and R. D. Saraiva, "A random-key genetic algorithm for solving the nesting problem," *Int J Comp Integ M*, vol. 29, no. 11, pp. 1159–1165, 2016.

[3] S. Wang, K. Tai, and M. Y. Wang, "An enhanced genetic algorithm for structural topology optimization," *Int J Numer Meth Eng*, vol. 65, no. 1, pp. 18–44, 2006.

[4] E. G. Castro and M. S. G. Tsuzuki, "Swarm intelligence applied in synthesis of hunting strategies in a three-dimensional environment," *Expert Syst Appl*, vol. 34, no. 3, pp. 1995–2003, 2008.

[5] S. Kirkpatrick, C. D. Gelatt, and M. P. Vecchi, "Optimization by simulated annealing," *Science*, vol. 220, no. 4598, pp. 671–680, 1983.

[6] A. Corana, M. Marchesi, C. martini, and S. Ridella, "Minimizing multimodal functions of continuous variables with the simulated annealing algorithm," *ACM Transactions on Mathematical Software*, vol. 13, pp. 262–280, 1987.

[7] L. Ingber, "Very fast simulated re-annealing," *Mathematical and Computer Modelling*, vol. 12, no. 8, pp. 967–973, 1989.

[8] T. C. Martins and M. S. G. Tsuzuki, "Simulated annealing applied to the rotational polygon packing," *IFAC Proc Vol (IFAC-PapersOnline)*, vol. 12, no. PART 1, 2006.

[9] T. C. Martins and M. S. G. Tsuzuki, "Rotational placement of irregular polygons over containers with fixed dimensions using simulated annealing and no-fit polygons," *J Braz Soc Mech Sci*, vol. 30, no. 3, pp. 205–212, 2008.

[10] T. C. Martins, A. K. Sato, and M. S. G. Tsuzuki, "Adaptive neighborhood heuristics for simulated annealing over continuous variables," in *Simulated Annealing - Advances, Applications and Hybridizations* (M. S. G. Tsuzuki, ed.), pp. 3–20, INTECHOpen, 2012.

[11] T. C. Martins, E. D. L. B. Camargo, R. G. Lima, M. B. P. Amato, and M. S. G. Tsuzuki, "Electrical impedance tomography reconstruction through simulated annealing with incomplete evaluation of the objective function," in *33rd IEEE EMBC*, pp. 7033–7036, 2011.

[12] T. C. Martins and M. S. G. Tsuzuki, "Electrical impedance tomography reconstruction through simulated annealing with total least square error as objective function," in *34th IEEE EMBC*, pp. 1518–1521, 2012.

[13] R. S. Tavares, T. C. Martins, and M. S. G. Tsuzuki, "Electrical impedance tomography reconstruction through simulated annealing using a new outside-in heuristic and GPU parallelization," *J Phys: Conf Series*, vol. 407, p. 012015, dec 2012.

[14] V. Černỳ, "Thermodynamical approach to the traveling salesman problem: An efficient simulation algorithm," *J Optimiz Theory App*, vol. 45, no. 1, pp. 41–51, 1985.

[15] E. H. L. Aarts and P. J. M. van Laarhoven, "Statistical cooling: a general approach to combinatorial optimization problems," *Philips J Res*, vol. 40, no. 4, pp. 193–226, 1985.

[16] T. C. Martins and M. S. G. Tsuzuki, "Placement over containers with fixed dimensions solved with adaptive neighborhood simulated annealing," *B

Pol Acad Sci-Tech, vol. 57, pp. 273–280, 2009.

[17] A. K. Sato, T. C. Martins, and M. S. G. Tsuzuki, "Rotational placement using simulated annealing and collision free region," *IFAC Proceedings Volumes*, vol. 43, no. 4, pp. 234–239, 2010.

[18] A. K. Sato, T. C. Martins, and M. S. G. Tsuzuki, "An algorithm for the strip packing problem using collision free region and exact fitting placement," *CAD*, vol. 44, no. 8, pp. 766–777, 2012.

[19] A. K. Sato, T. de Castro Martins, and M. S. G. Tsuzuki, "A pairwise exact placement algorithm for the irregular nesting problem," *Int J Comp Integ M*, vol. 29, no. 11, pp. 1177–1189, 2016.

[20] J. A. Bennell and J. F. Oliveira, "The geometry of nesting problems: A tutorial," *Eur J Oper Res*, vol. 184, no. 2, pp. 397 – 415, 2008.

[21] L. R. Mundim, M. Andretta, and T. A. de Queiroz, "A biased random key genetic algorithm for open dimension nesting problems using no-fit raster," *Expert Syst Appl*, vol. 81, pp. 358–371, 2017.

[22] J. Egeblad, B. K. Nielsen, and A. Odgaard, "Fast neighborhood search for two- and three–dimensional nesting problems," *Eur J Oper Res*, vol. 183, pp. 1249–1266, 2007.

[23] A. Elkeran, "A new approach for sheet nesting problem using guided cuckoo search and pairwise clustering," *Eur J Oper Res*, vol. 231, no. 3, pp. 757–769, 2013.

[24] A. K. Sato, T. C. Martins, A. M. Gomes, and M. S. G. Tsuzuki, "Raster penetration map applied to the irregular packing problem," *Eur J Oper Res*, vol. 279, no. 2, pp. 657–671, 2019.

[25] A. K. Sato, T. C. Martins, and M. S. G. Tsuzuki, "Collision free region

determination by modified polygonal boolean operations," *CAD*, vol. 45, no. 7, pp. 1029–1041, 2013.

[26] M. P. Bendsoe and O. Sigmund, *Topology optimization: theory, methods, and applications*. Springer Science & Business Media, 2013.

[27] T. R. Marchesi, R. D. Lahuerta, E. C. N. Silva, M. S. G. Tsuzuki, T. C. Martins, A. Barari, and I. Wood, "Topologically optimized diesel engine support manufactured with additive manufacturing," *IFAC-PapersOnLine*, vol. 28, no. 3, pp. 2333–2338, 2015.

[28] O. Sigmund and K. Maute, "Topology optimization approaches," *Struct Multidiscip O*, vol. 48, no. 6, pp. 1031–1055, 2013.

[29] G. I. Rozvany, "A critical review of established methods of structural topology optimization," *Struct Multidiscip O*, vol. 37, no. 3, pp. 217–237, 2009.

[30] A. Shukla and A. Misra, "Review of optimality criterion approach scope, limitation and development in topology optimization," *Int J Adv Eng Techn*, vol. 6, no. 4, p. 1886, 2013.

[31] S. Rojas-Labanda and M. Stolpe, "Benchmarking optimization solvers for structural topology optimization," *Struct Multidiscip O*, vol. 52, no. 3, pp. 527–547, 2015.

[32] F. A. Gomes and T. A. Senne, "An slp algorithm and its application to topology optimization," *Comput Appl Math*, vol. 30, no. 1, pp. 53–89, 2011.

[33] O. Sigmund, "On the usefulness of non-gradient approaches in topology optimization," *Struct Multidiscip O*, vol. 43, no. 5, pp. 589–596, 2011.

[34] D. Guirguis, W. W. Melek, and M. F. Aly, "High-resolution non-gradient topology optimization," *J Comput Phys*, vol. 372, pp. 107–125, 2018.

[35] N. Garcia-Lopez, M. Sanchez-Silva, A. Medaglia, and A. Chateauneuf, "A hybrid topology optimization methodology combining simulated annealing and simp," *Comput Struct*, vol. 89, no. 15-16, pp. 1512–1522, 2011.

[36] G. Y. Cui, K. Tai, and B. P. Wang, "Topology optimization for maximum natural frequency using simulated annealing and morphological representation," *AIAA J*, vol. 40, no. 3, pp. 586–589, 2002.

[37] M. Ohsaki, "Genetic algorithm for topology optimization of trusses," *Comput Struct*, vol. 57, no. 2, pp. 219–225, 1995.

[38] O. Sigmund, "A 99 line topology optimization code written in matlab," *Struct Multidiscip O*, vol. 21, no. 2, pp. 120–127, 2001.

[39] K. Suresh, "A 199-line matlab code for pareto-optimal tracing in topology optimization," *Struct Multidiscip O*, vol. 42, no. 5, pp. 665–679, 2010.

[40] T. C. Martins and M. S. G. Tsuzuki, "EIT image regularization by a new multi-objective simulated annealing algorithm," in *IEEE EMBC 2015*, pp. 4069–4072, IEEE, 2015.

[41] T. C. Martins, A. V. Fernandes, and M. S. G. Tsuzuki, "Image reconstruction by electrical impedance tomography using multi-objective simulated annealing," in *IEEE ISBI 2014*, pp. 185–188, jul 2014.

[42] H. R. Najafabadi, T. G. Goto, T. C. Martins, A. Barari, and M. S. G. Tsuzuki, "Multi-objective topology optimization using simulated annealing method," in *ICGG 2020*, pp. 343–353, Springer.

[43] A. Barari, "Automotive body inspection uncertainty associated with computational processes," *Int J Veh Des*, vol. 57. no. 2-3, pp. 230-241, 2011.

[44] A. Barari, H. A. ElMaraghy, and P. Orban, "NURBS representation of estimated surfaces resulting from machining errors," *Int J Comp Integ M*, vol. 22, no. 5, pp. 395-410, 2009.

[45] A. Lalehpour, C. Berry, and A. Barari, "Adaptive data reduction with neighbourhood search approach in coordinate measurement of planar surfaces," *J M Syst*, vol. 45. pp. 28-47, 2017.

[46] A. Barari, and S. Mordo, "Effect of sampling strategy on uncertainty and precision of flatness inspection studied by dynamic minimum deviation zone evaluation," *J Metrol Qual Eng*, vol. 4, no. 1, pp. 3-8, 2013.

[47] A. Lalehpour, and A. Barari, "Developing skin model in coordinate metrology using a finite element method," *Measurement*, vol. 109, pp. 149-159, 2017.

[48] E. K. Ueda, M. S. G. Tsuzuki, R. Y. Takimoto, A. K. Sato, T. C. Martins, P. E. Miyagi, and R. S. U. Rosso Jr, "Piecewise Bézier curve fitting by multiobjective simulated annealing," *IFAC-PapersOnLine*, vol. 49, pp. 49–54, jan 2016.

[49] E. K. Ueda, T. C. Martins, and M. S. G. Tsuzuki, "Planar curve fitting by simulated annealing with feature points determination," *IFAC-PapersOnLine*, vol. 51, pp. 290–295, jan 2018.

[50] E. K. Ueda, A. K. Sato, T. C. Martins, R. Y. Takimoto, R. S. U. Rosso Jr, and M. S. G. Tsuzuki, "Curve approximation by adaptive neighborhood simulated annealing and piecewise Bézier curves," *Soft Comput*, vol. 24, p. 18821–18839, 2020.

[51] E. K. Ueda, M. S. Tsuzuki, and A. Barari, "Piecewise Bézier curve fitting of a point cloud boundary by simulated

annealing," in *INDUSCON 2018*,
pp. 1335–1340, jan 2019.

[52] E. K. Ueda, A. Barari, A. K. Sato, and
M. S. G. Tsuzuki, "Detection of defected
zone using 3D scanning data to repair
worn turbine blades," in *21st IFAC
World Congress*, pp. 10666–10670, 2020.

[53] E. K. Ueda, A. Barari, and M. S. G.
Tsuzuki, "Determination of open
boundaries in point clouds with
symmetry," in *ICGG 2020*, pp. 332–342,
Springer.

Chapter 10

A Metaheuristic Tabu Search Optimization Algorithm: Applications to Chemical and Environmental Processes

Chimmiri Venkateswarlu

Abstract

Stochastic optimization methods are increasingly used for optimizing processes that are difficult to solve by conventional techniques. These methods are widely employed to optimize the processes which have higher dimensionality with severe nonlinearities. Different methods of this kind include the genetic algorithm (GA), simulated annealing (SA), differential evolution (DE), ant colony optimization (ACO), tabu search (TS), particle swarm optimization (PSO), artificial bee colony (ABC) algorithm, and cuckoo search (CS) algorithm. Among these methods, tabu search (TS) is a potential tool used to find a feasible optimal solution from a finite set of solutions. The memory used in TS will remember the current best solution and it also enables the TS to track the last solutions while guiding the search moves. The capability of memory and strategic adaptation features of TS enable it to make use of good solutions and also search for new feasible regions in the search space. TS has been successfully applied to solve a wide spectrum of optimization problems in different disciplines. This chapter describes the TS algorithm in detail and its applications to chemical and environmental processes, specifically, dynamic optimization of a copolymerization reactor and inverse modeling of a biofilm reactor. In dynamic optimization of copolymerization reactor, the meta heuristic Tabu search (TS) is designed and applied to determine the optimal control policies of a styrene–acrylonitrile (SAN) copolymerization reactor. In inverse modeling of biofilm reactor, the tabu search is designed and applied to determine the parameters of kinetic and film thickness models as consequence of the validation of the mathematical models of the process with the aid of measured data acquired from an experimental fixed bed anaerobic biofilm reactor used in the treatment of pharmaceutical industry wastewater. For both the cases, optimization by Tabu search is carried out by suitably formulating the desired objective functions and the problems are solved by encoding the variables and parameters using real floating point numbers. The results explain the efficacy of TS for optimal control of polymerization reactor and inverse modeling of biofilm reactor.

Keywords: Tabu search, Metaheuristic approach, Adaptive memory, Copolymerization reactor, Biofilm reactor

1. Introduction

Tabu search is a meta heuristic problem solving approach used to solve combinatorial optimization problems. It was first proposed by Glover [1] and further developed by Hansen [2]. TS has now become an established search procedure and has been successfully applied to solve a wide spectrum of optimization problems [3–9].

1.1 Basic principle

TS uses a memory which allows it to remember the current best solution and also enables it to explore the previous solutions and to direct the search moves. The features of memory adaptation and exploration facilitate TS to find better solutions and discover new potential regions in the search space. The memory adaptation feature helps to realize its course of action to exploit the search space efficiently. The ability of TS to explore good solutions enables it to assimilate the intelligent search mechanisms to search for good solutions and to discover new potential regions. The use of adaptive memory helps TS to learn and creates a more flexible and effective search strategy compared with memory less methods, such as simulated annealing (SA) and genetic algorithms (GA).

1.2 Components of TS

The components of TS are explained as follows.

1.2.1 Neighbors generations and neighborhood search

To optimize the function $f(x)$ globally from all the probable solutions $x \in X$ in the space X, it requires to specify a structure in the vicinity of the solution space and the staring solution. The search advances to alter the existing solution to create a set of promising solutions in the vicinity of solution space. During the search process, the number of solutions traversed by TS is the product of the number of solutions in the vicinity of the solution space, $N(x_i)$, and the number of iterations, k. The function at each iteration is evaluated for $N(x_i)$ solutions. The better move is chosen in the vicinity of the complete solution space. The search proceeds to the next iteration, to find a solution in the vicinity of the accepted move. Thus, the TS builds a set of viable solutions by using a history record of the search.

1.2.2 Tabu list

The tabu list in TS maintains the data of the previously visited solutions. The list is included with the latest moves and it is altered dynamically as the search proceeds. The data in the tabu list helps to direct the move from the present solution to the next solution. At each iteration, the search process is maintained by updating the tabu list. The tabu list also avoids re-visiting of recent neighbors recorded in the list and thus save computational time.

1.2.3 Short-term memory and long-term memory

The information is stored in tabu list as recency-based short-term memory (RSM) and a frequency-based long-term memory (FRM). When the search proceeds, the nearly better solution to the present solution is sorted as tabu, and added to the recency-based tabu list. As fresh solutions enter the list, older solutions are removed from the bottom. Long-term memory relies on the frequency of a solution that is visited. When the tabu list is filled with the highest number of elements in the frequency based tabu list, the solution with the smallest frequency index will be replaced.

1.2.4 Intensification and diversification

These strategies are used to create neighbors that have higher likelihood of finding optimal solutions based on the data in the tabu list. Intensification strategies are used to search potential areas in detail around the areas that are found good in the past. Intensification strategies are generally employed based on long term memory whose components are used to create neighbors for search intensification [4, 6, 10]. Diversification strategies are used to search the complete viable region, thus restricting the search getting trapped in local optima. These strategies promote probing unvisited regions by creating solutions radically different from those searched earlier [4]. A frequency-based tabu list is used to keep track of the search area.

During the generation of neighbor solutions, the difference between the present and fresh neighbor solution is managed by using a coefficient, α. The change from the current point is multiplied by α during the course of building new neighbors. The coefficient α is in the form of a sine function [10].

$$\alpha = \frac{1}{2}\left[1 + \sin\left[\frac{i\theta\pi}{N_{neigh}}\right]\right] \tag{1}$$

Here i is index of the neighbor, N_{neigh} is the total number of neighbor solutions generated at each iteration, and θ is a parameter that controls the oscillation period of α.

1.2.5 Aspiration criterion

The TS conditions at times prevent moves leading to unvisited solutions. Aspiration criterion is a condition that can override the tabu status of a certain move. To avoid certain missing solutions during the search, the aspiration criterion in certain cases may invalidate the tabu property and maintains an appropriate balance between diversification and intensification [4, 10, 11].

An aspiration criterion that is designed based on a sigmoid function as given by

$$S(k) = \frac{1}{1 + e^{-\sigma(k - k_{center} \times M)}} \tag{2}$$

where k_{center}, k, σ and M denote the tuning parameter, current iteration number, another tuning parameter and maximum number of iterations. The value of k_{center} can be in the range of 0.30–0.70, the value of σ can be in the range of 5/M to 10/M. A random number P that lies between 0 and 1 is generated from a uniform distribution at each iteration. If P is greater than $S(k)$, the tabu property is active and the best non-tabu neighbor is used as a fresh starting point. If P is less than or equal to $S(k)$, the aspiration criterion ignores the tabu property.

1.2.6 Stopping criteria

A stopping criterion is needed to terminate the search when the optimum is reached. The stopping criterion can be in the form of fixing the number of iterations or specifying a threshold for convergence of solution. The criteria like the maximum time termination [6] and termination-on convergence criteria [10] are also used as stopping criteria for the search process. The termination-on-convergence criteria is expressed by Lin and Miller [10] as.

$$\left| \frac{f_k(x) - f_{k-\Gamma}(x)}{f_{k-\Gamma}(x)} \right| < \delta \tag{3}$$

where δ is the ratio of the change in the objective function value, $\Gamma = \eta M$, and η is the fraction of the maximum iterations (M) by which the change in the objective function is compared. As per this stopping criterion, if the enhancement over Γ generations is no longer than a threshold (δ), continuation of further iterations can be ineffective, and the search should be discontinued.

1.3 TS implementation procedure

The flow chart of TS algorithm is shown in **Figure 1**. Tabu Search begins with an initial solution x_o. The neighbor solutions are created by altering the existing solution through a sequence of moves. The best new neighbor, x^*, is used as the starting point for the next iteration, unless it is in the tabu list. Thus, even if no neighbor solutions are better than the starting solution, the best solution is still selected as the starting point for the next iteration. A record of the best solutions ever found, x^*, is separately maintained. Also, the adaptive memory in tabu lists guides the search by utilizing the benefit of past information. This memory facilitates TS to make strategic choices and accomplish responsive exploration.

TS is implemented using the following steps.

1. Starts with an initial solution x_o

2. Find the best solution, x^* and define tabu list

3. Neighbor's generations and neighborhood search

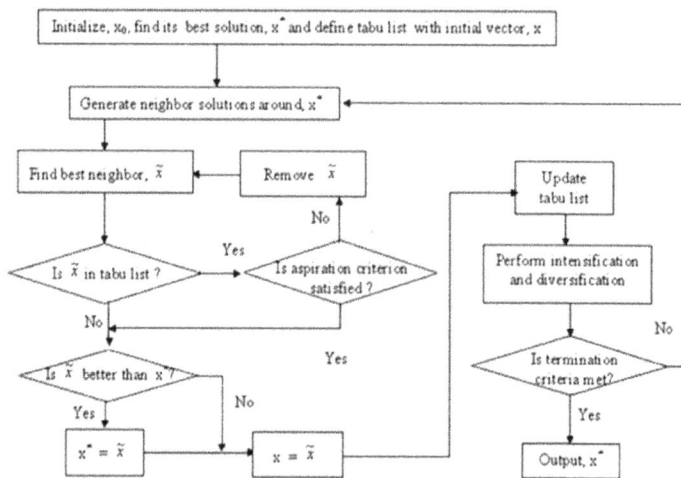

Figure 1.
Flow chart of the TS algorithm.

   a. Define a neighborhood structure with respect to the solution space

   b. Change the present solution to form a neighborhood of possible solutions

   c. TS explore neighbor solution at each iteration

   d. Searches entire neighborhood and select enhanced solution

4. Tabu list

   a. Majority of the recent solutions are in tabu list

   b. Older solutions are discarded

5. Short-term memory and long-term memory

   a. Recency-based short-term memory (RSM)

   b. Frequency-based long-term memory (FRM)

6. Intensification and Diversification

   a. To search potential areas

   b. To search whole feasible region

7. Aspiration criterion

   a. To explore unvisited solutions

   b. To evade feasible missing solutions

8. Stopping criteria

   a. According iterations

   b. Stopping on convergence

2. Application of Tabu search for optimal control of a polymerization reactor

The determination of open-loop time varying control policies that maximize or minimize a given performance index is referred as optimal control. The optimal control policies that guarantee the product property requirements and the operational constraints can be computed off-line, and are executed on-line in such a way that the process is operated in accordance with these control policies. Achievement of product quality is a major issue in polymerization processes since the molecular or morphological properties of a polymer product majorly influences its physical, chemical, thermal, rheological and mechanical properties as well as polymer applications. In a free radical copolymerization, composition drifts caused by variations in reactivities of comonomers can be mitigated by continuous addition of more reactive monomer while maintaining a constant mole ratio. The end use properties of the polymer product such as flexibility, strength and glass transition temperature are affected by the copolymer composition. Further, the molecular weight (MW) and molecular weight distribution (MWD) affects the important end use properties such as viscosity, elasticity, strength, toughness and solvent resistance. Hence, the determination of optimal control trajectories is important for the operation of polymerization reactors in order to produce a polymer with the desired product characteristics.

In the past, various methods have been reported for optimal control of polymerization reactors [12–16]. Most of the studies are based on classical methods of optimization such as Pontryagin's maximum principle [17], which has been applied to solve the optimal control problems of different polymerization reactors [18–23]. However, the classical methods have certain limitations for optimal control of polymerization reactors and these drawbacks have been discussed in literature [24]. The stochastic and evolutionary optimization methods are found beneficial over the conventional gradient-based search techniques because of their ability in locating the global optimum of multi-modal functions and searching design spaces with disjoint feasible regions.

2.1 Optimal control problem

The general open-loop optimal control problem of a lumped parameter batch/semi-batch process with fixed terminal time can be stated as follows. Find a control vector $u(t)$ over t_f $[t_o, t_f]$ to maximize (minimize) a performance index $J(x,u)$:

$$J(x, u) = \int_{t_0}^{t_f} \varphi[x(t), u(t), t] \, dt \tag{4}$$

Subject to

$$\dot{x}(t) = f(x(t), \ u(t), t), \quad x(t_0) = x_0 \tag{5}$$

$$h[x(t), u(t)] = 0 \tag{6}$$

$$g[x(t), u(t)] \leq 0 \tag{7}$$

$$x^L \leq x(t) \leq x^U \tag{8}$$

$$u^L \leq u(t) \leq u^U \tag{9}$$

In the above, J refers the performance index, x is the state variable vector, u is the control variable vector. Eq. (5) represents the system of ordinary differential equations with their initial conditions, Eqs. (6) and (7) specify the equality and inequality algebraic constraints and Eqs. (8) and (9) are the upper and lower bounds for the state and control variables.

2.2 Multistage dynamic optimization strategy

A multistage optimization results due to the natural extension of a single stage optimization. In a system involving multiple stages, the output from one stage becomes the input to the subsequent stage. The multistage dynamic optimization procedure can be referred elsewhere [24]. This type of optimization needs special techniques to split the problem into computationally manageable units. In multistage dynamic optimization, the optimal control problem of the entire batch duration is divided into finite number of time instants referred to as discrete stages. The control variables and the corresponding state variables that satisfy the objective function are evaluated in stage wise manner.

In this work, a multistage dynamic optimization strategy with sequential implementation procedure is presented for optimal control of polymerization reactors. The procedure for solving the optimal control problem is similar to that of the dynamic programming based on the principle of optimality [25]. The meta heuristic features of tabu search (TS) are exploited by implementing this strategy on a semi-batch styrene–acrylonitrile (SAN) copolymerization reactor. The procedure involves in discretizing the process into N stages, defining the objective function, f^i, the control vector, u^i and the state vector, x^i for stage i. This procedure is briefed in the following steps:

1. The optimum value of the objective function, $f^1[x^1]$ for stage 1 driven by the best control vector u^1 along with the state vector x^1 is represented as

$$f^1[x^1] = \min_{u^1} \ f_o^1[x^1, u^1] \tag{10}$$

2. The value of the objective function, $f^2[x^2]$ for stage 2 is determined based on the best control vector u^2 along with the state vector x^2 as given by

$$f^2[x^2] = f^1[x^1] + \min_{u^2} \ f_o^2[x^2, u^2] \tag{11}$$

3. Recursive generalization of the above procedure for the kth stage is represented by

$$f^k[x^k] = f^{k-1}[x^{k-1}] + \min_{u^k} \ f_o^k[x^k, u^k] \tag{12}$$

In this procedure, f_o^k represents the performance index of stage k.

2.3 The description of polymerization process and its mathematical representation

The solution copolymerization of styrene and acrylonitrile taking place in a semi-batch reactor is considered as a test bed for sequential implementation of the dynamic optimization strategy. The xylene is the solvent and AIBN is the initiator for the reaction. The feed which is a mixture of monomers, solvent, and initiator enters the reactor in semi-batch mode. The reactor initial volume is 1.01 L. The initially set design parameters are the solvent mole fraction $f_s = 0.25$ and initiator concentration $I_0 = 0.05$ mol/L. The mole ratio of monomers in the feed, M_1/M_2, is 1.5, where M_1 and M_2 are the molar concentrations of the unreacted monomers (styrene and acrylonitrile). The homogeneous solution free-radical copolymerization of styrene with acrylonitrile is described by the following reaction mechanism [26].
Initiation:

$$I \longrightarrow 2R$$

$$R + M_1 \xrightarrow{k_{i1}} P_{10} \tag{13}$$

$$R + M_2 \xrightarrow{k_{i2}} Q_{01}$$

Propagation:

$$P_{n,m} + M_1 \xrightarrow{k_{p11}} P_{n+1,m}$$

$$P_{n,m} + M_2 \xrightarrow{k_{p12}} Q_{n,m+1} \tag{14}$$

$$Q_{n,m} + M_1 \xrightarrow{k_{p21}} P_{n+1,m}$$

$$Q_{n,m} + M_2 \xrightarrow{k_{p22}} Q_{n,m+1}$$

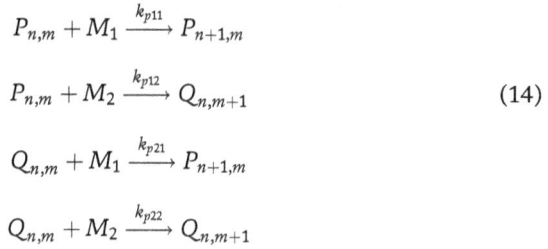

Combination termination:

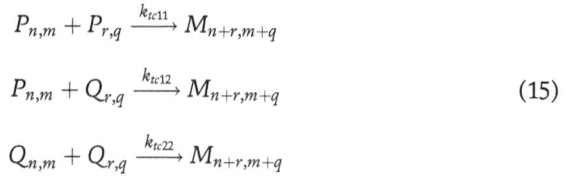

$$P_{n,m} + P_{r,q} \xrightarrow{k_{tc11}} M_{n+r,m+q}$$

$$P_{n,m} + Q_{r,q} \xrightarrow{k_{tc12}} M_{n+r,m+q} \tag{15}$$

$$Q_{n,m} + Q_{r,q} \xrightarrow{k_{tc22}} M_{n+r,m+q}$$

Disproportionation termination:

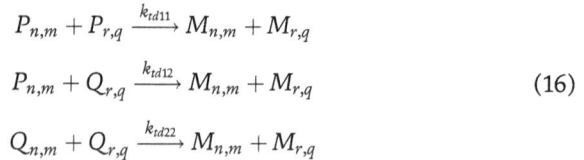

$$P_{n,m} + P_{r,q} \xrightarrow{k_{td11}} M_{n,m} + M_{r,q}$$

$$P_{n,m} + Q_{r,q} \xrightarrow{k_{td12}} M_{n,m} + M_{r,q} \tag{16}$$

$$Q_{n,m} + Q_{r,q} \xrightarrow{k_{td22}} M_{n,m} + M_{r,q}$$

Chain Transfer:

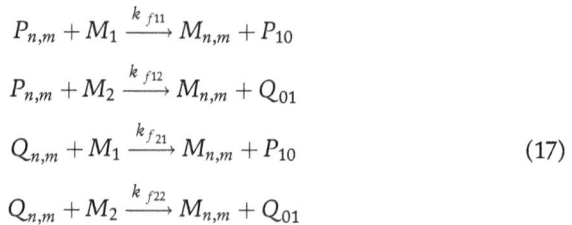

$$P_{n,m} + M_1 \xrightarrow{k_{f11}} M_{n,m} + P_{10}$$

$$P_{n,m} + M_2 \xrightarrow{k_{f12}} M_{n,m} + Q_{01}$$

$$Q_{n,m} + M_1 \xrightarrow{k_{f21}} M_{n,m} + P_{10} \tag{17}$$

$$Q_{n,m} + M_2 \xrightarrow{k_{f22}} M_{n,m} + Q_{01}$$

where $P_{n,m}$ stand for a growing polymer chain with n units of monomer 1 (styrene) and m units of monomer 2 (acrylonitrile) with monomer 1 on the chain end. Similarly, $Q_{n,m}$ refers to a growing copolymer chain with monomer 2 on the end. The $M_{n,m}$ denotes inactive or dead polymer.

The molecular weight (MW) and molecular weight distribution (MWD) of the copolymer are computed using three leading moments of the total number average copolymers. The instantaneous k^{th} moment is given by:

$$\lambda_k^d = \sum_{n=1}^{\infty}\sum_{m=1}^{\infty}(nw_1 + mw_2)^k M_{n,m} \ , \quad k = 0, 1, 2, \dots \dots \tag{18}$$

where w_1 is the molecular weight of styrene and w_2 is the molecular weights of acrylonitrile. The total number average chain length (X_n), the total weight average chain length (X_w) and the polydispersity index (PD) are defined as:

$$X_n = \lambda_1^d / \lambda_0^d$$
$$X_w = \lambda_2^d / \lambda_1^d \qquad (19)$$
$$PD = X_w / X_n$$

More details on modeling equations, reaction kinetics and numerical data pertaining to this system can be referred elsewhere [24].

The polymerization process model [24] is in the form of the general expression in Eq. (5). The set of state variables, X and the control vector U in the model are given by.

$$X(t) = [M_1(t), M_2(t), I(t), V(t), \lambda_0(t), \lambda_1(t), \lambda_2(t)]^T$$

and

$$U(t) = [T(t), u(t)]^T \qquad (20)$$

In the above, I is the concentration of the unreacted initiator, V is the reaction mixture volume, λ_k (k = 0, 1, ...) is the kth moment of the dead copolymer molecular weight distribution, T is the reaction mixture temperature, an u is the volumetric flow rate of the feed mixture to the reactor.

2.4 Control objectives

The desired values of copolymer composition (F_1D) and number average molecular weight (MWD) are chosen to be 0.58 and 30000, respectively. Minimizing the deviations of copolymer composition, F_1 and molecular weight, MW from their respective desired values during the entire span of the reaction are specified as the desired objectives. In order to attain these objectives, the control variables are set as monomer addition rate, u and reactor temperature, T. If one manipulative variable is used to control one polymer quality parameter, the uncontrolled property parameters may deviate from their desired values as the reaction proceeds. The optimal control problem involves in optimizing the single objectives as well as both the objectives simultaneously. The objectives of SAN copolymerization are specified as.

$$J_1 = [1 - MW(t)/MWD]^2 \qquad (21)$$
$$J_2 = [1 - F_1(t)/F_1D]^2 \qquad (22)$$
$$J_3 = [1 - F_1(t)/F_1D]^2 + [1 - MW(t)/MWD]^2 \qquad (23)$$

Here MW and F_1 are the molecular weight and copolymer composition, and MWD and F_1D are their respective desired values. The notation t here refers discrete time. The hard constraints are set as.

$$0 \leq u(t) \leq 0.07 \ (1/\min)$$
$$320 \leq T(t) \leq 368(k) \qquad (24)$$
$$V(t) \leq 4.0 \, l$$

These constraints on operating variables are selected by taking into consideration of reaction rate, heat transfer limitation and reactor safety. Determination of temperature policy, T, to satisfy J_1 (Eq. (21)) and monomer feed policy, u, to satisfy J_2 (Eq. (22)) are considered as single objective optimization problems.

Determination of both u and T policies that satisfy J_3 (Eq. (23)) is considered as a problem of simultaneous optimization.

2.5 Design and implementation

The design and implementation of tabu search for optimal control of copolymerization reactor is explained as follows. The total span of reaction time in SAN copolymerization reactor is split into finite number of time instants referred to discrete stages. The total time of reaction is fixed at 300 min. The duration of reaction time is divided into 19 stages with 20 time points, each stage having a duration of 15 min. Thus the discrete control sequences of feed flow rate, u and temperature, T are specified as.

$$u = [u_1, u_2, \ldots, u_{20}]^T \qquad (25)$$
$$T = [T_1, T_2, \ldots, T_{20}]^T \qquad (26)$$

The control sequences are located at equal distances. Initially, the control vector is specified as constant value at each of the discrete stages. The control input at the beginning of the first stage is chosen as the lower bound of the input space. The elements of tabu search for computing the optimal control policies are specified as follows. The sizes of both recency and frequency based tabu lists are set as 50. The sizes of these lists are chosen such that they prohibit revisiting of un-prosperous solutions in the search process. An intensification procedure in the form of a sine function is used. The oscillation period α of the sine function is controlled by a parameter (θ) whose value is 4.0001. A sigmoid function based aspiration criterion with the parameters as k_{center} = 0.3 and σ = 7/M with M as specified number of iterations is used. For optimizing MW, ten neighbors are formed at the end of the first stage by considering random changes in the search space of T with an incremental change of -0.4 to 0.4. The number of neighbors specified for each control input of each stage is 10. The integration of process model is performed for a time period of 15 min with a time step of 1 min from the starting to the end of the first stage and the objective function values are calculated for all the generated neighbors at the end of this stage. The iterative convergence of TS leads to establish the best control input (T) along with its objective function. This provides the optimal control point for the first stage, which is then used as a starting point for the second stage solution. For second stage solution, random neighbors are created at the end of the second stage around the optimal T of the first stage. The integration of the model is performed from the beginning to the end of the first stage based on the initial control point (T) and from the starting to the end of the second stage for each of the neighbors generated at the end of second stage. The optimal control inputs for successive stages are computed until the end of last stage in a similar manner. The control input values that are computed at the end of each stage thus represents the optimal control policy for T. For F_1 optimization, ten neighbors are generated at the end of first stage with an incremental variation of -1.0 x 10^{-6} to 1.0 x 10^{-6}. In analogous manner, optimal control policy for $u(F_1)$ is found by following the similar TS procedure as in T policy. For determining the dual control policies of T and u, multistage dynamic optimization by TS is performed by accounting incremental variations in neighbors generation of T and u within the ranges of -0.4 to 0.4, and -1.0 x 10^{-6} to 1.0 x 10^{-6}, respectively. This case of multistage dynamic optimization involves the computation of the objective function values for 100 neighbor combinations at each of the control point corresponding to T and u. **Figure 2** shows the implementation of TS strategy for optimal control of SAN copolymerization reactor.

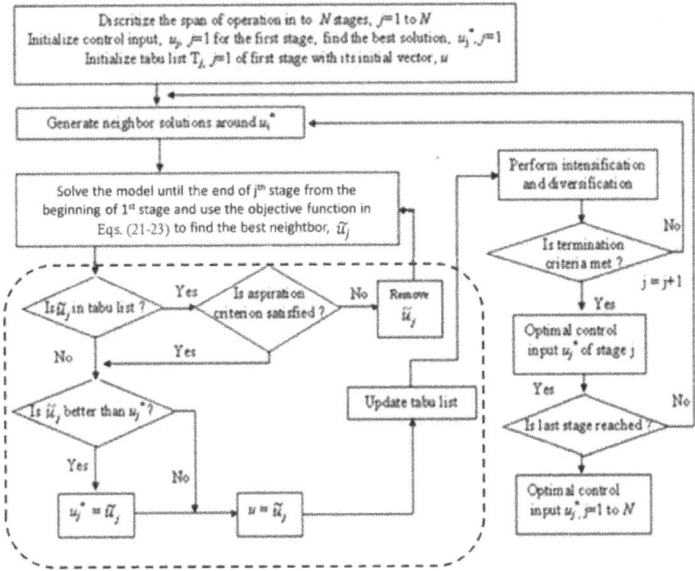

Figure 2.
Multistage dynamic optimization of copolymerization reactor using tabu search.

2.6 Analysis of results

Tabu search (TS) is designed and applied to compute the optimal control policies that meet the single and multiple objectives of SAN copolymerization reactor. The dual control policies of T and u computed by TS for multistage dynamic optimization of polymerization reactor along with the objective function values are shown

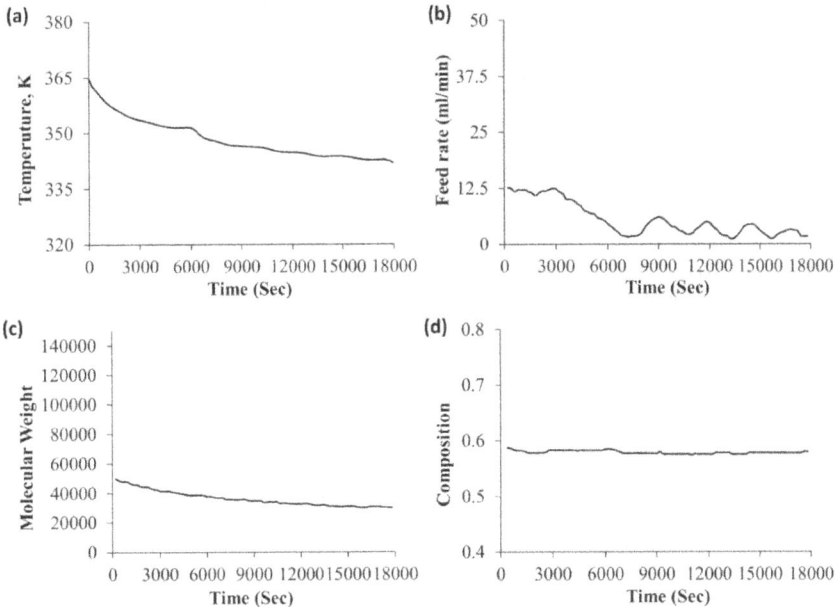

Figure 3.
Dual control policies and objectives: (a) Temperature policy, (b) Feed rate policy, (c) Molecular weight as an objective, and (d) Copolymer composition as an objective.

Strategy	Control Policy	Computational efficiency		Normalized absolute error	Memory storage, k
		Convergence time (sec)	Implementation time (sec)		
TS	T	1.89	16.83	0.000566	2224
	U	2.19	13.81	0.01379	2076
	T & u	4.25	19.21	0.007535	3084

Table 1.
Computational efficiencies of TS strategy.

in **Figure 3**. These results exhibit the efficiency of TS in computing both the temperature and feed rate policies for the reactor to maintain MW and F_1 almost near to their desired values. The computational efficiencies of TS are evaluated by means of execution times, normalized absolute error values and memory storage requirements as shown in **Table 1**.

3. Application of Tabu search for inverse modeling of anaerobic fixed bed biofilm reactor

Tabu search is applied to determine the parameters of kinetic and film thickness models through the validation of the mathematical models of the process with the help of measured data acquired from an experimental fixed bed anaerobic biofilm reactor used in the treatment of pharmaceutical industry wastewater.

3.1 Experimental biofilm reactor and its description

The schematic of a laboratory scale fixed bed biofilm reactor setup used for industry waste water treatment is shown in **Figure 4**. The description of the experimental unit, its auxiliary items and the packing specifications are given as follows.

The reactor consists of a QVF glass column of 1 m height and 0.1016 m internal diameter. A Teflon perforated plate of thickness 3 mm is provided at the base of the column to support the packing material. Two openings fitted with valves made of Teflon are provided at the bottom of the reactor so that one valve is used to control the flow rate of influent pumped into the reactor and the other valve is used to discharge the excess sludge accumulated. The top of the column has a provision to place a thermometer to measure the temperature. Three sampling ports are placed at equal distances along the side of the column to withdraw samples. The temperature of the column is maintained by varying the voltage through the 0.2 kW heating tape connected to a dimmer-stat. The column is insulated with 1.0 in. asbestos rope. The entire setup is supported by a 1.0 in. M.S. pipeline structure. The column and the packing specifications are given elsewhere [27].

3.2 Experiments and data generation

The experimental setup shown in **Figure 4** is used to conduct experiments for anaerobic treatment of pharmaceutical industry wastewater involving different packing materials with varying organic loading rates and feed rates, and different hydraulic retention times. The influent kept in a 2 *l* capacity vessel is fed to the column through an inlet connection located at the bottom using a peristaltic pump made of Watson Marlow. The effluent sample is collected from an outlet located at the top of the column. The column and the packing specifications are given

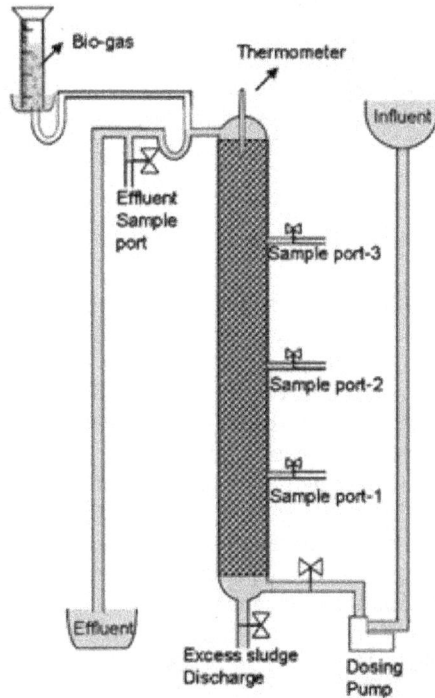

Figure 4.
Experimental biofilm reactor setup.

elsewhere [27]. The reactor is filled with the packing material and then seeded with 1.0 *l* of active anaerobic sludge brought from M/s. Alkabir, Hyderabad, India. To maintain the initial growth of microorganisms, the reactor is further added with 2.6 *l* of very dilute solution of synthetic glucose medium (1000 mg/l) with nutrients such as total nitrogen as urea (125 mg/l), total phosphorous as KH2PO4 (50 mg/l), NaCl (50 mg/l), KCl (40 mg/l), CaCl2 (15 mg/l), MgCl2 (10 mg/l) and FeCl3 (2 mg/l). The pH of the solution is maintained at 7.2 by adding 0.1 N NaOH. The reactor is made biologically active by keeping it undisturbed for about 20 days. The gas (CH4 + CO2) evolved at the top of the column confirms the biological activity of microorganisms.

The wastewater procured from a typical pharmaceutical industry is a complex medium and its composition is reported elsewhere [28]. Substrates of varying COD concentrations are prepared using the wastewater in order to use them as feed solutions for biofilm reactor experiments. Once the biological activity of the bed is detected, the diluted industry wastewater solution equivalent to three bed volumes is sent to the reactor by an electronic dosing pump. NaHCO3 is added to maintain the pH of the feeding solution at 7.2. Different experiments are conducted for treating the pharmaceutical industry wastewater using the feed solutions having the COD concentrations of 4,980 mg/l, 11,860 mg/l, 23,460 mg/l and 40,720 mg/l. The hydraulic retention time (HRT) of each experiment is varied from 2 to 12 days based on the substrate concentration of the feed solution. Each experimental run subsequent to attainment of its HRT is allowed to continue for 13 days so as the operation reaches stable state condition by that time.

Thirteen experiments are conducted using the different feed solutions prepared from the pharmaceutical industry wastewater. The reactor is provided with the sampling ports to withdraw samples at bed heights of 0.25 m, 0.5 m and 0.75 m from the bottom. Samples collected from the three sampling ports of the column as well as from the effluent water are analyzed for COD, BOD, TVA, TA, pH etc. The stable state condition of each experiment is observed with respect to the minimal variation in reactor effluent pH, alkalinity and COD concentrations. Thus 13 experiments are conducted under different feed stream conditions. The data corresponding to the influent and effluent COD concentrations, HRT and OLR of all experiments are given elsewhere [28].

3.3 Mathematical and kinetic models

The application of mathematical models to biofilm reactors suffers due to lack of availability accurate kinetic models and uncertainty in model parameters. Therefore, appropriate selection of kinetic model and accurate determination of its parameters is required for successful modeling of biofilm reactors. The biological reaction rates are generally represented by Monod kinetics and also occasionally by Contois and Haldane models. In the past, efforts have been made to determine the parameters of kinetic models of biofilm reactors, mostly experimentally [29–31]. Various researchers [32–35] have reported that numerical evaluation of kinetic parameters is an attractive alternative to the experimental methods for biofilm processes. By numerical approach, the parameters of the kinetic models of biofilm processes are determined as a consequence of the validation of their mathematical models with the aid of measured data. The approach of determining the model parameters such that the behavior of the process model approximates the observed process behavior is known as inverse modeling.

3.3.1 Mathematical models

Mathematical models of varying complexities involving different types of kinetic and film thickness models are used to represent the wastewater treatment process in the biofilm reactor.

3.3.1.1 One dimensional model

This model accounts the rate of mass transfer to be proportional to the concentration difference between the interface and the bulk fluid. This model assumes no accumulation of the component at the interface under steady state condition. The differential equation governing the fluid phase is given by.

$$-u\frac{dc}{dz} = k_g a_v \left(c - c_s^s\right) \tag{27}$$

with the boundary condition:

$$c = c_0 \text{ at } z = 0 \tag{28}$$

In this model, the substrate concentration in the bulk fluid varies only in the axial direction.

The differential equation governing the solid phase is given by.

$$\frac{D_f}{\xi^{a-1}}\frac{d}{d\xi}\left(\xi^{a-1}\frac{dc_s}{d\xi}\right) - r_s(c_s) = 0 \tag{29}$$

with the boundary conditions:

$$\frac{dc_s}{d\xi} = 0 \quad \text{at } \xi = 0$$

$$k_g\left(c_s^s - c\right) = -D_f \frac{dc_s}{d\xi} \quad \text{at } \xi = L_f \tag{30}$$

The superscript '*a*' in Eq. (29) refers to 1, 2, and 3 for planar, cylindrical and spherical geometries, respectively, and D_f denotes substrate diffusion coefficient in the biofilm (m^2/day). The notation for other terms in the above equations is denoted as k_g is mass transfer coefficient (m/s), a_v is specific surface area of particle (m^{-1}), c is substrate concentration in the bulk fluid (kg/m^3), c_s is substrate concentration in the bio-film (kg/m^3), c_s^s is substrate concentration on the bio-film surface (kg/m^3), ξ is space coordinate in the bio-film (m), z is axial coordinate (*m*) and L_f is thickness of the bio-film (*m*).

3.3.1.2 Two dimensional model

This model considers the variation of the substrate concentration in bulk fluid in both the axial and radial directions. The substrate transfer occurring through the biofilm is characterized by physical diffusion and reaction, and this process is described by second order differential equation. Pressure losses along the length of the reactor are neglected. The differential equation governing the bulk fluid phase is given by.

$$u\frac{dc}{dz} = \frac{\varepsilon D}{r}\left[\frac{\partial}{\partial r}\left(r\frac{\partial c}{\partial r}\right)\right] - k_g a_v\left(c - c_s^s\right) \tag{31}$$

where D is substrate diffusion coefficient in the bulk fluid (m^2/day), u is superficial velocity (m/s), r is radial coordinate (*m*) and ε is porosity of bed. The boundary condition for solving this equation is given in Eq. (28). The solid phase biofilm model for two dimensional model is same as in one dimensional model as in Eq. (29) and its boundary conditions are expressed by Eq. (30).

3.3.1.3 Mass transfer coefficient

The substrate mass transport occurring from the bulk fluid to the biofilm surface across the diffusion layer is defined by

$$j_D = \frac{k_g(sc)^{2/3}}{u} \tag{32}$$

where Sc is Schmidt number ($\mu/\rho D$). The j_D-factor is calculated using the following correlation:

$$j_D\varepsilon = \frac{0.765}{Re^{0.82}} + \frac{0.365}{Re^{0.386}} \tag{33}$$

$$\text{where} \quad Re = \frac{\rho u d_p}{\mu} \text{for } 0.01 < Re < 15,000$$

where d_p is equivalent particle diameter (m). The mass transfer coefficient, k_g obtained from Eq. (32) is used in bulk fluid phase equations, Eq. (27) and Eq. (31).

3.3.2 Kinetic models

Various kinetic models can be used to represent the bioprocess kinetics [28], of which the Haldane and Edward models are given as follows.

3.3.2.1 Haldane model

This rate expression is generally valid when the substrate concentration is high. According to this model, the substrate consumption rate, r_s in biofilm is given by the following equation:

$$r_s = (\mu_{max}\rho_s/Y) \; \frac{C_s}{K_s + C_s + \frac{C_s^2}{K_I}} \tag{34}$$

where C_s is substrate concentration, μ_{max} is maximum specific growth rate, ρ_s is density of biomass, Y is yield coefficient, K_s is half velocity constant, and K_I substrate inhibition constant.

3.3.2.2 Edward model

This model considers substrate inhibition, according to which the substrate consumption rate, r_s in biofilm is given by the equation:

$$r_s = \left(\frac{\mu_{max}\rho_s}{Y}\right) \left(\exp\left(\frac{-C_s}{K_I}\right) - \exp\left(\frac{-C_s}{K_s}\right)\right) \tag{35}$$

3.3.2.3 Film thickness

The biofilm thickness (L_f) is has significant influence on substrate conversion. A uniform film thickness with fixed value cannot represent the realistic situation and the film thickness L_f is expected to vary along the reactor [35]. The varying thickness of the biofilm is determined by the substrate flux from the bulk fluid into the biofilm as wellas the growth and decay rates of the bacteria. It is assumed that the biofilm is composed of a static portion and a variable portion. The variable portion is supposed to raise with the organic loading rate and lessen with the hydraulic loading rate. The above point of view leads to the following relations for estimating the biofilm thickness:

$$L_f = a + b \; OLR \tag{36}$$

$$L_f = a + b \; OLR - \frac{c}{HLR} \tag{37}$$

where OLR is organic loading rate (kg/m^3/day), HLR hydraulic loading rate (m/day), and a, b and c are constants to be determined.

3.4 Procedure for solving the model equations

To facilitate the solution of mathematical models, the height of the column (1 m) is divided into 50 equal steps, each step representing 0.02 m. In one dimensional model, the fluid phase equation (Eq. (27)) along with its boundary condition (Eq. (28)) is solved for each height step in the axial direction by using Runge–Kutta (RK) 4th order method. The solid phase equation for the biofilm (Eq. (29)) is solved for each segment using orthogonal collocation on finite elements (OCFE) [36]. Two finite elements, each with four internal collocation points are considered for OCFE implementation. In two dimensional model, the fluid phase equation (Eq. (31)) is transformed to ODE by finite difference technique and the resulting equation along with its boundary condition is solved by using 4th order RK method for each height step of the axial direction. The solution for solid phase equation of this model is same as in one dimensional model.

3.5 Tabu search for inverse modeling of biofilm reactor

The implementation procedure of tabu search is given in Section 1.3. In this work, tabu search is used to determine the parameters of kinetic and film thickness models in association with the validation of the mathematical models using the data of measurements obtained from an experimental fixed bed anaerobic biofilm reactor. The parameter estimation via inverse modeling is carried out by defining an objective function, J given by.

$$J = f(\alpha) = \sum_{i=1}^{l} (y_i - \hat{y}_i)^2 \tag{38}$$

where α is the vector of parameters, l is the number of observations, y_i is the measured value of the ith variable, and $\hat{y}_i$ is the corresponding predicted value. Tabu search is applied to estimate the parameters of kinetic and film thickness expressions of different modeling configurations with the support of one dimensional (1D) and two dimensional (2D) mathematical models of the biofilm reactor. Random variation in all the parameters is considered to generate neighbors so as to provide the maximum improvement to the current solution. Recency based tabu list with a length of 100 and frequency based tabu list with a length of 50 are employed. A sigmoid function based aspiration criterion, Eq. (2) is employed with k_{center} as 0.3 and σ as $7/M$, where M refers specified number of iterations. An intensification strategy with the coefficient α taking the format of a sine function, Eq. (1) is employed, in which the value of θ is assigned to be 4.0001. The maximum number of iterations are considered as 100. The termination-on-convergence criteria, Eq. (3) is used as stopping criterion.

3.6 Analysis of results

Parameter estimation by tabu search is performed by devising different modeling configurations to represent the biofilm reactor. These configurations involve the Edward kinetics and Haldane kinetics with substrate inhibition along with the two and three parameter film thickness expressions. The modeling configuration that is formed by 1D model along with Haldane kinetics and two parameter film thickness expression is referred to 1D-Haldane-2film. Other modeling configurations are formed and abbreviated in the same manner. The effectiveness of these modeling configurations are assessed by comparing the model predictions with experimental data and further by means of performance measures such as cost function (CF) and model efficiency (ME) [28]. A conventional optimization method called Nelder–Mead optimization (NMO) [37, 38] is also employed for comparison with the tabu search. The results are analyzed to find the better suitability of mathematical models using the quantitative performance measures (CF and ME). These results evaluated for 1D and 2D models with Haldane and Edward kinetics involving two film and three film expressions indicate the better suitability of 2D model over 1D model.

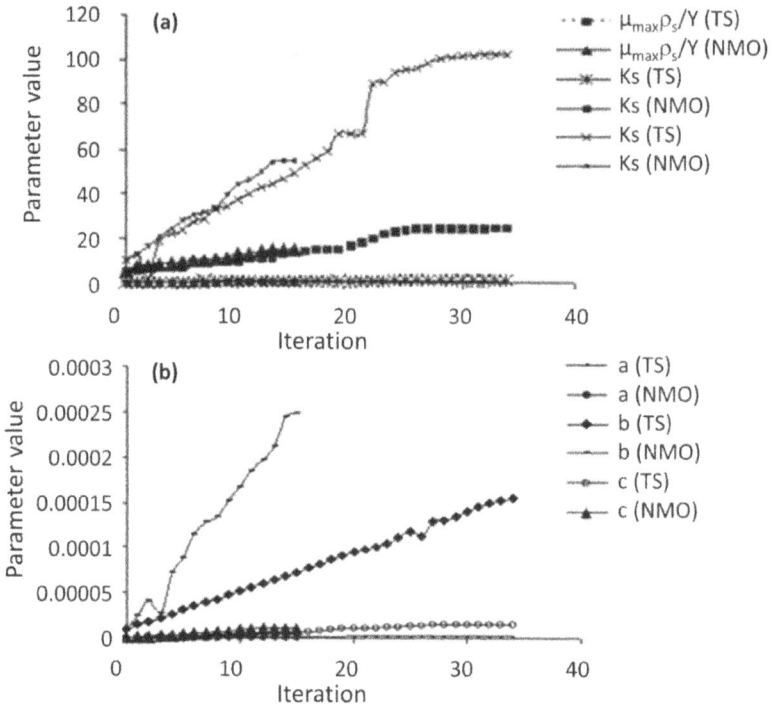

Figure 5.
Iterative convergence of parameter estimates by TS and NMO: (a) kinetic parameters, (b) film thickness parameters.

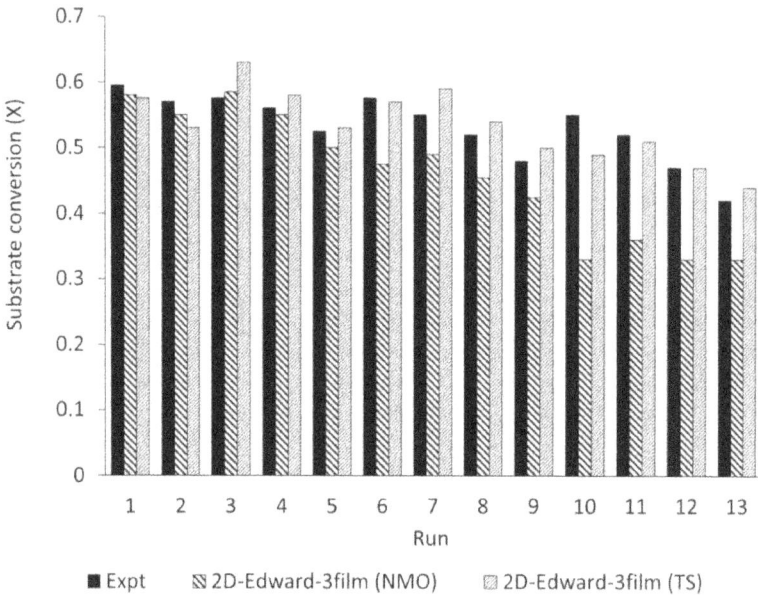

Figure 6.
Comparison of experimental results with model predicted substrate conversions of 2D model with Edward kinetics.

When the results are analyzed to assess the better suitability optimization algorithm, the analysis of results of different modeling configurations indicate the effectiveness of TS over NMO. The results evaluated to find the usefulness of kinetic models indicate the better suitability of Edward kinetics over Haldane kinetics. The results analyzed to assess the film thickness expressions indicate the appropriateness of the three parameter film thickness expression over two parameter expression. The iterative convergence of parameters estimated by TS and NMO are shown in **Figure 5**. The comparison of the prediction results of 2D model with Edward kinetics and three parameter film thickness expression evaluated using TS and NMO with those of experimental results are shown in **Figure 6**. The analysis of the results show the better performance of TS over NMO for inverse modeling of biofilm reactor. From these results, it is found that the TS involving 2D model, Edward kinetics and three parameter film thickness expression better represents the fixed bed biofilm reactor involved in the treatment of pharmaceutical industry wastewater.

4. Conclusions

Tabu search is a stochastic optimization method used to solve global optimization problems. The effectiveness of tabu search for modeling and optimization is explored with the applications concerning to chemical and environmental systems of varying complexities. The meta heuristic Tabu search (TS) is designed and applied to determine the optimal control policies of a SAN copolymerization reactor and inverse modeling of a biofilm reactor. The computational efficiencies of TS are evaluated in terms of execution times, normalized absolute error values and memory storage requirements. The results of polymerization reactor show the usefulness of TS in determining the optimal temperature and feed rate policies to maintain the objectives MW and F_1 near to their desired values. The results of biofilm reactor explain the efficacy of TS in appropriately configuring the kinetic and film thickness models as a consequence of validation of mathematical models.

Author details

Chimmiri Venkateswarlu
Indian Institute of Chemical Technology (CSIR-IICT), Hyderabad, India

*Address all correspondence to: chvenkat.iict@gmail.com

IntechOpen

References

[1] Glover F. Future paths for integer programming and links to artificial Intelligence. Comput. Operat. Res. 1986; 5:533-549

[2] Hansen P. The steepest ascent mildest descent heuristic for combinatorial programming, Numerical methods in combinatorial programming conference. Italy: Capri; 1986

[3] Cvijovic D, Klinowski J. Taboo search: An approach to the multiple minima problem. Science. 1995;**667**: 664-666

[4] Glover F, Laguna M. Tabu Search, Boston. Kluwer Academic Publishers; 1997

[5] Gendreau M, Laporte G, Semet F. A Tabu search heuristic for the undirected selective travelling salesman problem. Europ. J. Operat. Res. 1998;**106**:539-545

[6] Wang C, Quan H, Xu X. Optimal design of multi product batch chemical process using tabu search. Comput. Chem. Eng. 1999;**23**:427-437

[7] Mosat A, Hungerbuhler K. Batch process optimization in a multipurpose plant using tabu search with a design-space diversification. Comput. Chem. Eng. 2005;**29**:1770-1786

[8] Waligora G. Tabu search for discrete–continuous scheduling problems with heuristic continuous resource allocation. Eur. J. Oper. Res. 2009;**193**:849-856

[9] Fescioglu-Unver N, Kokar MM. Self controlling tabu search algorithm for the quadratic assignment problem. Comput. Ind. Eng. 2011;**60**:310-319

[10] Lin, B., D.C. Miller, D.C. Solving heat exchanger network synthesis problem with tabu search. Comput. Chem. Eng., 28: 1451–1464, 2004(a).

[11] Lin, B., Miller, D.C. Tabu search algorithm chemical process optimization, Comput. Chem. Eng., 28: 2287–2306, 2004(b).

[12] Wu GZA, Denton LA, Laurence RL. Batch polymerization of styrene-optimal temperature histories. Polym. Eng. Sci. 1982;**22**:1

[13] Choi KY, Butala DN. Synthesis of open loop controls for semi batch copolymerization reactors by inverse feedback control method. Automatica. 1989;**25**:917-923

[14] Arzamendi G, Asua JM. Monomer addition policies for copolymer composition control in semi-continuous emulsion polymerization. J. Appl. Poly. Sci. 1989;**38**:2019

[15] Gloor PE, Warner RJ. Developing feed policies to maximize productivity in emulsion polymerization processes. Thermochimica Acta. 1996; **289**:243

[16] Zavala VM, Tlacuahuac AF, Lima EV. Dynamic optimization of a semi-batch reactor for polyurethane production. Chem. Eng. Sci. 2005;**60**: 3061-3307

[17] Pontryagin LS, Boltyanski VG, Gamkrelidze RV, Mishchenko EF. The mathematical Theory of Optimal Processes. New York: John Wiley & Sons; 1962

[18] Thomas IM, Kiparissides C. Computation of the near optimal temperature and initiator policies for batch polymerization reactors. Can. J. Chem. Eng. 1984;**62**:284-291

[19] Ponnuswamy SR, Shah SL, Kiparissides CA. Computer optimal control of batch polymerization reactors. Ind. Eng. Chem. Res. 1987;**26**: 2229-2236

[20] Secchi AR, Lima EL, Pinto JC. Constrained optimal batch polymerization reactor control. Polym. Eng. Sci. 1990;**30**:1209-1219

[21] Ekpo EE, Mujtaba IM. Optimal control trajectories for a batch polymerization reactor. Int. J. Chem. React. Eng. 2007;**5**:1542-6580

[22] Chang JH, Lai JL. Computation of optimal temperature policy for molecular weight control in a batch polymerization reactor. Ind. Eng. Chem. Res. 1992;**31**:861-868

[23] Salhi D, Daroux M, Genetric C, Corriou JP, Pla F, Latifi MA. Optimal temperature-time programming in a batch copolymerization reactor. Ind. Eng. Chem. Res. 2004;**43**:7392-7400

[24] Anand P, BhagvanthRao M, Venkateswarlu C. Dynamic optimization of copolymerization reactor using tabu search. ISA Trans. 2015;**55**:13-26

[25] Bellman RE. Dynamic Programming. Princeton University Press, Princeton, NJ. 2003;**947-957**

[26] Butala D, Choi KY, Fan MKH. Multiobjective dynamic optimization of a semibatch Free-Radical copolymerization process with interactive CAD tools. Comp. Chem. Eng. 1988;**12**:1115-1127

[27] Rama Rao K, Srinivasan T, Venkateswarlu C. Mathematical and kinetic modeling of biofilm reactor based on ant colony optimization. Process Biochem. 2010;**45**:961-972

[28] Shiva Kumar B, Venkateswarlu C. Inverse modeling approach for evaluation of kinetic parameters of a biofilm reactor using Tabu Search. Water Env. Res. 2014;**86**:205

[29] Nguyen VT, Shieh WK. Evaluation of intrinsic and inhibition kinetics in biological fluidized bed reactors. Water Res. 1995;**29**:2520-2524

[30] Rittmann BE, McCarty PL. Model of steady state biofilm kinetics. Biotechnol. Bioeng. 1980;**22**:2343-2357

[31] Tsuneda, S., Auresenia, J., Morise, T., A. Hirata, A. Dynamic modeling and simulation of a three-phase fluidized bed batch process for wastewater treatment. Process. Biochem., 38: 599-604, 2002.

[32] Zhang, S., P.M. Huck, P.M. Parameter estimation for biofilm processes in biological water treatment. Water Res., 30: 456–464, 1996.

[33] Sarti A, Foresti E, Zaiat M. Evaluation of a mechanistic mathematical model of a packed bed anaerobic reactor treating waste water. Latin Am. Appl. Res. 2004;**34**:127-132

[34] Spigno, G., Zilli, M., Nicolella,C. Mathematical modeling and simulation of phenol degradation in biofilters, Biochem. Eng. J., 19, 267–275, 2004.

[35] Kiranmai, D., Jyothirmai, A., C.V.S. Murthy, C.V.S. Determination of kinetic parameters in fixed-film bio-reactors: an inverse problem approach. Biochem. Eng. J., 23: 73-83, 2005.

[36] Finlayson BA. Nonlinear Analysis in Chemical Engineering. McGraw-Hill Chemical Engineering Series. New York: McGraw-Hill; 1980

[37] Kuester JL, Mize JH. Optimization Techniques with Fortran. New York: McGraw-Hill; 1973

[38] Venkateswarlu C, Gangiah K. Dynamic modeling and optimal state estimation using extended kalman filter for a kraft pulping digester. Ind. Eng. Chem. Res. 1992;**31**:848-855

Uncertainty Management to Support Pollution Prevention and Control Decisions

Introductory Chapter: Uncertainty Management to Support Pollution Prevention and Control Decisions

Rehab O. Abdel Rahman and Yung-Tse Hung

1. Introduction

The progressive growth in industrialization and population caused severe environmental problems worldwide, these problems need to be analyzed, monitored, controlled and mitigated when appropriate to ensure the quality and sustainability of life [1]. Currently, there is growing international recognition for these problems and in particular environmental pollution is receiving considerable attention either on the international, regional, national and individual scales. To help controlling the existing pollution sources and preventing new pollution sources/areas, strengthen regulations have been issued and human and natural resources have been allocated all over the world [1]. The results of these efforts will be very helpful in supporting various sustainable development goals that were identified in the United Nation 2030 agenda [2]. Among these goals, the achievement of good health and well-being (Goal 2), clean water and sanitation (Goal 4), affordable and clean energy (Goal 5), industry, innovation and infrastructure (Goal 9), sustainable cities and communities (Goal 11), responsible consumption and production (Goal 12), climate action (Goal 13), life below the water and on land (Goals 14 and 15, respectively) are affected by the efforts to prevent and control the environmental pollution.

To ensure effective pollution prevention and control, there is a need to prove that each planned/operated human activity will not impose negative impacts on the human society and the environment. This situation is stressful for the decision makers, e.g. policy makers, designers, regulators, where the decisions must balance the benefits from this activity to the society and its potential negative impacts on the environment, their probabilities, and their consequences. Different assessment methodologies were ratified more than 5 decades ago and are used as tools to support the decision making process. These assessments aim to provide systematic procedures to study the impacts/risks of the human activities on their societies and on the environment. These assessments include life cycle assessment (LCA), life cycle sustainability assessments (LCSA), environmental impact assessment (EIA), strategic environmental assessments (SEA), and risk assessments (RA) [3, 4]. LCA is used to assess the environmental impacts associated with the life cycle stages of a product or service supply chain, e.g. raw material extraction processes, manufacturing and processing, transportation, usage and disposal. It includes goal definition and scoping, inventory assessment, impact assessment, and interpretation. LCSA aims to evaluate the impacts of a product or service on the environment (LCA), social life (social life cycle assessment S-LCA) and society's economic (life cycle costing, LCC) towards more sustainable products throughout their life cycle [5, 6]. EIAs are widely used for regulating human activities worldwide. EIAs focus on the evaluation of the impacts of specified project over its different life phases, i.e. construction, operation, and closure, on the ecological components of the environment. EIA performed by identifying the baseline project, assessing and mitigating the impacts, and monitoring planning. SEA aims to evaluate the environmental impacts of alternative visions and development intentions incorporated in policy, planning or program initiative [7, 8]. Finally, RA used to assess health risk assessment (HRA), hazard risk assessment (HZRA), and environmental (ecological) risk assessment (ERA).

To build confidence in these assessment's results and subsequently in the decisions to be taken based on them, there is a need to identify, present, and describe the uncertainties associated with data collection and analysis, scenario developments, and expert judgment. In this chapter, uncertainty management to support regulatory decision making process to prevent and control pollution will be presented. In this respect, it should be noted that basic elements for regulatory decision making process include clear identification of the applied laws, regulations and acceptance criteria, assessment of the safety significance, verification of collected data and information, assigning priorities, and clarification of the analysis/assessment to be performed [3]. Based on the safety significance of the activity, it might require a simple qualitative assessment, i.e. for activities of low safety significance, or it might need in-depth quantitative assessment, i.e. for activities of high safety significance. Depending on the existing regulations, this assessment can cover impacts (e.g. EIA) and/or risks (e.g. RA) to human, property, or/and the environment, and the results of the assessment are used to manage these impacts and/or risks, i.e. prioritize the efforts to minimize or mitigate the impacts and/or risk. The rest of this chapter is devoted to introduce the uncertainty management. This will be achieved by introducing the applications of risk assessment to support the regulatory decision making process, then elements of the uncertainty management will be overviewed.

2. Risk assessment to support regulatory decision making process

Early RA studies were limited to HRA and comprised health problem identification, dose–response assessment, exposure assessment, and risk characterization. Then HZRA studies were used as a tool to evaluate the risks of specific system or process. **Figure 1** illustrates the steps of HZRA that is used to support the decision making process for a system in the design phase, where the system's hazards are identified, accidents probabilities and consequences are evaluated, then risks are characterized. If the risk is acceptable, then the decision will support the construction and/or operation of the system, otherwise there will be a need to modify the system. Finally, integrated risk assessment (IRA) methodology was developed to estimate the health and ecological risks. It consists of three phases [9]:

- **Problem identification**: in which the hazard is identified and the assessment context is set up by identifying the goals, objectives, scope, and assessment activities.

- **Risk analysis**: aims to identify the exposures and their effects on human and the environment. In this step, assessment models are developed; required data are collected; and modeling results are analyzed to characterize the exposures and their effects. Detailed information about the development of assessment models are found elsewhere [1].

- **Risk characterization**: aims to estimate the risks based on the information of exposures and their effects.

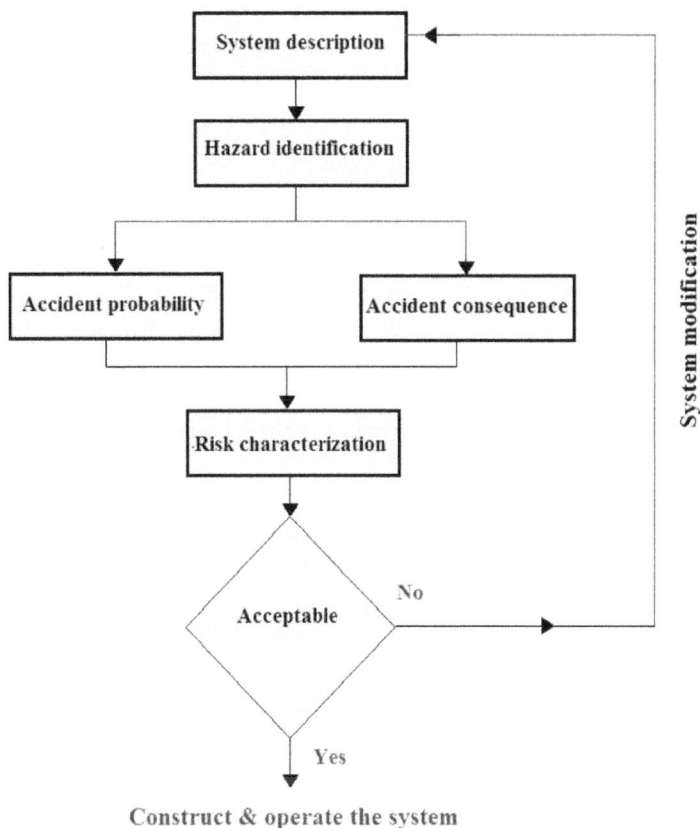

Figure 1.
Environmental risk assessment scheme for project during the design phase.

Each step in the risk assessment is associated with uncertainties that need to be identified, presented and described, and their effects needs to be quantified.

Risk assessment studies are applied to support the decision making process for policy and regulatory decision makers and for project decision makers, **Table 1** lists some of these applications and their examples [10]. Decision making process that relies on the risk assessments are classified as risk-based and risk-informed decision making processes [11]. On one hand, the first relies totally on the risk assessment results, thus allowing efficient risk management and ensure a defensible basis for the decision. On the other hand, the risk-informed decision making process consider other factors with the risk assessment results, (e.g. existing expert judgment, stakeholder involvement, and other engineering insights). The guidance for conducting the risk assessment is differed from country to another, where the level of acceptable risks, nature of the uncertainty analysis, and risk communication programs may be defined or not [12].

Risk assessments are classified based on the adopted technique to assess the risk into qualitative, semi-quantitative, and quantitative assessments. Qualitative assessments are widely used in chemical process industries to analyze potential equipment failure and human errors that can initiate incidents. They are applied throughout the facility life cycle to identify critical safety equipments for special maintenance, testing, or inspection, as a part of the facility management of change program, and to investigate possible causes of incidents [13]. Examples of qualitative hazard evaluation techniques include what if analysis, checklist analysis, and

Field	Applications	Example
Policy & regulatory	Design regulations	Determine acceptable risk level
	Prioritization of environmental risks	Identify regulated chemicals and products
	Provide basis for site-specific decision	Planning or sitting for certain installation
	Comparison of risk	Support substitution decision
Project	License application	Show compliance with legislation
	During design phase	Show the safe operation and product safety
	During sitting	Support Site selection study
	Prioritization and evaluation of risk reduction measures	

Table 1.
Application of risk assessment studies to support the decision making process.

Point of comparison	Deterministic approach	Probabilistic approach
Assumptions	Relies on conservative/bounding assumptions to address the uncertainties	Relies on best estimate, yet it uses conservative assumptions in determining the success criteria
Initiating events and hazards	Limited events are considered	Comprehensive set of events are selected including both DBA* & behind DBA
	Events frequencies & failure probabilities are treated approximately	Explicitly treatment of the initiating events frequency and failure probability
Consideration of accidents	Addressed separately	Integrate all the initiating events
Uncertainty management	Use of conservative assumptions, or best estimate codes and models with uncertainty analysis	Explicit uncertainty treatment in the models
Prioritization	Give rough indication on the relative importance of the system	Included in modern probabilistic safety assessment models

*DBA design basis accident.

Table 2.
Features of the deterministic and probabilistic risk assessment [16].

HAZOP. Examples of risk analysis tools include failure modes, effect and criticality analysis (FMECA), and layer of protection analysis (LOPA) [14]. Quantitative risk assessments originated in nuclear, aerospace, and electronic industries, they are further sub-classified into deterministic, probabilistic or combination of them. Traditionally, deterministic approaches were adopted by relying on the defense in depth strategy and appropriate safety margin, and following conservative requirements in the design, manufacturing and operation of the project. In this approach, design basis accident is identified during the problem identification phase, its consequences are determined within the risk analysis phase, finally safety barriers are designed to mitigate or prevent the accident consequence [3, 15, 16]. On the other hand, the probabilistic approach is used to analyze all feasible scenarios; where a broad spectrum of initiating events and their event frequency are addressed in the problem identification phase. Then the consequences of those events and weights are analyzed. **Table 2** summarizes the main features of both approaches [16]. An example of the risk-informed regulatory process in the nuclear industry is the USNRC risk-informed processes that consider compliance with regulations, consistency with defense in depth strategy, risk informed analysis, and performance monitoring. USNRC indicated that the application of risk-informed decision making process enhances the deterministic approach [15].

3. Uncertainty management in risk assessments

Uncertainty management is used to build confidence in the outcome of the RA results. Subsequently, the adaptation of well-developed uncertainty identification, classification, inventory, quantification and assessment, and combination schemes are essential for reliable decision making process. In the first step, identification of inherent uncertainty sources in the studied system is achieved. In the uncertainty combination step, the total uncertainty is obtained by aggregating all the quantified uncertainties, where different forms of uncertainties with different mathematical presentations are aggregated to produce a confidence sentence in the system performance. In this section, approaches to classify, inventory and quantify uncertainties will be introduced.

3.1 Uncertainty classification

In general, different sources of uncertainty associate the problem identification and risk analysis phases in the risk assessment methodology. These uncertainties might be related to the system variability and randomness, the presence of errors, either in the measurements, or modeling and analysis, scenarios or data insignificance, and lack of knowledge, indeterminacy, judgment, and linguistic imprecision in decision making [17–25]. Some of these uncertainties could be reduced and others are irreducible.

Two uncertainty classification systems are used; the first is based on the ability to reduce these uncertainties and the second is based on their sources [18–24]. The first consists of two classes, i.e. Epistemic & Aleatory, and this system is effectively used in building confidence in the uncertainty management outcomes, where:

- **Irreducible uncertainties (Type I)** are aleatory, i.e. related to the randomness/stochastic nature of the system, and cannot be reduced but they could be better characterized. Examples include uncertainties associated with natural hazard identification, and those associated with the system heterogeneity [18, 22].

- **Reducible uncertainties (type II)** are epistemic, i.e. arose due to the lack of knowledge, and could be reduced by gaining additional information or data. Epistemic uncertainties are associated with the nature of some mechanisms at specified conditions, e.g. radiological health effects at low doses [23].

It should be noted that during uncertainty management it is important to differentiate between these types and justify the consideration of certain type that associates the features, events, or processes (FEP) of the studied system towards reliable uncertainty quantification.

Uncertainty classification based on the uncertainty sources includes the following classes, where each class includes both epistemic and aleatory uncertainties [23, 24]:

- **Natural variation in the system properties/features**: the spatial and temporal heterogeneity of the system properties or features is associated with uncertainty. This heterogeneity is inherent in the system and needs to be identified during the problem definition phase. In modeling the fate and transport of pollutants in the environment or within the engineered systems that prevent and control the migration of these pollutants, this variability act as a source for two types of uncertainties. The first is aleatory due to the randomness of the system properties/features, e.g. the spatial distribution of permeability of the geological formation. The second is epistemic due to the lack of knowledge about the temporal evolution of this randomness, e.g. how the permeability will be changed with time.

- **Measurement errors**: these errors associate the data collection process and include random and systematic errors. The first is relatively easy to be detected and quantified, where they associate the reduced tool/device sensitivity, presence of noise, and imprecise definition. Uncertainties due to random errors are addressed using probabilistic methods [23, 24]. The systematic errors are harder to be quantified, they resulted from a bias in the sampling and they need a perfect calibration procedure to account for them.

- **Conceptual model development**. Models are abstractions of the real system; during conceptual model development there is a need to optimize the studied system. In this step, the less important FEPs are excluded [1]. This is a source of epistemic uncertainty that is reduced by performing a sensitivity analysis to optimize the selected FEPs. This source of uncertainty is more prominent in deterministic approaches.

- **Computational model errors**: there are several types of errors that associate the computational modeling of the data. Regardless the type of the used mathematical models, e.g., empirical, mechanistic, or black box, its application is associated with errors that arise from the validity of the model to represent the studied system and the accuracy and stability of the numerical model [25–27]. Uncertainties associated with model validity might be aleatory or epistemic, whereas those due to the accuracy and stability of the numerical model are epistemic and are reduced by adopting a systematic verification procedures.

- **Subjective judgment**: during the analysis of the data, expert judgment is required especially in the following cases; lack of data, and lack of knowledge; this will lead to subjective uncertainty. Examples of these cases are the need for extrapolation or interpolation of the data, and assignment of parameter distribution [28, 29].

It should be noted that both types of uncertainty classifications are used to quantify, assess and minimize the uncertainty in the decision making process. Skinner et al. developed a classification system based on the ability of the uncertainty to be reduced and their location, in the system, data, model and the subjective uncertainty in the form or language, extrapolation and decision, and their associated sub-location as illustrated in **Figure 2** [19].

3.2 Uncertainty inventory

Uncertainty inventory includes all the information and questions relating to the identified and classified uncertainty. It is developed to obtain traceable, updatable and defensible record of uncertainty assessment and quantification, where

- It includes multiple quantities of interest and permits their categorization at different levels of information, e.g. system vs. component level,

- It supports the determination of the nature of the total uncertainty and prioritizes the efforts towards efficient uncertainty reduction,

- It focuses on the information necessary for decision making, and risk/reliability estimation,

3.3 Uncertainty quantification

Both linguistic and numerical uncertainty quantifications approaches are used to analyze and assess the uncertainty in a given system. The quantification methods depend on the propagation of uncertainties in the system model and then assess the model output response due to this uncertainty propagation. The mathematical representations of the uncertainty are based on the use of probability, imprecise probability and possibility theories. For deterministic risk assessment, the uncertainty might be quantified either using a one factor at a time (OFAT) or multi- variant techniques. OFAT allows the change of one uncertain factor or parameter within a specified range with keeping the rest of the factors or parameters fixed [27]. This allows the examination of the effect of the factor variability/randomness/presence on a single process output or multi outputs. **Figure 3** illustrates the application of OFAT in assessing the risk of a system, where a single valued specified factor is propagated through the system, and the model outputs are quantified (**Figure 3a** and **b**). To quantify the uncertainty in the risk estimate of that system, discrete values or probabilistic uncertainty information of the uncertain parameter are propagated through the system which generates statistical information in the risk values for the uncertain parameter (**Figure 3a** and **c**). Different sampling methods could be used to represent the probabilistic information in that parameter, i.e. Latin hypercube sampling. OFAT does not allow the investigation of the interaction between uncertain parameters and their effect on the system output (s), nor allowing the determination of the outputs dependence [24, 26]. To overcome the latter, the parameters are often selected based on their ability to produce a conservative decision.

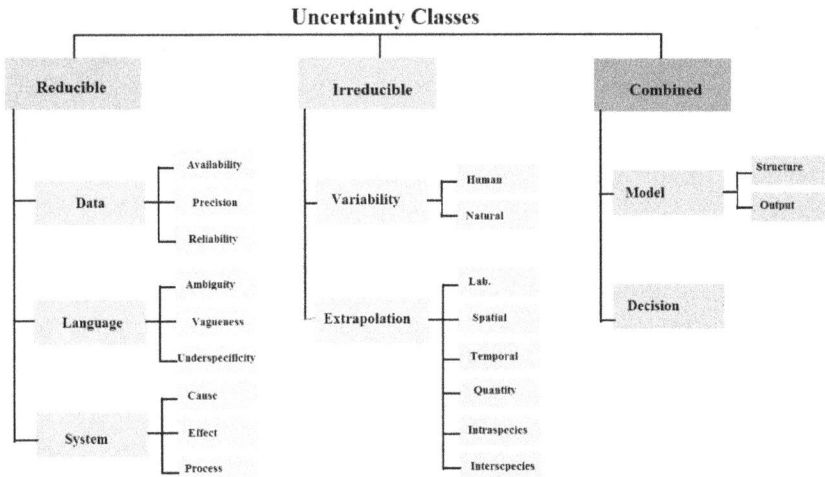

Figure 2.
Uncertainty classification according to Skinner et al. [19].

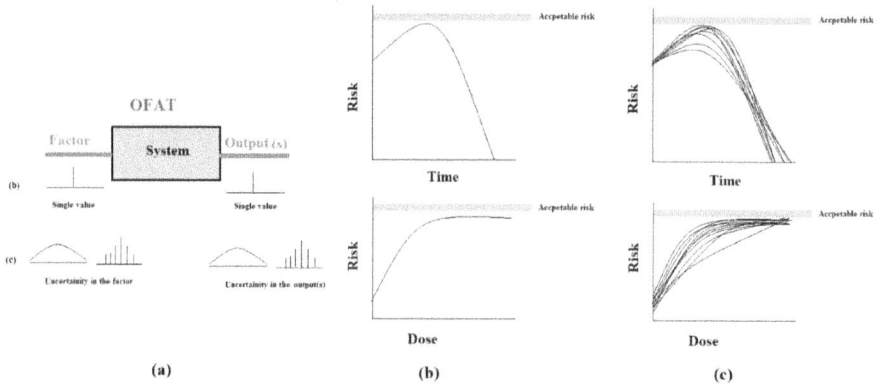

Figure 3.
Application of the OFAT approach in uncertainty quantification, (A) uncertainty propagation in the model, (b, c) modeling outputs for single parameter value (b) and uncertain parameter (c).

The use of the multi-variant approach is adopted by varying the factors or parameters simultaneously and investigating their individual and combined effects on the process output. This approach is applied using statistical experimental design, (e.g. response surface methodology, Taguchi) which allows the development of regression models that correlate between the multi variant inputs and the process outcomes either for multi-variant – single objective or multi-variant – multi objective problem based [25, 27, 30–33]. Integrated tools were developed to quantify and assess the uncertainty in RA, an example of these tools is the Quantifying Margin and Uncertainty, which used to support the certification of the reliability and safety for a physical system and quantify the performance thresholds and their margins and the associated uncertainty in their evaluation. This tool widely used to quantify uncertainties that are dominated by lack of knowledge in risk-informed decision analysis [34].

3.4 Sensitivity analysis to support the uncertainty management

Sensitivity analyses are used as tools to reduce the uncertainty, where it is used to prioritize the research efforts to reduce uncertainty associated with the scenario, conceptual model, input data, modeling process, and the designed system [35]. Differential and probabilistic sensitivity analyses are used to support the uncertainty quantification and reduction. Differential sensitivity analysis is used when exact risk formula exists, this technique is computationally efficient; however, it is only valid in vicinity of the base case and might require intensive efforts to drive the sensitivity coefficients [35]. Probabilistic sensitivity analyses are conducted by assign probability density functions to each input parameter, generate an input matrix using suitable sampling method, calculate the outputs, and assess the influences and relative importance of each input/output relationship [36]. In probabilistic risk analysis, the marginal distributions of the studied parameters and the dependence between them need to be specified [36]. In this case, interval probability, Dempster-Shafe structure, and probability boxes are widely used approaches.

4. Conclusion

In this chapter the approaches to manage uncertainty within the risk assessment framework to support the decision making process for pollution prevention and control systems are introduced. In this respect, the risk assessment, its need, and approaches were introduced and discussed. The classification of sources and reducibility of uncertainty is presented. The approaches to quantify the uncertainty were overviewed with special reference to the role of the sensitivity analysis in uncertainty management.

Author details

Rehab O. Abdel Rahman[1*] and Yung-Tse Hung[2]

1 Hot Lab. Center, Egyptian Atomic Energy Authority, Cairo, Egypt

2 Department of Civil and Environmental Engineering, Cleveland State University, Cleveland, USA

*Address all correspondence to: alaarehab@yahoo.com

IntechOpen

References

[1] Rehab O Abdel Rahman, Introductory chapter: Development of assessment models to support pollution preventive and control decisions, In: R.O. Abdel Rahman (Ed.) Kinetic Modeling for Environmental Systems, IntechOpen, 2019, DOI: 10.5772/intechopen.83822

[2] United nation, the sustainable development agenda https://www.un.org/sustainabledevelopment/development-agenda/ (last accessed 29 January, 2021)

[3] R.O. Abdel Rahman, M.W. Kozak, Y.T. Hung, Radioactive pollution and control, Ch (16) In: Y.T. Hung, L.K. Wang, N.K. Shammas (Eds.), Handbook of Environment and Waste Management, World Scientific Publishing Co, Singapore, (2014) 949-1027, http://dx.doi.org/10.1142/9789814449175_0016

[4] K. Zhang, Y. Pei, C. Lin, An investigation of correlations between different environmental assessments and risk assessment, Procedia Environ. Sci. (2010) 643-649

[5] S. Valdivia, A. Ciroth, M. Finkbeiner, J. Hildenbrand, W. Klöpffer, B. Mazijn, S. Prakash, G. Sonnemann, M.; C.M.L. Traverso, Ugaya, et al. Towards a Life Cycle Sustainability Assessment: Making Informed Choices on Products; UNEP/SETAC Life Cycle Initiative: Paris, France, 2011

[6] C. Wulf, J. Werker, C. Ball, P. Zapp, W. Kuckshinrichs, Review of Sustainability Assessment Approaches Based on Life Cycles, Sustainability 11 (2019) 5717.

[7] K. Swangjang, Development of Conceptual Model for Eco-Based Strategic Environmental Assessment, in Rehab O. Abdel Rahman, Kinetic Modeling for Environmental Systems,

IntechOpen, (2019), ISBN: 978-1-78984-727-7 DOI: 10.5772/intechopen.79240

[8] M.R. Partidário, 1998. Significance and the Future of Strategic Environmental Assessment. International Workshop on Strategic Environmental Assessment, Tokyo.

[9] G. Suter, T. Vermeire, W. Munns, J. Sekizawa, Framework for the integration of health and ecological risk assessment, WHO, 2012 WHO/IPCS/IRA/01/12, https://www.who.int/ipcs/publications/en/ch_2.pdf?ua=1 (Last assessed in 29 January 2021)

[10] EEA, Environmental Risk Assessment - Approaches, Experiences and Information Sources, 1998, Environmental issue report No 4, European environmental agency (EEA), ISBN: 92-9167-080-4

[11] E. Zio, N. Pedroni, Risk informed decision making process: an overview (2012). Numéro 2012-10 des Cahiers de la SécuritéIndustrielle, Fondation pour une Culture de SécuritéIndustrielle, Toulouse, France (ISSN 2100-3874).

[12] I. Linkov, J. P.-Oliveira, Assessment and Management of Environmental Risks Cost-efficient Methods and Applications, Kluwer Academic Publishers (2000)

[13] CCPS, Guidelines for hazard evaluation procedures, 3rd ed, (2008) Center for chemical process safety, John Wiley & Sons, Inc., Hoboken, New Jersey

[14] CCPS, Guidelines for risk based process safety, 3rd ed, (2007) Center for chemical process safety, John Wiley & Sons, Inc., Hoboken, New Jersey

[15] USNRC, white paper on risk-informed and performance-based

regulation, 1999, USNRC, https://www.nrc.gov/reading-rm/doc-collections/commission/srm/1998/1998-144srm.pdf (Last assessed in 29 January 2021)

[16] IAEA, Risk informed regulation of nuclear facilities: Overview of the current status IAEA-TECDOC-1436, International Atomic Energy Agency, Vienna (2005)

[17] J.M. Booker, T.J. Ross, An evolution of uncertainty assessment and quantification, Scientia Iranica D (2011) 18 (3), 669-676

[18] M.E. Pate-Cornell, Uncertainties in risk analysis: Six levels of Treatment, Reliab. Eng. Syst. Safe. 54 (1996) 95-111

[19] D.JC. Skinner, S.A. Rocks, S.JT. Pollard, G.H. Drew, Identifying uncertainty in environmental risk assessments: the development of a novel typology and its implications for risk characterization. Hum. Ecol. Risk Assess. 20 (3)(2014) 607-640.

[20] S.F. Wojtkiewicz, M. S. Eldred, R.V. Field, Jr., A. Urbina, J.R. Red-Horse, Uncertainty Quantification In Large Computational Engineering Models, AIAA-2001-1455 American Institute of Aeronautics and Astronautics, 1-11

[21] K.A. Notarianni, G.W. Parry (2016) Uncertainty. In: Hurley M.J. et al. (eds) SFPE Handbook of Fire Protection Engineering. Springer, New York, NY. https://doi.org/10.1007/978-1-4939-2565-0_76

[22] P. Grossi, Sources, nature, and impact of uncertainties on catastrophe modeling, 13th World Conference on Earthquake Engineering Vancouver, B.C., Canada August 1-6, 2004 Paper No. 1635

[23] L. Uusitalo, A. Lehikoinen, I. Helle, K. Myrberg, An overview of methods to evaluate uncertainty of deterministic models in decision support, Environ. Model. Softw. 63(2015)24-31.

[24] H.M. Regan, M. Colyvan, M.A. Burgman, A taxonomy and treatment of uncertainty for ecology and conservation biology. Ecol. Appl. 12 (2002) 618-628.

[25] IAEA, Safety Assessment Methodologies for Near Surface Disposal Facilities: Volume 1 Review and enhancement of safety assessment approaches and tools. International Atomic Energy Authority, Vienna, 2004

[26] A.M. El-Kamash, R.O. Mohamed, M.E. Nagy, and M.Y. Khalill, Modeling and validation of radionuclides releases from an engineered disposal facility, Radioactive Waste Management and Environmental Restoration. 22(4), pp. 373- 393 (2002)

[27] R.O. Abdel Rahman, O.A. Abdel Moamen, N. Abdelmonem, I.M. Ismail. Optimizing the removal of strontium and cesium ions from binary solutions on magnetic nano-zeolite using response surface methodology (RSM) and artificial neural network (ANN), Environ. Res. 173 (2019) 397-410

[28] R.O. Abdel Rahman, Preliminary assessment of continuous atmospheric discharge from the low active waste incinerator, Int. J. Environ. Sci, 1, No 2 (2010), 111-122

[29] R.O. Abdel Rahman, A.A. Zaki, Comparative study of leaching conceptual models: Cs leaching from different ILW cement based matrices, Chem. Eng. J., 173 (2011) 722- 736. doi:10.1016/j.cej.2011.08.038

[30] O.A. Abdel Moamen, I.M. Ismail, N.M. Abdel Monem, R.O. Abdel Rahman, Factorial design analysis for optimizing the removal of cesium and strontium ions on synthetic nano-sized zeolite, J. Taiwan Inst. Chem. E. 55 (2015) 133-144

[31] M.S. Gasser, H.S. Mekhamer, R.O. Abdel Rahman, Optimization of the utilization of Mg/Fe hydrotalcite like Compounds in the removal of Sr(II) from aqueous solution, J Environ. Chem. Eng., 4 (2016) 4619-4630

[32] M.S. Gasser, E. El Sherif, R.O. Abdel Rahman, Modification of Mg-Fe hydrotalcite using Cyanex 272 for lanthanides separation, Chem. Eng. J., 316C (2017) 758-769.

[33] M.S. Gasser, E. El Sherif, H.S. Mekhamer, R.O. Abdel Rahman, Assessment of Cyanex 301 impregnated resin for its potential use to remove cobalt from aqueous solutions, Environ. Res. 185 (2020) 109402

[34] A. Urbina, S. Mahadevan, T. L. Paez, Quantification of margins and uncertainties of complex systems in the presence of aleatoric and epistemic uncertainty, Reliab. Eng. Syst. Safe. 96 (9)(2011)1114-1125.

[35] D.M. Hamby, Review of techniques for parameter sensitivity Analysis of environmental models, Environ. Monit. Assess. 32(1994) 135-154.

[36] S. Ferson, R.B. Nelsen, J. Hajagos, D.J. Berleant, J. Zhang, W.T. Tucker, L.R. Ginzburg, W.L. Oberkampf, Dependence in probabilistic modeling, Dempster-Shafer theory, and probability bounds analysis, SAND2004-3072. (2004)

Chapter 12

Uncertainty Management in Engineering: A Model for the Simulation and Evaluation of the Operations Effectiveness in Land Use and Planning

Ermelinda Serena Sanseviero

"Uncertainty is the natural habitat of human life, although the hope of escaping it is the engine of human activities"

Zygmunt Bauman

Abstract

The quality of the environment is essential for our health, our economy and our well-being. However, it faces a number of major challenges, not least those related to climate change, unsustainable consumption and production, and various types of pollution.Spatial planning policies (and EU legislation) protect natural habitats, keep water and air clean, ensure adequate waste disposal, improve knowledge of toxic substances and support the transition of businesses towards a sustainable economyThe goal of the work is to develop a standardized methodology for the monitoring and management of spatial information as the basis for spatial planning. The present work makes use of data analysis methods in spatial planning, where the proposed "mathematical" model is of help in supporting decision making. In fact, certain decisions often arise only from the evaluation of certain parameters, which are always small; it is necessary to consider them all, even in a disaggregated way, and give the right weight to each one. The proposed model describes the territorial system as an interaction between the physical system and the social system; it interpret needs and identify problems concerning the physical system and the social system; and formulate purposes and deduce objectives expressed in quantitative magnitude; formulate forecasts on the consequences of decisions to change the uses of the physical system, through electronic processing. The model could be used to evaluate alternative guidelines for change; andto choose from among the possible alternatives the one that is believed to contribute most to the pursuit of the objectives.

Keywords: Uncertainty, decision making, land use, planning, scenario

1. Introduction

Environmental policies and EU legislation protect natural habitats, keep water and air clean, ensure adequate waste disposal, improve knowledge of toxic substances and support the transition of businesses to a sustainable economy. While there is not a single Community planning policy because the "planning" has more weight, but in essence it is left to the responsibility of national Governments; on the other hand certainly you can have recourse to an important series of strategies, in fact, the European Union has moved so far mainly at the level of standards (Directives and Regulations) and less at the level of plans and programs. There are "strategic" guidelines of political value and certainly not a "European environmental planning", because the U. E. sets only a reference framework in the various sectors, indicates priorities, suggests the most appropriate means and instruments, but does not interfere with the role of the Member States in terms of planning, except to suggest a general criterion: "integrate planning sustainable in community policy"[1]. This point is important for two reasons:

- because planning is considered a necessary and useful political tool as long as it is "sustainable";

- because national planning cannot ignore the community framework.

One of the strategies, for example, concerns the importance of soil protection and was the subject of attention by the European Commission in 2002. Following the publication in 2001 of the sixth European Environment Action Plan, the European Commission adopted the COM [1] "Towards a Thematic Strategy for Soil Protection", building on a specific political commitment to this issue. Soil performs many environmental key-functions which are: biomass production, storage, filtering, buffering and transformation and plays a central role in water protection and the exchange of gases with the atmosphere. It is also a habitat and gene pool, an element of the landscape and cultural heritage, and a provider of raw materials. In order to perform its many functions, it is necessary to maintain soil condition[2]. The soil, however, is an integral part of the territory and the territorial dimension is particularly important in the "planning" processes. So land assessment as a spatial *planning* tool promotes *sustainable development* of space and settlements; it serves to defend the soil resource, to protect it from pollution, contributing to the conservation and improvement of the quality of the environment. The assessment of soils is of particular importance in areas that are already heavily contaminated and ecologically sensitive in which there continues to be a huge demand for spaces intended for construction and expansion.

The present work makes use of data analysis methods in spatial planning. The problem faced in first appeal sets as objective to define the social and economic orders of the area by the point of view of the situation of the state of fact. In these cases, the task is really difficult as it is necessary to take into account a multiplicity of aspects of different nature (economic, social, historical, urban and environmental) and of various dimensions, moreover it is necessary to summarize them to obtain useful information for make assessments that meet community goals. Far this aim, I follow a methodology based on the analysis of the data and in particular on the method of the principal components, on the cluster analysis and on coverings and partitions with fuzzy sets [2]. In such an approach, I have ascertained that a lot of importance it's to attribute to the changes of the general system on which one

---

[1] Comment taken from "Law and environmental management - II Ed." by Stefano Maglia and AmedeoPostiglione - IrnerioEditore

[2] APAT Agency for the Protection of the Environment and for Technical Services - A "SOIL DEFENSE - EUROPEAN STRATEGY"

intervenes, kept account that the means that we have to disposition often result inadequate to gather the entity of the aforesaid changes.

The method is generally adaptable; however, to achieve satisfactory results, it should be used directly to related information with the insights that you pursue.

It is considered that a correct methodological approach to the problems of decision resides in multidimensional analysis. In fact in such type of analysis all the technical, economic, social and environmental aspects, are opportunely valued and balanced.

It is recognized by many people that any intervention on the system baits a complex connection of phenomena, necessity emerges then to favor the orientation and the control of the scales through the use of a model that aim at rationalizing the complexity of the situations and to reduce uncertainty that permeates the planning process. Of it the planning and the evaluation of the interventions have a lot of importance. The evaluation is a tool of control (in progress and ex post) of uncertainty on the evolution of the several elements constituent the general context, kept account of all the different demands expressed from consumers, decision men and technicians, and considering the multiplicity of the preset objectives, the potentialities of the available technologies and the constraints [3].

The government policy of territorial planning implemented in our country as well as in other European and non-European nations has made the problem of finding more valid tools for forecasting and decision-making urgent, with the necessary deepening of the study of methodologies linked to urban culture. The goal then is to implement a methodology adapted for the control and management of information concerning the territory in order to provide a scientific setting of the work steps that precede the action of spatial planning. Such an approach starts after the Second World War, especially in the Anglo-Saxon and US world with the study of urban systems, activity patterns and founding principles of their balance, spatial organization and environmental design models. A stop in the search for patterns of systematic and ordered study has had prevailed when the reductive opinion of mathematical studies to be able to exclusively deal with solving quantitative problems.

In this respect, starting from the eighties onwards.

Explains and justifies the denial explicated by certain architects towards mathematical methods, attitude further motivated error exchange for *schematismo* previously established the need for systematic, order, organic and its symbolism of a scientific theory.

The main role of the scientific disciplines that coordinate and assist the construction of such a setting, it must identify a methodology to build up the logic of abstraction and verification process. In conclusion, the method is always applicable; however, the results must be interpreted in relation to the particular circumstances in which the investigation takes place.

Nowadays the new sciences of complexity, the increasingly heated debate among the philosophers of science, from which relativistic and localistic positions seem to emerge, do not leave much room for the revival of scientific paradigms with models of absolute validity."Local scientific approaches do not cooperate harmoniously with an image, with a theory of knowledge and the universe, but on the contrary they intersect, overlap, ignoreeachother, contrast, integrate, split".

Complexity arises from the awareness that "the type of problems that contemporary society faces cannot be standardized, like the problems faced par excellence by strong disciplines. As already mentioned, the use of procedural standards is as of little use as proposing physical standards. In reality, this awareness is not new, but can be traced back to the 1980s, when the weakening of the nation-state form, the crisis of political units of great territorial dimension and of the concept of territory as a univocal entity, make the "gravitational" spatial paradigm obsolete, which presupposes a hierarchical structuring of space.

The "restitution" of complexity is strongly linked to the processing of information: moving in a field of consolidated knowledge practices, one is forced to discard everything that is not compatible or can be tamed with knownmeans, reducing the complexity of the phenomena. It's necessary to abandon the old certainties to take new paths, such as those traced by the complexity of relationships and continuous interactions between animate and inanimate components of the same world [4]. The theory of complexity, first of all, is not a scientific theory in the strict sense. It would be better to speak (and indeed some authors do) of "complexity challenge" or "complexity thought" or, better still, "complexity epistemology". It is precisely as an epistemological perspective, in fact, that complexity plays a crucial role in contemporary thought. This is because complexity involves three equally elevant epistemological innovations: a new alliance between philosophy and science, a new way of doing science, a new conception of natural evolution [5]. This chapter makes use of data analysis methods in spatial planning. The exploratory methods of multivariate analysis make it possible to arrive at territorial types (at different scales) that are also significant at an environmental level. The methods of *data analysis* and the *main components* in the work aim to describe the territorial system as an interaction between the physical system and the social system; interpreting needs, evaluating alternative directions of change; support the choice between the alternatives of what you think may contribute most to achieving the objectives of preservation [6].

2. Models for the mathematical representation of the real situations: the territory and the complexity

I have cited of complexity as a structuring character of our reality and in particular the territory. It is therefore obvious that the "management of complexity" is the fundamental problem for those involved in reading, representation or knowledge of the city and the territory. Hence the need to question those methods consolidated by practice but certainly not by results, with which we are used to working because they are equipped with tested tools capable of selecting and modeling a reality that at the end of the process "must" be verifiable. A real situation usually presents itself in a confusing, complex, vague way. It is not immediately clear how we can formulate a mathematical model to represent the phenomena observed. However after a first process of abstraction, using logical principles and common sense, you can try to get a '"acceptable" mathematical representation of the situation to be studied. From the point of view of the study of the territory, the difficulty lies precisely in transferring a perception that belongs to our mind into a model and therefore into artificial symbols. Multiple internal and external factors exert their influence, one on the other, in a continuous process of dependencies and reciprocal relations thus configuring what can be defined as the "system" [7]. The system is in a symbolic language, the area on which it operates. "At this point it is worthwhile to include the definition already indicated that the models are abstract representations of reality, helping us to perceive meaningful relationships in the real world, to manipulate them and then predict other. In this sense, the model while qualifying as a design tool, as part of a rational methodology should not be confused with the project; the project subtends a model but in support to this penetrates into the elements and relationships in a much more analytical (..) consequently, the model belongs to the moment meta projectual of the design process." [8]. "The elements, relationships and interrelationships to be searched, analyzed and interpreted involve an enormous amount of work [9], since it seems superfluous to assert it, the territory, the city represents the" impact "of the community structures, the projection plane, and the organization of social, economic, administrative, cultural, residential activities etc." [2].

Meanwhile, following this procedure, the following organization chart can be adopted:

a. **Definition of the collective or universe "U".** It must be chosen in such a way as to be suitable to the purposes of the research; in this specific case it seemed appropriate to choose, as the statistical units for the phenomenon to be studied, small parts of the territory (municipalities), either because they have bureaucratic and organizational structures, either because they are compared to known statistical summaries. In general, the statistical universe that is going to be investigated and which you want to investigate the interrelationships and structure, is the set $U = \{O1, O2, O3, ... Om\}$ of "statistical units" or "Objects".

b. **Definition of the phenomenon.** A phenomenon is what is observed in the elements of a collective. For example in the collective "group of municipalities" the hierarchical structure, the socio-economic profile, etc., are phenomena And the variables selected for analysis are intended to clarify what should be observed to study the evolution of the phenomenon.

c. **Choice of variables.** In this regard, it introduces the concept of a statistical nature. If "U" is the collective statistics that you consider, in the case study is the set of municipalities, and V is any set, is said character or statistical variable defined in U and set of values or V mode, each X function defined in U and values in V. If V is contained in the set R of real numbers the statistical variable is said real. In the case study, we have been taken into consideration 10 variables represented by the following territorial real variables:

$X1$ = number of inhabitants.
$X2$ = variation of population divided by the total population;
$X3$ = percentage divided by total active population;
$X4$ = occupied housing divided by the total housing;
$X5$ = agricultural area divided by the municipal land area;
$X6$ = Industry insiders divided by the total number of employees;
$X7$ = active in agriculture divided by total assets;
$X8$ = Percentage tertiary sector divided by the total number of employees;
$X9$ = Active public administration divided by total assets;
$X10$ = trade workers divided the total number of employees;

In general, given the phenomenon, to describe it is considered the 'ordered set $X = (X1, X2, ...Xn)$ of real variables that you think will adequately represent the phenomenon, and it is called performance of the set of phenomena values assumed by these variables in the various objects of the collective. For each object Oi set U and for each variable Xj is determined so zij the value assumed by the object Oi in the variable Xj. In order to better understand the phenomenon variables Xj are replaced by variables centered $Zj = aij (Xj-mj)$, where I is the average of Xj and aij different from 0 is an appropriate multiplier.

The statistical survey result is thus represented by means of a Eq. (1)

$$\text{"objects-characters"} : OC1 = \begin{bmatrix} Z_1 & Z_2 & Z_3 & \dots & Z_N \\ z_{11} & z_{12} & z_{13} & \dots & \dots \\ \dots & \dots & \dots & \dots & \dots \\ \dots & \dots & \dots & \dots & \dots \\ z_{m1} & z_{m2} & z_{m3} & \dots & z_{mn} \end{bmatrix} \begin{matrix} O_1 \\ O_2 \\ O_3 \\ \dots \\ O_m \end{matrix} \qquad (1)$$

In the case study, it was thus obtained a base matrix of 10 to 46 objects indicators (Municipalities). The MM1 = (U tern, X, OC1) is defined mathematical model of the phenomenon.

Once created the MM1 model, starting from the foregoing considerations, we finally have the means to manage information globally taking into account all significant relationships between the variables.

The MM1 model contains easily visible information such as the values of the objects in the collective U Xj variables and other information hidden or latent. In order to highlight all the information, latent or apparent, contained in the model we introduce the concept of "equivalent models".

Two models MM1 = (U1, Z, OC1) and MM2 = (U2, Y, OC2) are equivalent if that of OC2 derives from the knowledge of the OC1 object-mode matrix and vice versa. MM1 MM2 implies if the knowledge of OC1 to OC2 is deduced.

In this chapter, we consider only pairs of models (MM1, MM2) wherein U1 = also U2 and, once assigned the MM1 model, suppose that MM2 is such that each yj component of the vector Y is a linear function of those of the vector Z, namely that there is a matrix T, the general term tij, such that Y = TZ. In this case we say that MM1 and MM2 linearly implies that T is the transformation matrix from MM1 to MM2.

If $Y = [Y1, \dots Yr]$, $Z = [Z1, \dots Zn]$ the Y = TZ is written in full:

$$TZ = \begin{bmatrix} y_1 & = & t_{11}z_1 & \dots & t_{1n}z_n \\ \dots & = & \dots & \dots & \dots \\ y_r & = & t_{r2}z_1 & \dots & t_{rn}z_n \end{bmatrix} \qquad (2)$$

If the information matrix T is square and invertible then also implies MM1 MM2 linearly and is said to MM1 and MM2 are linearly equivalent in this case by:

$$Y = TZ \qquad (3)$$

follows (the transpose inverse):

$$Z = T^{(-1)} Y \qquad (4)$$

From (3) is obtained, in particular the relationship

$$(OC2)^t = T(OC1)^t \qquad (5)$$

and (4) the

$$(OC1)^t = T^{(-1)} (OC2)^t \qquad (6)$$

By suitably selecting T and then passing from the model MM1 to MM2 linearly equivalent model occurs in general that some properties contained in MM1 but latent, become evident in MM2.

If the matrix T is orthogonal then it is invertible and is

$$T^{(-1)} = T^t \tag{7}$$

so that (5) and (6) reduce respectively to

$$(OC_2) = (OC_1)T^t, (OC_1) = (OC_2)T \tag{8}$$

In this case the models MM1 and MM2 call them orthogonally equivalent.

The transition from one model to an equivalent orthogonally can be useful because they are preserved to the particular mathematical properties that allow to better interpret the urban phenomenon.

Consequently, I will adopt some statistical techniques essentially consisting in the passage from the initial MM1 model to specific equivalent and in particular orthogonally equivalent models. These techniques are called factorial analysis.

The techniques mentioned above that contribute to the research of the relationships between variables, have in particular the advantage of highlighting that for some new variables y1, y2, ... yn-called "factors" such that only some of them y1, y2, y3, ... ys, (with s < < n), are relevant for the explanation of the phenomenon. This, in fact, allows a reduction in the number of variables and thus a simplification of the model.

3. An application of factorial analysis methods for the analysis of an urban system

One could refer to one of the detected information on statistical units study outline, there is the method of principal components if T is orthogonal and the matrix of variances and covariances, is a diagonal matrix, = diag (l1 ... ln) with them, elements of the diagonal, arranged in descending order. The Yi obtained by the relationship Y = TZ in this case are called "main components". They are factors not related and equipped with various mathematical properties.

The principal component analysis is basically a theory for the study of a phenomenon represented by many random variables X1, X2, ...Xn, from the point of view of its variability. It is proposed to represent the same phenomenon with new centered variables Y1, Y2, ...Yn said main components, not related to each other, with decreasing variability, such that the sum of the variability of Yi is equal to that of Xj and so that already with a few variables explain a large proportion of the variability of the phenomenon.

These variables allow to obtain, inter alia,:

a. simple geometric representations of the phenomenon under examination.

b. simplification of the procedure of multivariate regression of new variables considered.

c. decomposition of the variability in a manner easily understandable phenomenon.

d. formation of homogeneous groups with more simple procedures which includes immediacy with the practical significance.

e. ability to solve problems of multi-objective programming choices.

In our research we studied the "socio-economic balance of the municipalities of the province of Pescara" phenomenon. It was represented by 10 random variables {X1 .... X10} and 46 objects {O1 ... O46} with O1, O2 etc. we indicate each of the 46 Municipalities of the Province of Pescara.

For each variable Xj j = 1, ..10, was calculated the average mj and the vector X = (X1 .... X10) has been replaced by S = scraps (S1 ... S10); with Sj = Xj -mj; for each j = {1 ... 10} was taken as the *multiplier value* aj = $\frac{1}{\sigma_j}$, with σj *standard deviation* of Xj.

It is thus obtained a vector Z = (Z1 ... Z10), and then there was obtained OC1 matrix. If Mx is object-matrix characters relating to the variables Xj and M it is the matrix that has as its j-th column vector with all components equal to mj, you get

$$OC_1 = (M_x\text{-}M)D^{-1} \tag{9}$$

with D = diag (s1 ... sn), diagonal matrix of the standard deviations. The matrix of variances and covariances of Z is then

$$R = \frac{1}{n}(OC_1)^t(OC_1), \tag{10}$$

generic element

$$r_{hk} = \text{cov}(Z_h, Z_k) = \frac{1}{n}\sum_{r=1}^{n} Z_{rh}Z_{rk} \tag{11}$$

The R is evidently also correlation matrix for both the vector X and the vector Z.

In our work, we were prepared using the formulas (9) (10) calculated with the program Mathematica 2.2.

It is a verification by calculating instead of the R of the variances and covariances S matrix of Xj is also performed given by the formula

$$S = (M_x\text{-}M)^t(M_x\text{-}M) \tag{12}$$

and obtaining

$$R = D^{-1}\Sigma D^{-1} \tag{13}$$

The eigenvalues and eigenvectors of R were then calculated.

Said L = diag (l1 ... l10) the matrix of the eigenvalues, in ascending order and called A the matrix that has the corresponding eigenvectors for columns is

$$RA = A\Lambda \tag{14}$$

that is

$$A^tRA = \Lambda \tag{15}$$

Then the matrix T = At is the orthogonal transformation matrix that is passed from Zj to new variables Y equivalent orthogonally variables and uncorrelated.

In fact from Y = TZ it is obtained.

OC2 = OC1T^t = (OC1)A and therefore the variance and covariance matrix of Y is the

$$\Delta = \frac{1}{n}(OC_2)^t(OC_2) = \frac{1}{n}A^t(OC_1)^t(OC_1)A = A^tRA = \Lambda, \tag{16}$$

which is a diagonal matrix

The OC2 matrix provides the values assumed in correspondence with the objects Oi from the main components Yj. It was found that the total variability was practically absorbed by the first three main components, so that the phenomenon observed in the urban system can be sufficiently described by the OC2 submatrix formed by the first three columns.

The profiles show the coordinates of the municipalities and variables with respect to the first three factorial axes. On the basis of these results, a hierarchical classification of the area reported in the following images was also obtained (**Figures 1** and **2**).

As evident, the analysis outputs are the principal components that cut the cloud of points (municipalities) in the direction of greatest inertia. The coordinates of the objects (always the municipalities) and the coordinates of the characters (the variables considered)on the factorial axes (always on the same axis system) are important for determining the results (**Figures 3–5**).

The analysis (not all reported for the sake of brevity) made it possible to obtain "artificial" variables (from 10 to 3) which, unlike the originals, are not correlated but equally provide information on the area. The procedure is valid when the variables to be analyzed are many 40, 50 etc. The first three main components explain respectively 27%, 21.4% and 13.5% for a total of 61.9% of the total system inertia. Each of them gives rise to a particular composition of the municipalities on the factorial axes as can be clearly seen by analyzing the thermometer graphs (**Figures 3–5**).

The first component explains more the commercial aspect of the areas and determines a different position of the municipalities depending on whether they are negatively or positively correlated with it. The value assumed by it, allows to

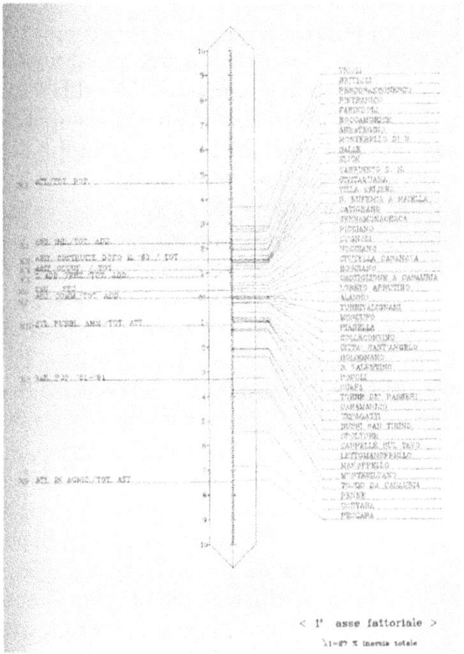

Figure 1.
An hierarchical classification of the area.

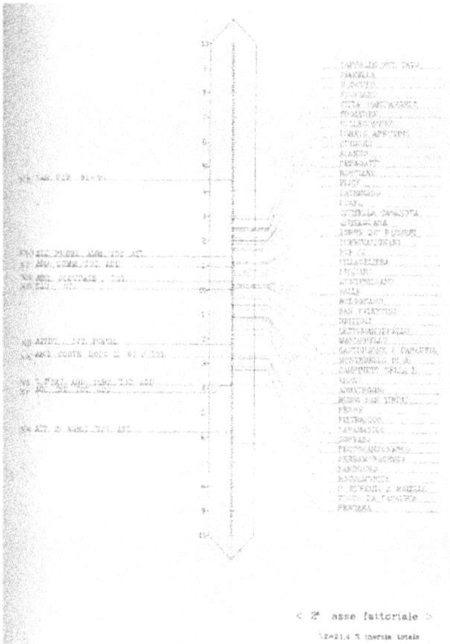

Figure 2.
An hierarchical classification of the area.

classify the municipalities in the ranking and gradually group those municipalities with similar values up to constitute.

Homogeneous classes, or at least similar in relation to this aspect, with common characteristics. The same interpretation applies to the variables explained on the second and third factorial axes. But a better representation is obtained with the observation of the factorial plans (**Figures 1** and **2**).The origin of the axes represents the baricenter of the system of masses that gravitate around them. Urban areas are characterized as places in the territory where various and multiple functions (as well as activities) are concentrated and interacted. The different weight and the different location that define the presence and organization of these functions in the different areas represent the synthesis, the landing point, of all the transformation processes of human activities in the long term.

The first factor explains the eighty 83.4% of the total inertia. The greatest contribution to the formation of territorial morphologies given by the position of the municipalities on this axis is given by groups of motion or variable correlated with each other and linked in complex form to the social structure of the resident populations. It is a group of variables that articulates the tertiary sector in a satisfactory way, useful for recognizing urban and non-urban social forms. The second factor, with a similar distribution of municipalities, explains the aspect linked to industry to the extent of 4.54% of total inertia; the most significant variables are those that sufficiently articulate industry and the production of consumer goods. The representation of the variables on the third factorial axis highlights that the indicative variable of the mechanical industry employees alone explains about 3% of the variability of the system based on this specialization, a hierarchical but sectoral classification of municipalities is created. It is in fact a specific variable that cannot take into account the urban reinforcement of the province but that at least highlights a characterizing aspect: the high concentration of industrial activities in mechanical processing.

In the graphs 4 and 5 the position of Pescara (the largest city of the considered area) is that of the vertex of a tetrahedron, or of a hypertetrahedron, thinking of a

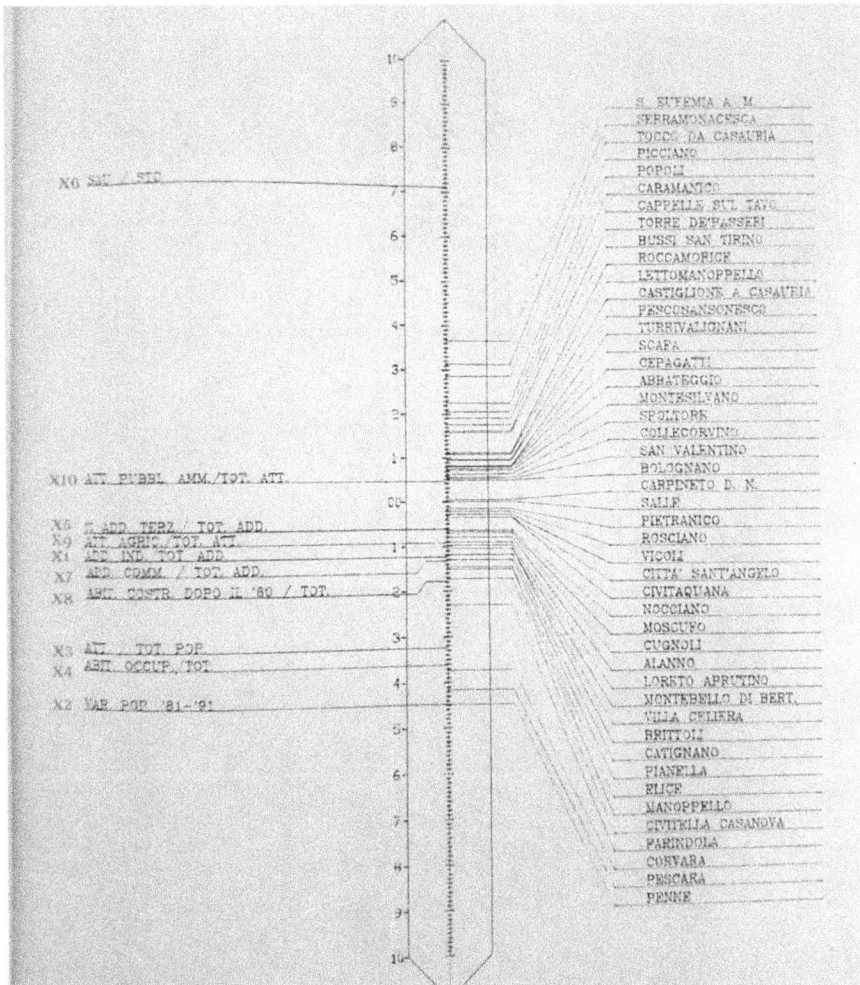

Figure 3.
The first three main components explain respectively 27%, 21.4% and 13.5% for a total of 61.9% of the total system inertia. Each of them gives rise to a particular composition of the municipalities on the factorial axes: first factorial axis.

space relative to the variables that is not three-dimensional but ten-dimensional (are the variables).

The other municipalities that occupy positions increasingly close to the center of gravity are those that represent characteristics that are increasingly closer to the morphology and less to the characteristics of rural morphologies. The purpose of this analysis is to define a map of the Province that is able to grasp, at a fine territorial scale, the incidence and location of the functions that take place in this territorial area.

The interaction of the various functions examined (residential, industrial, commercial, exception, cc) defines the typology of the different territorial morphologies. From this descriptive approach it is possible to go back, by interpretative way, to synthetic results (the map in fact) which expose the prevailing ways of the interaction between functions (land use) and territorial extension and their mutual influence. In areas with a greater concentration of industrial functions, a greater

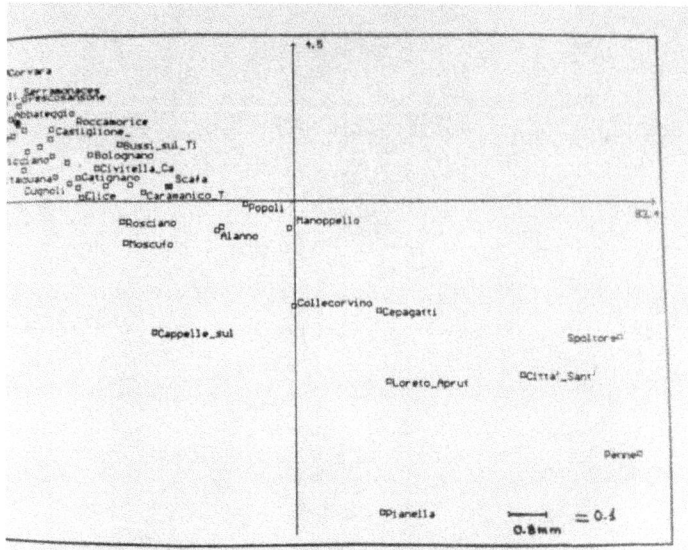

Position of the Municipalities on the plane formed by the I and II factorial axis

Figure 4.
Second factorial axis.

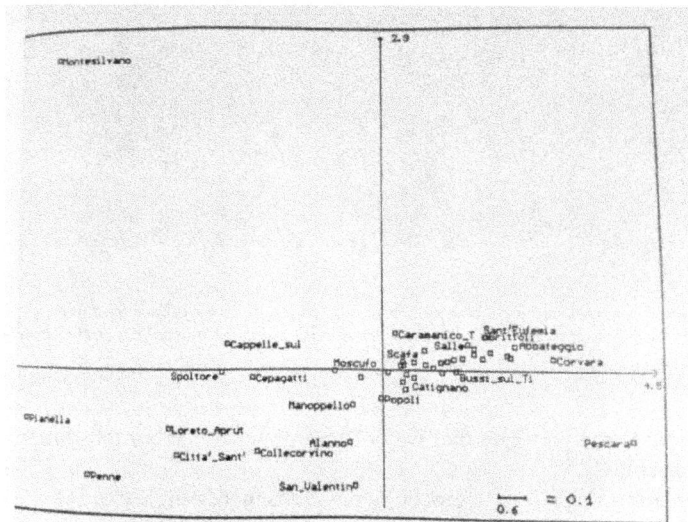

Position of the Municipalities on the plane formed by the II and III factorial axis

Figure 5.
Third factorial axis.

concentration of pollution of various types (soil, air, water, etc.) may be seen and it will be necessary to intervene with targeted policies to reduce negative emissions. In **Figure 6** the concentric circles representing different categories of territorial (spatial) morphologies, and on them you can insert a spatial system in which the centers or the analyzed unit is positioned with its coordinates (derived from the position on the factorial axes); evaluating such a graphical representation, in relation to a first

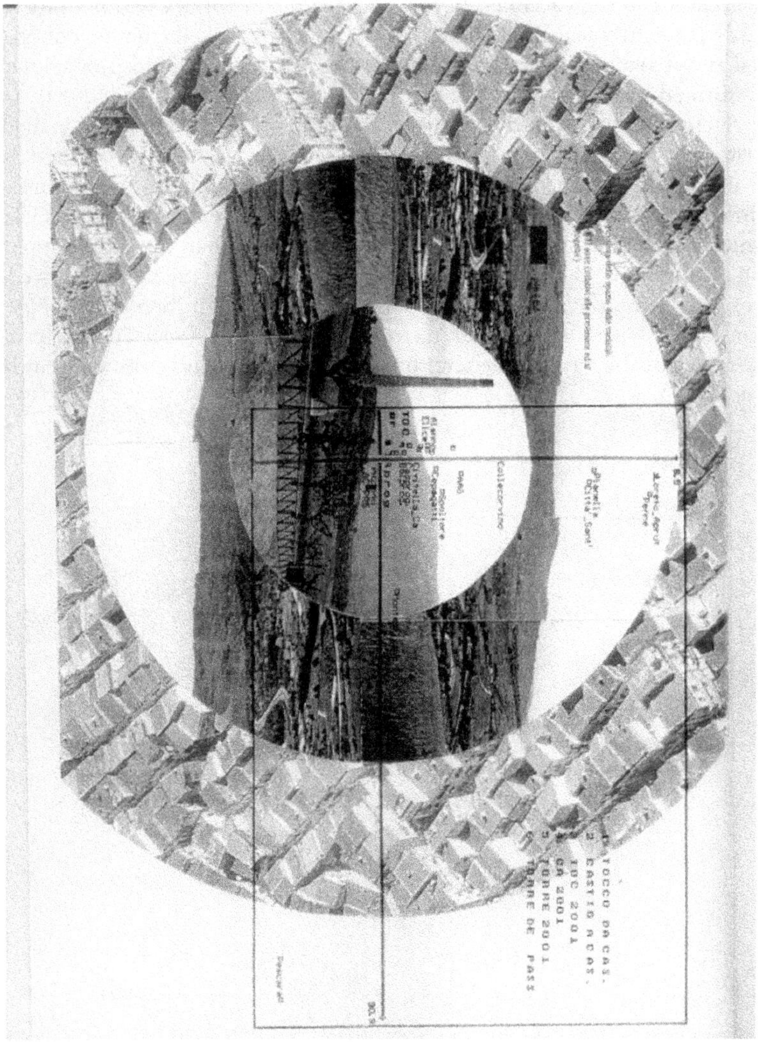

Figure 6.
The concentric circles representing different categories of territorial (spatial) morphologies: urban, productive, agricultural.

phase (the state of fact analysis) and in a second phase (project) you could observe the impact of a possible intervention on the part of analyzed territory (diachronic representation).

4. Conclusions

Finally, the usefulness of such an application lies in the consideration that complex variables can hardly be simplified and assumed uniquely in a model; instead of them we prefer an indicator (factorial axis) composed of different fractions of several variables and therefore more consistent with reality and more representative of the single variables. The profiles that can be outlined starting from the proposed model show the positions (coordinates) of the territories considered (small agglomerated, municipalities or parts of larger and more widespread urbanizations) and the variables, with respect to the factorial axes about these results, a *soft* hierarchical classification of the area is obtained (that is, non-rigorous nuanced and sharp outlines). You can therefore obtain spatial morphologies classes that are formed with the passage of territorial units considered from one to another cluster. Such shifts are caused perhaps by the simulation of interventions made on the territory or from variation (oscillation) of some input in the initial models.

The urban context, therefore, as a complex system, cannot be simply broken down into its constitutive dimensions (spatial, temporal, environmental, social, and economic), being the sum of the same, something more than the simple superimposition of the single dimensions, in a Gestalt vision which cannot be ignored. It becomes, therefore, the subject of analysis through the definition of the object "relationship between the size", taking into account that the social characterizes the very existence of the city (of the territory), since there are no cities without human presence.

Author details

Ermelinda Serena Sanseviero
D'Annunzio University of Chieti-Pescara, Italy

*Address all correspondence to: sansevi@unich.it

IntechOpen

References

[1] COMMISSION OF THE EUROPEAN COMMUNITIES - Brussels, 16.4.2002 COM(2002) 179 final - COMMUNICATION FROM THE COMMISSION TO THE COUNCIL, THE EUROPEAN PARLIAMENT, THE ECONOMIC AND SOCIAL COMMITTEE AND THE COMMITTEE OF THE REGIONS Towards a Thematic Strategy for Soil Protection

[2] S. Sanseviero mathematical statistic models for the analysis of land managementpag 395–407 in in Atti del CongressoNazionaledella MATHESIS (Italian Society of Mathematical and Physical Sciences), Verona, 28/30 novembre, 1996, entitled The foundations of mathematics for its teaching and their links with contemporary society, 1997.

[3] Maturo A., Sanseviero S. A Mathematical Model for a Complex Evaluation and Decision Making Process. In: Franchino R., Maturo A., Ventre A. G. S., Violano A., a cura di. Strategies, processes and decision-making models for environmental management, Edizioni Goliardiche, Trieste, p. 228–237. Naples 2004

[4] Sanseviero S, The representation of the contemporary urban territory, Aracne Ed 2016 ISBN 978–88–548-8926-2

[5] Sanseviero S, The learning society and the new frontiers of knowledge on the web Springer Nature 2020 ISBN 978-3-030-65273-9

[6] J. P. Benzecri & C. L'analyse des donnes. DUNOD 1980 Paris

[7] C Bertuglia, G. Rabino: Model for a district organization Guida Editori, Napoli

[8] F. Forte (edited by) Urban planning methodology, operational research, urban modeling. Napoli Guida copyr. 1972

[9] A. Fadini Introduzione alla teoria degliin siemis focati, 1979 Liguori Editore.

Chapter 13

Machine Learning in Estimating CO_2 Emissions from Electricity Generation

Marco Rao

Abstract

In the last decades, there has been an outstanding rise in the advancement and application of various types of Machine learning (ML) approaches and techniques in the modeling, design and prediction for energy systems. This work presents a simple but significant application of a ML approach, the Support Vector Machine (SVM) to the estimation of CO_2 emission from electricity generation. The CO_2 emission was estimate in a framework of Cost-Effectiveness Analysis between two competing technologies in electricity generation using data for Combined Cycle Gas Turbine Plant (CCGT) provided by IEA for Italy in 2020. Respect to other application of ML techniques, usually developed to address engineering issues in energy generation, this work is intended to provide useful insights in support decision for energy policy.

Keywords: CO_2 emissions, energy systems, machine learning, support vector machines, cost-effectiveness analysis, forecasting

1. Introduction

The science of decision support is foundational for every type of policy, and this work offer a proposal to analyze its role in energy policy.

An example of application of a particular machine learning (ML) technique to an energy policy problem is presented. It is important to understand the role of ML in energy and environmental analysis, for two solid reasons.

The first concerns the need to process large volumes of data and to elaborate and model complex relationships, typical of the energy analysis and of the environmental analysis. In this context, the use of AI (Artificial Intelligence) and machine learning is almost mandatory.

The second concerns the need to a concerted effort to identify how these tools may best be applied to tackle major problems of recent years, like climate change [1]: about this, CO_2 emissions is key variable that we must control to achieve the global objective of mitigating damage for humanity.

This work has a specific goal. Using known tools from the scientific literature on energy generation costs, we intend to show how the use of a machine learning technique (the support vector machines, SVM) can produce a more accurate modeling of these costs.

The link with CO_2 emissions is provided by the possibility of using the cost model in a cost-effectiveness analysis (C-E A), in which the cost is represented by the Levelised Cost of Energy (LCOE) and the effectiveness is represented by the CO_2 emissions of the technologies considered per unit of energy produced.

The CO_2 estimation is then obtained by selecting the best generation options according to the C-E A results.

The meaning of this work is the following.

Imagine that you are an energy analyst, in the public or private sector, and you need to use only one or just few variable/s (such as a forecast on the cost of natural gas), to estimate the costs of an electricity generation technology.

This task can be accomplished using a cost model of electricity generation in which a single piece of information can vary, leaving everything else unchanged (or imposing a certain trend on it).

The metric used is the indicator LCOE (Levelised Cost of Energy) provided by IEA (International Energy Agency), using 2020 data.

Once you have obtained a certain level of accuracy in estimate of energy cost, it is possible to move into a context of cost-effectiveness analysis, in which the best energy option in terms of Incremental Cost-Effectiveness Ratio (ICER) was selected to produce energy and, finally, provide a certain level of CO_2 emissions for the time horizon in which such a technology is still the "best option".

In other words, the estimate of energy cost and the cost-effectiveness analysis, allow us to trace the scenarios for electricity generation mix and, finally, calculate a quantitative forecast of the CO_2 emitted.

The proposed work just intends to show the application of one of the existing machine learning techniques to the estimation of the LCOE, starting from some explanatory variables.

A linear model (LM) and an SVM are compared in the prediction of the LCOE value for a combined cycle gas plant (CCGT) with a focus on the fuel cost, Operation and Maintenance (O&M) cost and CO_2 price using IEA data for Italy in 2020.

The work carried out intends to highlight the possibilities of applying machine learning techniques not only in the purely engineering aspects of energy systems, but also in the statistical-economic ones at a higher level of abstraction.

Some words about why to focus on power generation systems.

As countries work towards a low carbon world, it is crucial that policymakers, modelers, and experts have at their disposal reliable information on the cost of generation.

IEA [2] reports that the levelised costs of electricity generation of low-carbon generation technologies are more and more low the costs of conventional fossil fuel generation. Renewable energy costs continue their descent in recent years and their costs are now competitive with dispatchable fossil fuel-based electricity generation for many countries.

2. Methodology

This section presents the main tools used in this work: the LCOE methodology provided by IEA and the SVM, the used machine learning technique. Just before SVM presentations a very brief remind about ML and its use in energy systems and CO_2 emissions estimates will be provided.

2.1 Levelised cost of energy

The Levelised Cost of Energy (LCOE) is the selected tool to measure the cost of an energy unit produced by the considered technologies. LCOE is a methodology described in the joint report by the International Energy Agency and the OECD

(Organization for Economic Co-operation and Development) Nuclear Energy Agency (NEA) (now at the ninth edition in a series of studies on electricity generating costs) [1]. This report includes cost data on power generation from natural gas, coal, nuclear, and a broad range of renewable technologies.

The metric for plant-level cost chosen is the well-known levelised cost of electricity (LCOE) (IEA are now considering system effects and system costs with the help of the broader value-adjusted LCOE, or Levelised Cost of Value-Adjusted LCOE, VALCOE metric, here not considered).

The LCOE is widely considered as the principal tool for comparing the plant-level unit costs of different base load technologies over their operating lifetimes since indicates the economic costs of a technology family, not the financial costs of a certain projects in a certain market. Due to the equality between discounted average costs and the stable remuneration over lifetime electricity production LCOE recall the costs of electricity production in regulated electricity markets with stable tariffs than to the variable prices in deregulated markets.

Despite many limitations, LCOE has maintained its utility and appeal since it is a uniquely straightforward, transparent, comparable, and well understood metrics remaining a widely used tool for modeling, policy making and public debate.

The calculation of the LCOE is based on the equivalence of the present value of the sum of discounted revenues and the present value of the sum of discounted costs. Another way on the left-hand side one finds the discounted sum of benefits and on the right-hand side the discounted sum of costs:

$$LCOE = P_{MWh} = \frac{\sum (Capital_t + O\&M_t + Fuel_t + Carbon_t + D_t) * (1+r)^{-t}}{\sum MWh * (1+r)^{-t}} \quad (1)$$

where:

P_{MWh}	The constant lifetime remuneration to the supplier for electricity;
MWh	The amount of electricity produced annually in MWh;
$(1+r)^{-t}$	The real discount rate corresponding to the cost of capital;
$Capital_t$	Total capital construction costs in year t;
$O\&M_t$	Operation and maintenance costs in year t;
$Fuel_t$	Fuel costs in year t;
$Carbon_t$	Carbon costs in year t;
D_t	Decommissioning and waste management costs in year t
P_{MWh}	is equal to levelised cost of electricity (LCOE).

Eq. (1) is the formula used here to calculate average lifetime levelized costs based on the costs for investment, operation and maintenance, fuel, carbon emissions and decommissioning and dismantling provided by OECD countries and selected non-member countries.

2.2 Machine learning

Machine learning (ML) is the field of artificial intelligence (AI) that provide methods to learn from data over time creating algorithms not being programmed to do so.

The literature about ML is relatively recent but is so vast that only some hint to review works can be made here, as an access point to this world[1].

Machine learning approaches are normally categorized as in the follows.

---

[1] Here we just remind a recent review of the state of art in machine learning techniques [3].

Supervised machine learning, that trains itself on a labeled data set; **unsupervised machine learning** that uses unlabeled data with algorithms to extract the features required to label, sort, and classify the data in real-time, without human intervention; **semi-supervised learning** (SsL) namely a medium between supervised and unsupervised learning: SsL uses a smaller labeled data set during training and make classification and feature extraction from a larger, unlabeled data set; **reinforcement machine learning** is like supervised learning, but do not requires sample data for training (since using "trial and error" mode).

About the machine learning algorithms for use with labeled data the **regression algorithms** (as linear and logistic regression); **decision trees** (based on a set of decision rules to perform classification); **instance-based algorithms**: it uses classification to estimate how likely a data point is to be a member of one group, or another based on its proximity to other data points.

Methods based for use with on unlabeled data are: **clustering algorithms**: (like K-means, TwoStep, and Kohonen clustering); **association algorithms**: (that find patterns in data by identifying 'if-then' relationships namely association rules); **neural networks**: (that create a layered network of calculations featuring an input layer, when data in; one or more hidden layer, where calculations are performed; and an output layer. Where each conclusion is assigned a probability); **deep neural network** that uses multiple hidden layers, each of which successively refines the results of the previous layer. Deep learning models are typically unsupervised or semi-supervised. Certain types of deep learning models—including convolutional neural networks (CNNs) and recurrent neural networks (RNNs)—are driving progress in areas such as computer vision, natural language processing (including speech recognition), and self-driving cars.

In this work, the machine learning approach used is the SVM one.

SVMs[2] are machine learning algorithms built on statistical learning theory for structural risk minimization. In pattern recognition, classification, and analysis of regression, SVMs outperform other methodologies. The significant range of SVM applications in the field of load forecasting is due to its ability to generalize (also, local minima lead to no problems in SVM).

SVM was chosen, in this work, for the sake of simplicity, since the performed Support Vector Regression (SVR) [5], extremely easy to understand in comparing a traditional statistical tool with a competing machine learning based one.

Often, the available applications of SVM in the energy sector are oriented on the engineering side[3] while in this work the approach is oriented in support decisions for energy policy field.

Using one of the possibilities offered by SVMs, namely the SVR, the follows show how it is possible to obtain more accurate forecasts of costs per unit of energy produced, using LCOE as a metric.

The best available accuracy is then used in a context of cost-effectiveness analysis.

In the following, a method to select among competing options (options that can be differ even for slight changes in some significant LCOE parameters), the one characterized by the best Incremental Cost-Effectiveness Ratio (ICER) is presented.

The possibility of making this choice during the lifetime of the plant leads to the possibility of identifying the best technology available, year by year, to get the corresponding profile of the associated CO_2 emissions.

---

[2] For a good introduction to this topic see [4].
[3] See, for example [6].

2.2.1 Machine learning for energy systems and CO$_2$ emission estimation

The growing utilization of data collectors in energy systems has resulted in a massive amount of data accumulated (an increasing mass of mart sensors are now extensively used in energy production and energy consumption) leading to a continuous production of big data and, consequently, to a massive number of opportunities and challenges in decision support science.

Today, ML models in energy systems are essential for predictive modeling of production, consumption, and demand analysis due to their accuracy, efficacy, and speed or to provide an understanding on energy system functionality in the context of complex human interactions.

Salimi et al. [7] propose a comprehensive review of essential ML to present the state of the art of ML models in energy systems and discuss their likely future trends.

Machine learning was used for estimate CO$_2$ emission from energy systems in several context, using different approach. It is possible to recall, among an increasing number of works in recent years:

Leerbeck et al. [8] about flexibility of the electricity demand, a machine learning algorithm developed to forecast the CO$_2$ emission intensities in European electrical power grids distinguishing between average and marginal emissions in Danish bidding zone DK2;

Magazzino et al. [9] an investigation on the causal relationship among solar and wind energy production, coal consumption, economic growth, and CO$_2$ emissions for these three countries;

Cogoljević et al. [10] on the linkage between energy resources and economic development the focus of that work is to develop and apply the machine learning approach to predict gross domestic product (GDP) based on the mix of energy resources with a higher predictive accuracy;

Wu et al. [11] about proposing a standardized framework for estimating the indirect building carbon emissions within the boundaries of various types of Local Climate Zones (LCZs using a random forest machine learning method);

Mele and Magazzino [12] on the relationship among iron and steel industries, air pollution and economic growth in China (using a Long Short Term Memory, LSTM, approach);

Li et al. [13] on the forecasting of energy consumption related carbon emissions for the Beijing-Tianjin-Hebei region.

Huang et al. [14] on the uses of gray relational analysis to identify the factors that have a strong correlation with carbon emissions for China to reduce carbon emissions by studying prediction of carbon emissions (using LSTM).

Csillik and Asner [15] on the creation of an automated, high-resolution forest carbon emission monitoring system that will track near real-time changes and will support actions to reduce the environmental impacts of gold mining and other destructive forest activities for the Peruvian Amazon (using deep learning models).

Csillik et al. [16] on the use of a random forest machine learning regression workflow to map country of Peru by combining 6.7 million hectares of airborne LiDAR measurements of top-of-canopy height with thousands of Planet Dove satellite images into, to create a cost-effective and spatially explicit indicators of aboveground carbon stocks and emissions for tropical countries as a transformative tool to quantify the climate change mitigation services that forests provide.

Niu et al. [17] to determine whether China can achieve the commitment of reducing carbon emission intensity in 2030, through a general regression neural network (GRNN) forecasting model based on improved fireworks algorithm (IFWA) optimization is constructed to forecast total carbon emissions (TCE) and carbon emissions intensity (CEI) in 2016–2040.

2.3 Our methodology

The present work reports an experiment performed using a simple LCOE model, built according to basic methodology proposed by IEA. The performed experiment is simple and straightforward. Two energy scenarios were produced, one based on a certain hypothesis of change in the fuel cost, the other based on a hypothesis of change in fuel cost, O&M cost, and CO_2 price, for the CCGT type plant, over a period of 30 years.

In each scenario, a certain LCOE profile is obtained for the time horizon considered. A simple regression analysis is then performed on this variable, using as explanatory variables, first the cost of fuel, and then the operating costs.

The analysis is carried out both using a LM and the SVM, with further manual tuning of the last to improve its performance. The manual tuning for SVR was used for the sake of simplicity since the main goal of the study is to suggest the application of this ML technique to gain forecasting accuracy to use in the following phase, the cost-effectiveness analysis.[4]

To evaluate the accuracy of the forecast, the Root Mean Square Error (RMSE), the Mean Average Error (MAE) and the Mean Average Percentage Error (MAPE) were used.[5]

This simple test was performed to show the accuracy of the fuel cost and O&M cost as a predictor of CCGT LCOE.

Once established the best technique, the data from the two scenarios in a third scenario are modified, under certain hypothesis explained in the follows, to made a C-E A between a technology represented by IEA data and another of the same type with little changes in O&M costs. Using ICER as a winning criterion, it is possible to select the best energy generation option and, finally, to trace the corresponding CO_2 emission estimate trend over the plant's lifetime.

All the data coming from IEA [2].

The LCOE model.

First, a LCOE model based on IEA Eq. (1), with the following level of detail, was built.

The basic relationships of the model are:

$$PF = Power * 8760 * AVLF * \frac{AAF}{100} * (1 - AuxP) \tag{2}$$

$$ws = 1 - wd \tag{3}$$

$$ks = krft + EMRP * B \tag{4}$$

$$i = wd * kd + ws * ks \tag{5}$$

$$d = i/(1+i) \tag{6}$$

$$dfi = \sum_j 1/(1+i)^j \tag{7}$$

$$icfinal = icfinal + \left(\frac{ic}{CnsT}\right) * dfi \tag{8}$$

$$df = \sum_j 1/(1+d)^j \tag{9}$$

---

[4] Indeed, manual tuning is often considered as one of the most significant choice [18].
[5] See [19] for a complete discussion about the used metrics.

$$\text{icfinal} = \text{icfinal} + \left(\frac{\text{ic}}{\text{CnsT}}\right) * \text{dfi1} \tag{10}$$

$$\text{dfi} = \sum_j 1/(1+i)^j \tag{11}$$

$$Pro = Pro + PF * df \tag{12}$$

$$OM = ((FOM + VOM) * PF) * df \tag{13}$$

$$Fue = ((CFue) * PF) * df \tag{14}$$

$$CO2 = ((PCO2) * PF) * df \tag{15}$$

$$Cost = \sum_j (OM + Fue + CO2) \tag{16}$$

$$Decom = n * Decom * Pro \tag{17}$$

$$LCOE = (Power * \text{icfinal} * 1000 + Cost + Decom)/Pro \tag{18}$$

Where:

CC	Cost of Capital (USD/MWh)
Power	net capacity (MWe)
AVLFmin	AVerage Load Factor min value (%)
AVLFmax	AVerage Load Factor max value (%)
AAF	Average Availability Factor (%)
AuxP	Auxiliary Power (%)
Lifetime	Time horizon of plant (years).
wdmin	min weight of cost of debt on total cost (%)
wdmax	max weight of cost of debt on total cost (%)
kdmin	min value of debt rate (%)
kdmax	max value of debt rate (%)
tmin	min value of taxation (%)
tmax	max value of taxation (%)
krftmin	min value of free risk rate (%)
krftmax	max value of free risk rate (%)
EMRPmin	min value of Expected Market Risk Premium (%)
EMRPmax	max value of Expected Market Risk Premium (%)
Bmin	min value of Beta (%)
Bmax	max value of Beta (%)
CnsTmin	min value of Construction Time (years)
CnsTmax	max value of Construction Time (years)
FOMmin	Fixed Operation and Maintenance Costs min (USD*MWh)
FOMmax	Fixed Operation and Maintenance Costs max (USD*MWh)
VOMmin	Variable Operation and Maintenance Costs min (USD*MWh)
VOMmax	Variable Operation and Maintenance Costs max (USD*MWh)
Cfuemin	min value of Costs of Fuel (USD*MWh)
Cfuemax	max value of Costs of Fuel (USD*MWh)
Effmin	min value of Efficiency (%)
Effmax	max value of Efficiency (%)
PCO₂min	min value of CO₂ price (USD*MWh)
PCO₂max	max value of CO₂ price (USD*MWh)
Decommin	min value of Decommissioning (USD*MWh)
Decommax	max value of Decommissioning (USD*MWh)

All other parameters are settled using the IEA values.

	Fuel Cost (baseline 45.5 USD/MWh)	O&MCost (baseline: 6.99 USD/MWh)	CO₂ price (10.1 USD/MWh)
Scenario 1	Linear decreasing of 2% per year except every 6 years	constant	constant
Scenario 2	Linear decreasing of 2% per year except every 6 years	Linear decreasing of 2% per year except every 6 years	Linear decreasing of 2% per year except every 6 years

Table 1.
Scenarios used for the regression of LCOE on fuel cost and O&M cost

We have set two type of scenario, basing on the following assumptions about certain variables of the model. The basic hypothesis is a constant decreasing of 2% for every variable changed, except every 6 years (a totally arbitrary choice), simulating an increasing amplification of this cycle (every 6 years, the percentage variation of the cost respect to the previous value is double than it and then is multiplied for the number of the occurring, so the first time at year 6, this value is roughly 4, namely 2% multiplied by 2 and then multiplied per variation 1).
Table 1 describes the hypothesis used in this first step of the analysis.

3. Results

Figure 1 shows the results obtained by performing a SVR about the data from IEA [1] for the first scenario considered (**Figure 2**).
The values of RMSE for the Linear Model (LM), the SVM Model Before Tuning (SVMBT) and the SVM Model After Tuning (SVMAT) are:

	RMSE	MAE	MAPE
Linear Model	1,30E-14	8,39E-15	8,39E-17
SVM	5,25E-01	4,01E-01	4,01E-03
Tuned SVM	1,74E-03	1,54E-03	1,54E-05

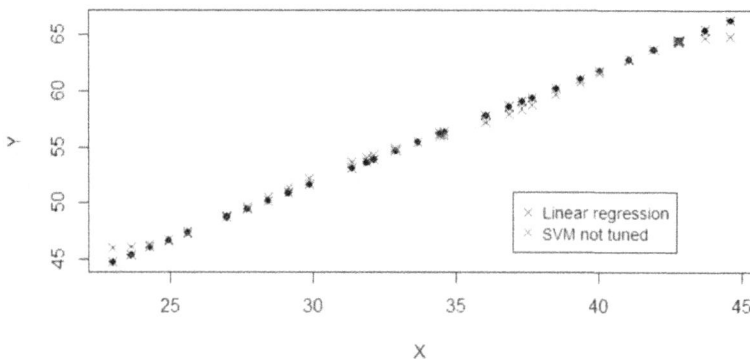

Figure 1.
Comparison between LM and SVMBT in predicting LCOE of CCGT technology for Italy (simulating data over lifetime of the plant - base data: Italy, 2020 - sources: IEA) - scenario 1 - Y = LCOE (USD/MWh), X = fuel cost (USD/MWh).

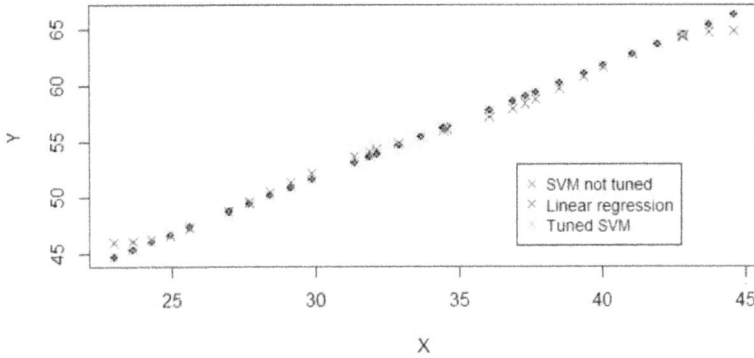

Figure 2.
Comparison between LM and SVMAT in predicting LCOE of CCGT technology for Italy after tuning (simulating data over lifetime of the plant - base data: Italy, 2020 - sources: IEA) - scenario 1 - Y = LCOE (USD/MWh), X = fuel cost (USD/MWh).

with a clear improvement of performance of the SVM after tuning. The linear model since the strong relationships between the fuel cost and the LCOE is clearly preferable respect to the SVM (**Figures 1–4**).

The values of RMSE for the Linear Model (LM), the SVM Model Before Tuning (SVMBT) and the SVM Model After Tuning (SVMAT) are:

	RMSE	MAE	MAPE
Linear Model	3.87E+00	2.70E+00	2.70E-02
SVM	2.77E+00	1.59E+00	1.59E-02
Tuned SVM	2.61E+00	1.45E+00	1.45E-02

Recalling that in the second case the O&M cost was used as a predictor, we can more appreciate the gain in terms of RMSE obtained by using the SVM.

The increasing accuracy of the SVR respect to the LM, can be used to perform a CO_2 emission estimation in a cost-effectiveness analysis.

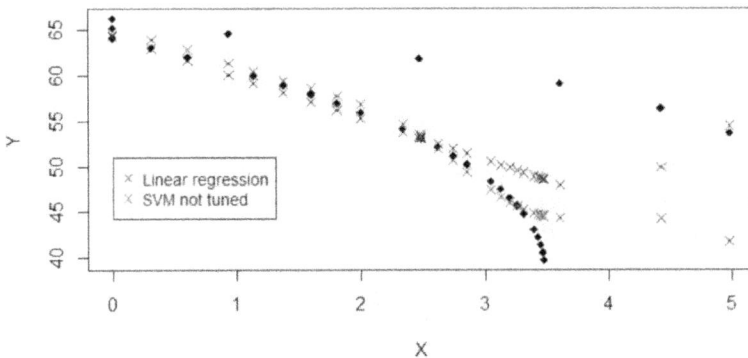

Figure 3.
Comparison between LM and SVMBT in predicting LCOE of CCGT technology for Italy (simulating data over lifetime of the plant - base data: Italy, 2020 - sources: IEA) - scenario 2 - Y = LCOE, X = O&M Cost.

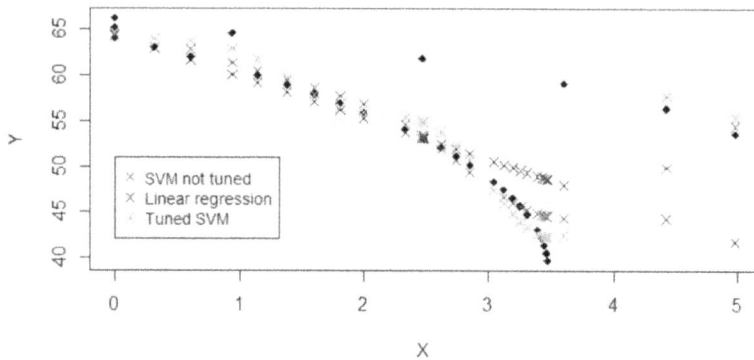

Figure 4.
Comparison between LM and SVMAT in predicting LCOE of CCGT technology for Italy after tuning (simulating data over lifetime of the plant - base data: Italy, 2020 - sources: IEA) - scenario 2 - Y = LCOE, X = O&M Cost.

Let us look at a simple and plain experiment based on IEA data [2] for Italy, 2020 in the following scenario:

	Fuel Cost (baseline 45.5 USD/MWh)	O&MCost (baseline: 6.99 USD/MWh)	CO_2 price (10.1 USD/MWh)
Scenario 3	Decreasing of 15% at 15th year then linear decreasing of 1% until rest of the lifetime.	Decreasing of 15% at 15th year then linear decreasing of 1% until rest of the lifetime.	Decreasing of 15% at 15th year then linear decreasing of 1% until rest of the lifetime.

In scenario 3 we made a simulation basing on the hypothesis of a sudden shock for the three variables above reported in the 15th year, immediately followed by a linear decrease of them until end of the lifetime, starting from IEA 2020 data as a baseline value.

For scenario 3 the errors in predicting LCOE using O&M Cost over the considered time horizon are:

	RMSE	MAE	MAPE
Linear Model	4.25878	3.49147	0.03491
SVM	2.70117	1.52912	0.01529
Tuned SVM	2.58541	1.52378	0.01524

In Cost-Effectiveness Analysis it is possible to calculate the Incremental Cost-Effectiveness Ratio (ICER), used as a measure of cost the LCOE and used as a measure of effectiveness through the quantity of CO_2 emitted. The ICER can be used as a selection criterion between different options then, the winning options will be producing a certain level of emissions.

Now, let us imagine comparing two types of plants of the same technological family, in this case the CCGT. In this hypothetical exercise, the second type of plant is characterized by higher operating costs (+5% of the IEA base value).

In addition to this, let us imagine that the second type of plant has an average load factor of 94%.

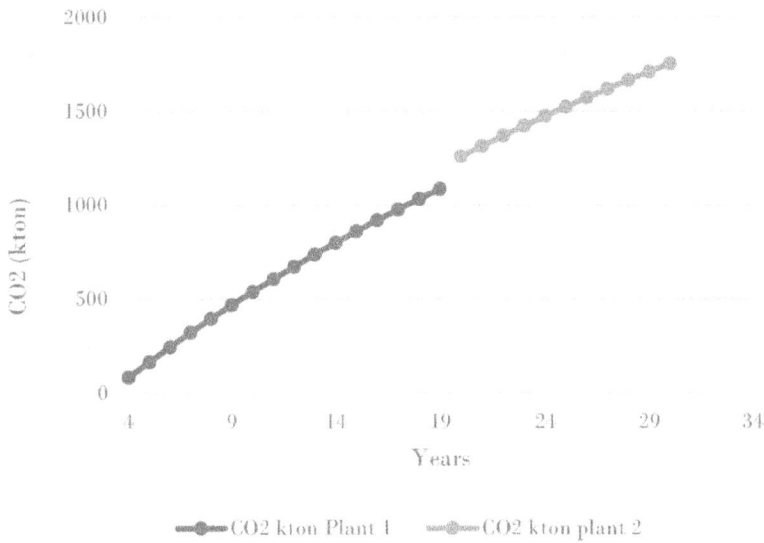

Figure 5.
CO_2 emissions from different kind of CCGT plants in scenario 3 (sources: IEA, 2020 + imaginary data).

Now, let us repeat the simulation performed for scenario 3 for the first type of CCGT plant (the real one), but only from the 20th year.

The meaning of this operation is as follows:

- to use systems with different characteristics (in this case we have changed the O&M costs and the load factor of a single technology family);

- to calculate the ICER corresponding to each plant in a defined time interval (in this case, from when the LCOE starts to vary);

- to calculate the degree of uncertainty on the value of the ICER thanks to the MAPE of the SVR, defining the variation range for the ICER[6];

- to select the technology that has the lowest ICER and then we calculate the corresponding emissions over the time horizon considered;

- finally, to calculate the emissions profile corresponding to the winning technology, year by year.

The results are shown in **Figure 5**.

Figure 5 illustrates what happens using the ICER criterion as a selector of the winning generation option. For the first 20 years, the first type of installation is selected, and the corresponding emissions are those of the blue line. From 20 years of age onwards, using the ICER as a criterion means choosing the second type of plant and the curve that shows the new profile of the emissions is the orange one.

4. Conclusions

ML can help in providing accurate forecasts of CO_2 emissions from power generation, especially when we face simultaneous variation of major driver (like

---

[6] Namely, ICER max/min = ICER +/− ICER*MAPE.

fuel cost, operating cost of the plant and so on); only a little piece of the possible comparisons between traditional techniques and a particular ML method was shown, focusing on the better performance of the ML one (SVM) respect to the traditional one (the LM).

In our case, the performed step was:

1. improving LCOE forecasting performance,

2. comparing multiple competing options by use of the ICER in Cost-Effectiveness Analysis;

3. consider the uncertainty about ICER using the MAPE (in this case, but is just an option) calculated by SVM;

4. choosing the best technology and calculating the CO_2 emissions for it;

5. defining the trend of the CO_2 emissions in the lifetime of the plant by step 4.

Recalling that a basic LCOE model can be brought to a great level of granularity, it is easy to imagine how this type of analysis could gain in depth and significance if the required data are available. Indeed, also in case of missing data, significant simulation can be provided by using each available piece of information on energy costs.

The experiment performed was conducted at the highest level of simplicity to better focus on the reasons that suggest ML integration not only about the engineering features of electricity generation field but also in support decision tools about energy policy.

Conflict of interest

The authors declare no conflict of interest.

Author details

Marco Rao
ENEA, Italian National Agency for New Technologies, Energy and Sustainable Economic Development, Rome, Italy

*Address all correspondence to: marco.rao@enea.it

References

[1] Rolnick, D. and Donti, P. and H. Kaack, L.H., and Kochanski, K. and Lacoste, A. and Sankaran, K. and Ross, A. and Milojevic-Dupont, N., and Jaques, N., and Waldman-Brown, A., and Luccioni, A., and Maharaj, T., and Sherwin, E., and Mukkavilli, S.K., and Konrad P. K, and Carla Gomes, K., and Ng, A., and Hassabis, D., and C. Platt, J. C., and Creutzig, F., and Chayes, J. and Bengio, Y., Tackling Climate Change with Machine Learning, 2019

[2] IEA, Projected Cost of Generating Electricity 2020, IEA, 2020

[3] Saravanan and Sujatha, P., "A State of Art Techniques on Machine Learning Algorithms: A Perspective of Supervised Learning Approaches in Data Classification," *2018 Second International Conference on Intelligent Computing and Control Systems (ICICCS)*, Madurai, India, 2018, pp. 945–949, doi: 10.1109/ICCONS.2018.8663155.

[4] Wang, L. Support vector machines: theory and applications, Springer-Verlag, Berlin, 2005

[5] Awad M., Khanna R. (2015) Support Vector Regression. In: Efficient Learning Machines. Apress, Berkeley, CA. https://doi.org/10.1007/978-1-4302-5990-9_4

[6] Zendehboudi, A., Baseer, M.A., Saidur, R. Application of support vector machine models for forecasting solar and wind energy resources: A review, Journal of Cleaner Production, Volume 199, 2018, Pages 272-285, DOI: 10.1016/j.jclepro.2018.07.164

[7] Salimi, M., Mosavi, A., Faizollahzadeh, A.S., Amidpour, M., Rabczuk, T., and Shamshirband, S., State of the Art of Machine Learning Models in Energy Systems, a Systematic Review, Energies 2019, 12, 1301; doi: 10.3390/en12071301

[8] Leerbeck, K., Bacher, P., Grønborg Junker, R., Goranović, G., Corradi, O., Ebrahimy, R., Tveit, A., Madsen, H., Short-term forecasting of CO$_2$ emission intensity in power grids by machine learning, Applied Energy, Volume 277, 2020, DOI:10.1016/j.apenergy.2020.115527

[9] Magazzino, C., Mele, M., Schneider, N. A machine learning approach on the relationship among solar and wind energy production, coal consumption, GDP, and CO$_2$ emissions, Renewable Energy, Volume 167, 2021, Pages 99–115 DOI: 10.1016/j.renene.2020.11.050.

[10] Cogoljević, D., Alizamir, M., Piljan, I., Piljan, T., Prljić, K., Zimonjić, S. A machine learning approach for predicting the relationship between energy resources and economic development, Physica A: Statistical Mechanics and its Applications, Volume 495, 2018, Pages 211-214, DOI: 10.1016/j.physa.2017.12.082.

[11] Wu, Y., Sharifi, A., Yang, P., Borjigin, H., Murakami, D., Yamagata, Y. Mapping building carbon emissions within local climate zones in Shanghai, Energy Procedia, Volume 152, 2018, Pages 815-822, DOI: 10.1016/j.egypro.2018.09.195.

[12] Mele, M., Magazzino, C. A Machine Learning analysis of the relationship among iron and steel industries, air pollution, and economic growth in China, Journal of Cleaner Production, Volume 277, 2020, 123293, DOI: 10.1016/j.jclepro.2020.123293.

[13] Li, M.; Wang, W.; De, G.; Ji, X.; Tan, Z. Forecasting Carbon Emissions Related to Energy Consumption in Beijing-Tianjin-Hebei Region Based on Grey Prediction Theory and Extreme Learning Machine Optimized by Support Vector Machine Algorithm.

Energies 2018, *11*, 2475. https://doi.org/
10.3390/en11092475

[14] Huang, Y., Shen, L., Liu, H. Grey
relational analysis, principal component
analysis and forecasting of carbon
emissions based on long short-term
memory in China, Journal of Cleaner
Production, Volume 209, 2019, Pages
415-423, DOI: 10.1016/j.
jclepro.2018.10.128.

[15] Csillik, O. and Asner, G.P. 2020
Environ. Res. Lett. **15** 014006

[16] Csillik, O., Kumar, P., Mascaro, J.
et al. Monitoring tropical forest carbon
stocks and emissions using Planet
satellite data. Sci Rep **9**, 17831 (2019).
https://doi.org/10.1038/s41598-019-
54386-6

[17] Niu, D. Wang, K., Wu, J., Sun, L., Yi
Liang, Y., Xu, X., Yang, X. Can China
achieve its 2030 carbon emissions
commitment? Scenario analysis based
on an improved general regression
neural network, Journal of Cleaner
Production, Volume 243, 2020, 118558,
DOI: 10.1016/j.jclepro.2019.118558.

[18] Korovkinas, K., Danènas, P.,
Garsva, G., Support vector machine
parameter tuning based on particle
swarm optimization metaheuristic,
Nonlinear Analysis: Modelling and
Control, Vol. 25, No. 2, 266–281, DOI=
10.15388/namc.2020.25.16517

[19] Botchkarev, A. A New Typology
Design of Performance Metrics to
Measure Errors in Machine Learning
Regression Algorithms,
Interdisciplinary Journal of Information,
Knowledge, and Management,
Volume 14, pp. 045–076, 2019, DOI=
10.28945/4184.

Chapter 14

Coastal Water Quality: Hydrometeorological Impact of River Overflow and High-resolution Mapping from Sentinel-2 Satellite

Annalina Lombardi, Maria Paola Manzi, Federica Di Giacinto, Valentina Colaiuda, Barbara Tomassetti, Mario Papa, Carla Ippoliti, Carla Giansante, Nicola Ferri and Frank Silvio Marzano

Abstract

The increase of human settlements and activities in coastal areas is causing a significant impact on coastal water quality. Predicting and monitoring the latter is of fundamental importance for assessing sustainable coastal engineering and ecosystem health. This trend is strongly influenced by the presence of rivers' mouths, acting as critical links between inland and sea. Forecasting river discharges and overflows, using hydrometeorological modelling, can provide a quantitative estimate of the excessive supply of sea nutrients, favouring algal proliferation and eutrophication phenomena. The river overflow contributes to the increase of the coastal bacterial concentration, contaminating marine bioindicators, such as bivalve molluscs. Coastal water status can be monitored by satellite high-resolution optical spectroradiometers, such as Sentinel-2 constellation, capable to retrieve Chlorophyll-a concentration as well as total suspended sediments, at the resolution of about 10 meters. This remote mapping is complementary to *in situ* samplings, both essential for supporting decisions on the management of coastal mollusc farming and fishing. In this work, we report the recent advancements in hydrological model-based prediction of river surges and remote sensing techniques exploiting Sentinel-2 imagery as well as their implications on coastal water quality management. As a pilot area, we select the central Adriatic Sea in the Mediterranean basin and the Abruzzo region coastline in Italy.

Keywords: sustainable coastal engineering, ecosystem health, remote sensing techniques, hydrometeorological modelling, faecal indicator organism, bivalve, Sentinel-2

1. Introduction

In recent decades, the increase of human settlements and activities in coastal areas is causing a significant impact on the resilience of the world's coastal and marine natural capital. The coastal environment is a dynamic ecosystem where natural and anthropogenic processes add up and interact, modifying their geomorphological, physical, and biological characteristics. Coastal areas are also defined as ecotones, which are very important from an ecological point of view as they are a natural transition zone between two different and adjacent ecological systems.

The human pressures are different and include climate change, overfishing, offshore commerce, and land-based activities. The several pressures on the coastal ecosystem and the possible overlapping pressures can cause cumulative adverse effects [1–3]. Land-based stressors link coastal marine systems to terrestrial human activities and represent dominant stressors in coastal ecosystems [1, 4–6]. Nutrient and chemical pollution run-off create coastal eutrophication, harmful algae blooms, or hypoxic or anoxic dead zones [4, 6–8], and these impacts are able, not only to harm coastal species and ecosystems [9–11] but also affect human health [12, 13] and economic activities.

A few research is available, where the impacts of human wastewater on coastal ecosystems and community health [14–17] are assessed. The combined effects from multiple pressures are not still considered in management or planning processes and this reduces the overall resilience of marine ecosystems.

According to the Water Framework Directive [18], 93% of the European marine area is under different pressures from human activities and about 28% of its coastline is affected by pressures causing changes in hydrographic conditions, for example, in seawater movement, temperature, and salinity. According to the hydromorphological pressure assessments made in coastal waters, the main sources are atmospheric deposition and discharges from urban wastewater treatment plants on the coast, or further in the catchment area [14, 19–21].

Coastal developments modify natural hydrological conditions and impact habitats where the pressure at the catchment scale is the highest on the coastline of the Mediterranean Sea. Intense human activities in regions surrounding enclosed and semi-enclosed seas, such as the Mediterranean, always produce, a strong environmental impact causing increasing coastal and marine degradation, in the long term. The sustainable development in the Mediterranean area is influenced by diverse factors, such as i) the rapid growth of the urbanisation rate; ii) the increase in tourism; iii) the rapid development that determines the degradation of coastal areas; iv) water scarcity; and v) commercial activities. This condition highlights the need to define mitigation strategies, using timely and action-oriented information.

Due to its morphology, the Italian Peninsula can be divided into two main basins that can be considered semi-enclosed. The first includes the western Mediterranean, limited eastward by the Sicilian channel, and characterised by wide abyssal plains. The second, the eastern Mediterranean, dominated by the Mediterranean ridge system, is characterised by more complex morphology. In Italy, populated areas are mainly concentrated along with coastal areas than the rest of the territory; according to the Corine Land cover data [22], the Italian coast has a length of about 8300 km: more than the 9% of the littoral is now artificially bordered by works grazing the shore (3.7%), ports (3%) and partially superimposed structures on the coast (2.4%). The artificialisation of housing and transport structures in coastal areas is gradually increasing. It has been estimated that a relative increase of the 5% in the area 10 km away from the shore was generally recorded in European countries between 2000 and 2006 [23].

The assessment of the sea and coastal systems and their interaction, based on scientific knowledge, are the indispensable basis for the management of human activities, in view of promoting the sustainable use of the seas and coasts and conserving marine ecosystems and their sustainable development.

In 1975, 16 Mediterranean states and the European Community under the auspices of the United Nations Environment Programme (UNEP) defined the Action Plan for the Mediterranean (MAP) [24, 25], aimed at protecting the environment and promoting sustainable development in the Mediterranean basin. The 19th Meeting of the Contracting Parties in 2016 agreed on the Integrated Monitoring and Evaluation Programme of the Mediterranean Sea and Coast and related evaluation criteria (IMAP) [26] which establishes the principles of integrated monitoring: for the first time, biodiversity and non-native species, pollution and marine, coastal, and hydrographic litter will be considered in an integrated way. The IMAP implementation defines 27 common indicators, foreseen in the Integrated Monitoring and Evaluation Programme, in line with the UNEP/MAP Barcelona Convention. The prediction and monitoring of water quality are among the main activities to be carried out for the protection of coastal ecosystems.

As for the prediction, water quality is strongly influenced by atmospheric events that could affect the pollution management systems, such as rainfall-dependent sewage drains and tributary river flow. For this reason, river mouths act as critical links between the hinterland and the sea. The prediction of river discharges and overflows using hydrometeorological models can be fundamental for indirect estimation of water quality, given that the drainage network runoff is closely related to the supply of marine nutrients, favouring algal proliferation and eutrophication phenomena. It also contributes to the increase in the concentration of faecal bacteria, such as *Escherichia coli* recognised as a faecal indicator organism in the European legislation, which contaminates marine bioindicators, such as bivalve molluscs. This condition has a relevant socio-economic impact; for example, it limits the consumption of bivalve molluscs collected from contaminated waters. The reductions in water quality after high precipitation events and the subsequent increase in river discharges lead local authorities to close shellfish harvesting areas after large events. But the inability of local authorities to accurately predict these events or to immediately assess the water quality exacerbates the losses of fishing economies.

From the above premises, it is clear that knowing the ecological status of water bodies is of critical importance to monitor how human activities are impacting or, the other way around, impacting by the coastal ecosystem. Monitoring coastal waters, indeed, is fundamental for both the evaluation of ecosystem health and as a support to local fishing economies, in terms of sustainability and site selection. Member States of the European Union are required, by the EU Marine Strategy Framework Directive [6] and the Water Framework Directive [18], to preserve territorial waters within the first nautical mile and achieve good ecological status. According to the European Environment Agency water assessment [20], only 46% of water bodies are actively monitored, 23% of monitoring did not include *in situ* water sampling and 4% still had unknown ecological status [27]. The proportion of water bodies without observation data is much larger than the ones for which monitoring is granted and for those monitored, in most cases, the status of surface waters was not classified as "good" [28]. Satellite observations offer a solution to the current limits shown by conventional water sampling methods—they allow to achieve much wider spatial and temporal coverage and larger water bodies. Remote mapping, therefore, is complementary to *in situ* sampling, both essential for supporting decisions on coastal aquaculture operations. It can also provide support in quantifying elements of environmental status that are currently not reported, such as phytoplankton blooms. In this context, the European Union together with the European Space Agency has boosted the development of the most advanced satellite-based instruments to observe optical water quality. Through the Copernicus framework, the spatial sector has had significant investment in recent years, and this enhances the cost-benefit of using satellite-based technologies for monitoring surface waters, also being satellite data freely available.

As for the environmental surveillance, coastal water status can be monitored by satellite high-resolution optical spectroradiometers, capable to retrieve suspended sediments or algae presence, at spatial resolutions of up to 10 meters [23, 29].

2. Hydrometeorological impact of river mouth on coastal water quality

Most human activities are related to water and take place in coastal areas or along riverbanks. The correct management of these areas is therefore primarily referred to as the basin-scale [18] where the territorial planning activities need to consider the aggregate effects of hydrographic changes caused by human activities at sea and on land [18, 30–32].

The Marine Strategy Framework Directive [6] highlights the importance of the assessment of the hydrographical conditions through seawater physical-chemical parameters, that is, temperature, salinity, depth, currents, waves, turbulence, turbidity (from a load of suspended particulate matter), upwelling, wave exposure, mixing characteristics, residence times, the spatial and temporal distribution of nutrients, oxygen, and acidification [30, 33–35]. All these variables are essential for understanding the dynamics of marine ecosystems that can be altered by anthropic presence.

Hydrographical conditions are site-specific and depend on landscape features, morphology, and lithology and are often conditioned by large-scale forcings, such as tide, general ocean circulation and climate. Small-scale features, such as land use and human-induced pressures, are also relevant to the river dynamics, especially in coastal areas.

The offshore waters of the Mediterranean Sea are extremely oligotrophic, and the coastal areas have been historically known to be influenced by natural and anthropogenic inputs of nutrients, mainly concentrated in the Adriatic basin [36].

Monitoring of contamination in different mollusc species is a well-known methodology, applied to assess the level of sea-water contamination, which exploits the bivalve capability to accumulate and retain contaminants. Moreover, mollusc edible species contamination is constantly monitored to ensure the introduction of safe products in the food market.

The European Regulation No 627/2019 [37] assesses the official control programmes for bivalve molluscs and provides the classification of mollusc production areas based on microbiological monitoring for the bacterium *E. coli* in the mollusc flesh and intervalvular liquid, used as a faecal indicator organism (FIO). Since rivers are routes for the transfer of organic matter including faecal bacteria from inland to the sea, the contamination is strongly affected by the inland drainage network, which collects also the most important abiotic factors affecting bacterial contamination of molluscs [38].

Even if the influence of FIO on the quality of coastal waters is studied since the XVIII century, few studies exist that attempt to evaluate the relationship between fluvial transport and shellfish hygiene in the sea [39, 40].

The land-sea-river system is extremely complex; therefore, it is not straightforward to establish a relationship between runoff, precipitation, and contamination levels. Connections are site-specific and dependent on the physiographical characteristics of each catchment and, in addition to the existing, local human pressures. Nevertheless, contamination decay is also connected to local environmental parameters which affect the bacterial dilution and the self-purification process of the bivalves.

The analysis of mollusc contamination has a socio-economic value; therefore, it is doubly important to evaluate in terms of both monitoring contamination levels and attempting to forecast possible pollution events. Nevertheless, different competencies may be required to achieve this purpose; on one hand, prediction of environmental processes requires deep knowledge of earth system modelling and data interpretation. On the other hand, environmental implications on food security monitoring are requested to adequately assess the design of useful tools or instruments able to really ameliorate the bivalves' productions. A wide number of capabilities should be connected to work together to find ICT solutions: biologists, physicists, engineers, and economists, for example. A virtuous example of this collaboration was the CapRadNet project (http://cetemps.aquila.infn.it/capradnet/), which originated a fruitful collaboration between different institutional levels and different actors usually involved in diverse activities. The outcome of the presented work is born from the project collaboration. The main aim of the present research was to investigate the relationships between two main environmental variables (flow discharge and precipitation) and the contamination level of molluscs harvested areas in a target site. The feasibility study resulted from the capitalisation of other two previous projects, funded by the IPA-Adriatic CBC Programme. The proposed analysis has taken advantage of *E. coli* concentration data analysed in the framework of the CAPS2 project (www.caps2.eu) and the hydrological modelling system developed in the same area to predict possible flood events due to severe meteorological events, as an outcome of the AdriaRadNet project (http://cetemps. aquila.infn.it/adriaradnet/).

The good practices defined in the CapRadNet project are being tested in a new project financed at the regional level, which intends to create the Early Warning System as a final product to improve the economic and production efficiency of the plants through the environmental information made available to aquaculture producers, which will be described below.

2.1 Early Warning System for sanitary risk: Hydrometeorological operational forecasted tool

The outcome of the feasibility study carried out during the CapRadNet project [41] had shown that an Early Warning System (EWS) for the sanitary risk assessment could be set up operationally, given the existing relationship between discharge overflow and *E. coli* concentration increases in bivalves' harvested area, in the Pescara River mouth. In general, rainfall is the most referred environmental factor influencing microbial contamination in coasts and estuaries affected by stormwater runoff and long sea and shoreline outfalls [42–44]. Rainfall-induced contamination could persist in molluscs as much as 6 days after the rainfall event [45, 46], even if a shorter lag time (<3 days) has been reported in the literature [43, 47]. The differences are often associated with the hydrogeology of the catchment (i.e., concentration time) and residence times in the receiving water [46]. Being recognised as an important predictor of microbial contamination of bivalve harvesting areas [48], river flows are explicitly mentioned in the EU legislation on official controls of bivalve molluscs intended for human consumption. In the study proposed by Campos et al. [46], the highest levels of *E. coli* were detected when total rainfall exceeded 2 mm and water levels in the main tributaries exceeded the mean flow.

Few published studies have considered the contribution of river flows in informing official public health controls for bivalve mollusc fisheries [48, 49], but the scientific literature has shown that catchment-scale microbial dynamics determining bathing water compliance are often determined by hydrological events [50–52].

2.2 Pilot Study

A numerical experiment was carried out, using 6 months of *E. coli* concentration data in the mollusc flesh and intravalvular liquid, detected in three pilot areas around the Pescara River mouth (**Figure 1**). Since official discharge data from hydrological annals are available until 2010, for the same 6-month period, a hydrological simulation was performed by using the CHyM distributed hydrological model [53–55], forced with observed rainfall data. The model has been extensively used in Abruzzo Region for flood forecast activities [56, 57]. The hydrometeorological conditions preceding each *E. coli* concentration exceedance were investigated, in terms of accumulated rainfall in the coastal area and runoff in the Pescara River mouth. A quick overview of obtained results is here given and discussed; more detailed information is available in Colaiuda et al. [41].

The analysed catchment originates in the inner, northern part of the Abruzzo region, draining an area of about 3147 km^2 before flowing into the Adriatic Sea. It is characterised by a very complex orography with altitudes spanning from zero up to almost 3000 m a.s.l. in the range of 150 km. The last ten kilometres along the river path are strongly urbanised, with a relevant solid transport amount, estimated at 106 tons/year, considering only the Pescara city urban area.

Outcomes of hydrometeorological investigations linked to the *E. coli* concentration peaks suggested that *i) E. coli* concentrations appeared to be most linked to the discharge peaks with respect to the precipitation values and *ii) E. coli* peaks

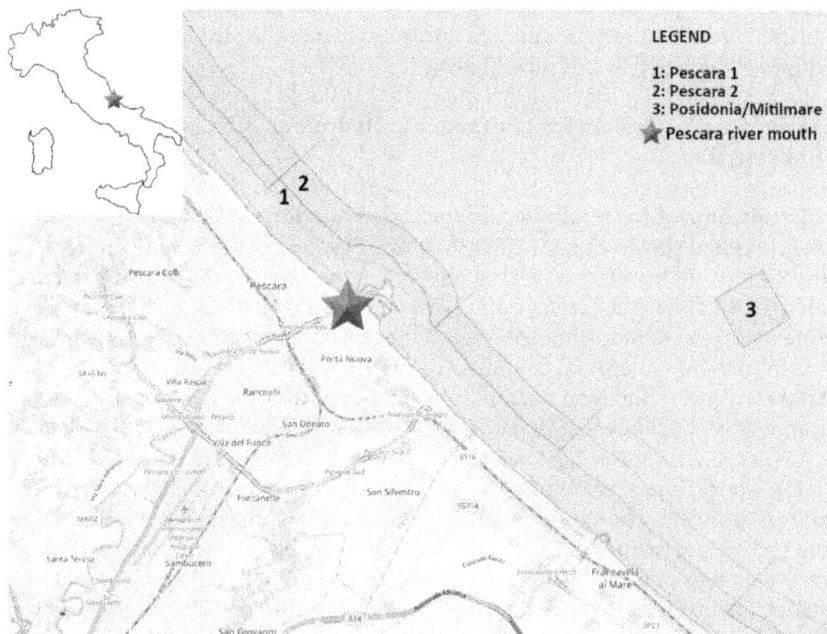

Figure 1.
Three pilot areas around the Pescara River mouth.

exceeding the reported threshold occurred after 2 or 3 days after the Pescara River discharge peak, in most cases.

In more detail, 29 samplings were analysed, and exceeding *E. coli* concentrations were linked to a runoff overflow in 83% of cases and a rainfall event in 50% of cases. As for the mussel farm sampling location, in the open sea at ~5km south-east the Pescara mouth, the 100% of exceeding *E. coli* concentrations were linked to a river overflow, while only the 33% were preceded by rainfall in the coastal area. The case study that occurred on March 8, 2016, revealed a high peak of *E. coli* without any river runoff increase a few days before the event. This case study was then deepened and the hydrometeorological analysis revealed that a huge discharge peak, reaching about 400 m^3/s, affected the Pescara River 7 days before, suggesting a longer river effect on bacterial transport for this case. Moreover, in some cases, a precipitation event over coastal areas and a river discharge increase occurred at the same time and the contribution of the two forcings cannot be discriminated at a first glance. The rainfall effect may also include the presence of sewer overflows (CSOs), direct land-runoff into the estuary, and re-suspension of contaminated sediments within the estuary itself. Finally, increased levels of *E. coli* in bivalves from all monitoring points under high river flow conditions suggest that stormwater runoff is contributing to a significant proportion of *E. coli* accumulation in bivalves [46]. Nevertheless, due to the catchment extension and geographical location, the coastal rainfall does not represent an environmental descriptor indicative of possible faecal contamination related to weather events. The discharge overflow estimation is indeed more representative of the hydrometeorological precursor (**Figure 2**).

A significant association between *E. coli* concentrations and the magnitude of the antecedent discharge peak has been carried out [41]. The Spearman's correlation coefficient r_D calculated was 0.69, and the associated *p*-value was low (~4.5 × 10 5), confirming the correlation hypothesis. The correlation between rainfall maxima and *E. coli* concentrations resulted in a lower correlation coefficient (r_R ¼ 0.35) and the associated *p*-value was high (~0.065), not confirming the correlation hypothesis.

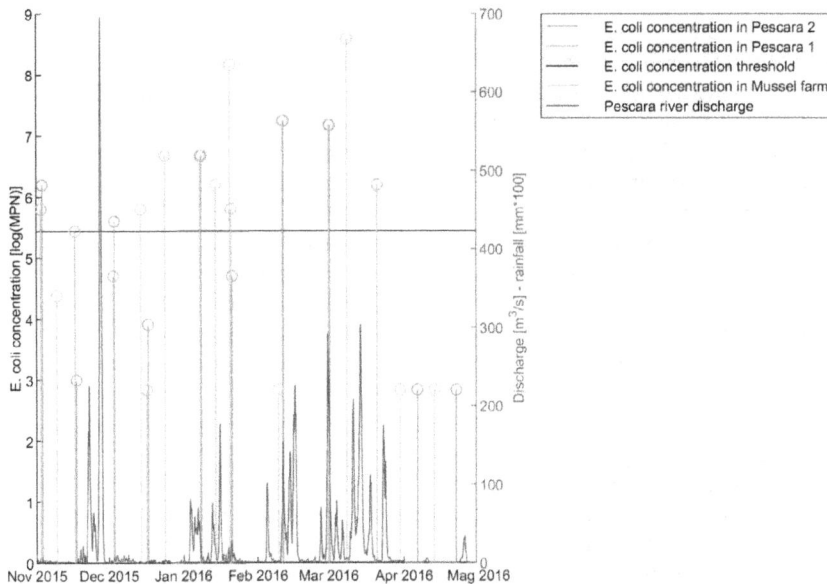

Figure 2.
Time series showing E. coli concentrations at P1, P2, and Mussel Farm, the discharge at the mouth of the Pescara River from November 2015 to May 2016.

Hydrological conditions prior to river flow peaks, such as heavy rainfall, are important in determining the presence of *E. coli* in seawater, but it cannot be ruled out that even low rainfall events could cause significant increases in concentrations when they follow a dry period.

Local regulations for monitoring water and molluscs have been planned regardless of weather conditions, the river flows, or other abiotic factors that can affect the concentration of FIO, and sampling intervals to detect the potential microbial contamination may, therefore, not be representative of variations in these conditions. For this reason, it is reasonable to assume that the data underestimate the strength of the correlations between bacterial concentration, precipitation, and river flow.

To overcome these limits, a holistic approach based on the correlations between data of precipitation (intensity and position) and variation in the river flow discharge is essential. This strategy is useful for predicting times and places of exposure to microbiological contamination. The combined assessment of abiotic factors (physical and chemical), hydrometeorological components and biotic factors also provides holistic information on the health of the ecosystem.

3. High-resolution coastal mapping of Chl-a concentration from Sentinel-2 satellite

As previously introduced, the use of satellite data is reaching constantly wider applicability in the earth monitoring sector. The availability of free satellite data throughout the globe, characterised by a great variety of accessible sensed data types, has sped up this process. In Europe, the raise of programmes, such as Copernicus (European Union's Earth Observation Programme) founded by the European Commission, allowed the birth of new value-adding activities and studies on earth monitoring and services for disaster prevention [58, 59]. The whole programme is composed of seven missions, with a fleet of around ten satellites among future and operational ones. The most important thematic streams of Copernicus services are dedicated to land and marine environments. An example of marine applications is Sentinel-3 satellites that have a push-broom imaging spectrometer called OLCI (Ocean Land Color Instrument). This instrument measures solar radiation reflected by the Earth in 21 spectral bands and has a ground spatial resolution of 300 m [60].

For what concerns the marine environment, as above extensively analysed, its health status is strictly connected to many human activities, especially for coastal waters which represent a vital asset. The possibility to monitor it constantly and rapidly, covering large portions of territory, plays a crucial role and perfectly fits the enhancement introduced using satellite data for monitoring.

The symptoms to be monitored for marine environment's health status evaluation are several and an example is the detection of algae presence. Under certain conditions, indeed, algae can reproduce in an accelerated way, giving birth to what is called an "algal bloom". Some kinds of algal bloom can be toxic, may cause skin rashes or illness in humans as well as can be poisoning for some marine species (e.g., shellfish). A possible parameter to measure algae presence in a water body is chlorophyll-a (Chl-a), the pigment that is used by algae for photosynthesis which constitutes the part that mostly interacts with solar radiation. The concentration of Chl-a contributes to the so-called particulate organic matter present in a water body. On the other hand, water clarity is also connected to all the possible suspended matter that contributes, when in high concentrations, to increase its turbidity [61, 62]. The concentration of all possible suspended particles in a water body is defined as Total Suspended Matter (TSM) concentration. Coastal waters, moreover, operate as a link between land and ocean systems. Rivers physically allow this connection, acting as a conduit for delivering significant amounts of dissolved and particulate materials from terrestrial environments to the coastal ocean, increasing TSM concentrations. In some cases, part of this TSM can be composed of soil particles detached from the coastline and dragged away from waterpower [63].

Several approaches have been followed for years to perform marine monitoring through satellite observations, depending on the specific parameter to be estimated (Sea Surface Height, Wind Speed, Sea Ice, just to mention a few) [64]. For what concerns the detection of water quality, the estimated parameters are connected to its bio-optical properties. At some specific wavelengths, indeed, the suspended particles inside water can interact with radiation incoming from the atmosphere, giving back in return an upwelling radiant flux that has some characteristic responses (it can be absorbed in certain wavelengths more than at others). Optical sensors, therefore, have been widely used to identify the so-called spectral marine inherent optical properties (IOPs, e.g., absorption and scattering) [65]. Usually, the water-leaving signal is quite low (sometimes 1% or less of downwelling irradiance) and requires the sensors to work in a set of narrow, sensitive spectral channels and to remove the atmospheric effects. Those sensors are usually referred to as "ocean color" sensors. The spectral signals received can be used to estimate phytoplankton abundance and other radiatively active constituents.

Usually, the main approaches followed to retrieve IOPs from satellite measurements are two [66]—the first applies atmospheric correction (AC) algorithms to remove the contribution of the atmosphere from the signal received at the top of the atmosphere (TOA) by the sensor and leads to the estimate of the bottom of atmosphere (BOA) reflectance (calculated as the ratio of water-leaving radiance to downwelling irradiance just above the air-sea interface). Different kinds of algorithms can be then applied to AC reflectance values to produce estimates of geophysical properties (e.g., inversion model [67, 68]). The second approach tries to find a direct relationship between the spectral radiance at the top of the atmosphere and IOPs [69, 70]. This one is more immediate and removes the AC step that can sometimes lead to misinterpretation of the atmospheric contribution in presence of optically complex water masses.

However, the resolution of the satellite images used remains the main constraint for accuracy and precision obtained through monitoring developed solutions. It is important to note that coastal areas are also spatially and optically complex and would require more frequent spatial and spectral sampling to enhance the monitoring capability [71].

3.1 Advantages of Sentinel-2 satellite data for coastal water remote sensing

As said, satellite data can boost the realisation of more effective environment monitoring algorithms. Among the most used satellite data for coastal water studies, we can find several studies that employ high-resolution optical data obtained from the OLI (Operational Land Imager) sensor on board of Landsat-8 satellite and MERIS (Medium Resolution Imaging Spectroradiometer) on the Envisat satellite [72, 73]. Additionally, to the OLI sensor, Landsat 8 satellite payload is also made of Thermal Infrared Sensor (TIRS). These two sensors are characterised by a spatial resolution of 30 meters (visible, NIR, SWIR); 100 meters (thermal); and 15 meters (panchromatic). MERIS, which is a push broom radiometer, reaches a spatial resolution of 300 m at nadir (for full resolution products) and 1200 m for reduced resolution data and its spectral range varies from 390 nm to 1040 nm.

In parallel with the aforementioned missions, and despite being built mainly as a land monitoring mission, also Sentinel-2 satellite has gained popularity for marine applications. Indeed, thanks to its high spatial resolutions together with a high revisit frequency, Sentinel-2 allowed to overcome several limitations of existing missions. Sentinel-2 is equipped with a MultiSpectral Instrument (MSI) with 13 spectral bands from the visible and near-infrared to the short-wave infrared (from 443 to 2190 nm). The spatial resolution varies from 10 m to 60 m, depending on the spectral band, with a 290 km field of view [74]. The MSI sensor is made of a three-mirror, 150 mm aperture telescope which collects light and focuses it into two separate focal planes—one for visible (VIS) and near-infrared (NIR), and the other for short-wave infrared (SWIR) wavelengths, respectively. Each focal plane is composed of 12 detectors staggered in two rows.

Its revisit frequency is increased with respect to other missions thanks to the simultaneous operations of two identical satellites: Sentinel-2A and Sentinel-2B, launched in 2015 and 2017, respectively. This more frequent data availability offers several benefits—from a higher probability of finding imagery clear of cloud and sun glint; to more effective applicability of change detection algorithms [75]. Moreover, the free availability of its data allowed to facilitate the spread of their usage.

3.2 Retrieval algorithms for Chl-a concentrations through Sentinel-2 data

In the studies conducted by Marzano et al. [76], Sentinel-2 data played a crucial role in the detection of water quality. Their study was focused on Case-II waters, as per Morel and Prieur water classification [77]. In coastal areas, indeed, water quality is mainly conditioned by Chl-a and TSM concentrations variations and their study was focused on different retrieval approaches to these quantities.

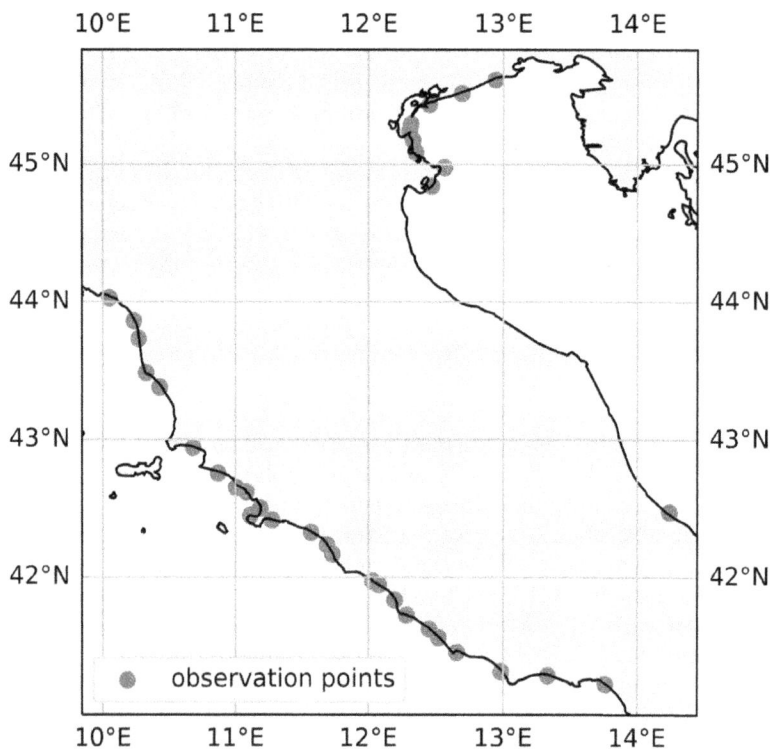

Figure 3.
In situ observation points locations of ARPAs' dataset.

The region of interest in this study is central-northern Italy over Tyrrhenian and the Adriatic Sea. The *in situ* observations of Chl-a concentrations were provided by Italian regional environmental agencies named ARPA (Agenzia Regionale per la Protezione Ambientale), for the regions of Tuscany, Lazio, Abruzzo, and Veneto, and covered the time period between 2016-08-04 and 2018-04-19. In **Figure 3**, the spatial distribution of the dataset along the Italian coasts is shown.

They analysed the use of both empirical and model-based regressive algorithms to retrieve IOPs. More in detail:

- the first method is based on the use of atmospheric correction for the retrieval of BOA reflectance from TOA radiances; values obtained using maximum band ratio (MBR) model on satellite data were put in comparison with observations provided by ARPA's for the evaluation of the empirical regressive algorithm, realized through models defined in literature.

- in the latter was developed a radiative transfer equation (that uses observations to determine absorption and scattering coefficients) through which synthetic reflectance values are retrieved. Those synthetic values are then used to evaluate the model-based regressive algorithm.

The atmospheric correction software used is ACOLITE, with the Dark Spectrum Fitting (DSF) enabled. **Figure 4** reports an example of RGB composite images for Top Of Atmosphere and Bottom Of Atmosphere reflectance before and after atmospheric correction, respectively.

Focusing on the Empirical Regressive algorithm (EmpReg), developed in [76], the retrieval of Chl-a concentrations was defined through the following:

$$r_{MBR} = \frac{\max\left(R_{wlB1}, R_{wlB2}\right)}{R_{wlB3}}, \tag{1}$$

where the numerator is the maximum between B1 and B2 (blue bands) water-leaving reflectance and the denominator is the water-leaving reflectance for B3 (green band). Indeed, the bands more sensitive to chlorophyll presence were

S2A/MSI 2016-08-04 10:06:13
ρ_t RGB

S2A/MSI 2016-08-04 10:06:13
ρ_s RGB

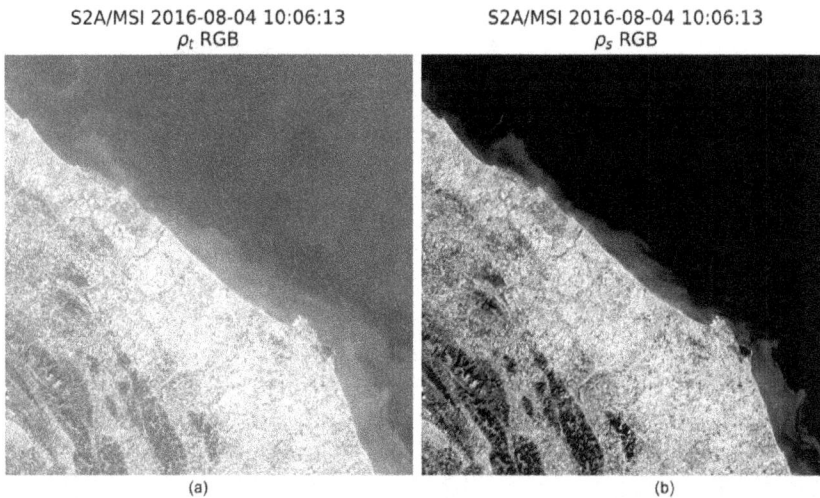

(a)

(b)

Figure 4.
RGB Sentinel-2 remote-sensing reflectance images over the Adriatic coast in the Marche region before (a) and after (b) the atmospheric correction using the ACOLITE software.

considered to retrieve Chl-a concentrations. This blue-to-green reflectance maximum band ratio (MBR) model is among the most used ones in literature [78, 79].

The empirical regressive retrieval algorithm, which is the optimal regressive formula found with respect to the area of analysis and dataset used in the paper, is defined as:

$$\hat{C}_{Chla} = a_1 \exp(-a_2 r_{MBR}) \tag{2}$$

where a_1=59.795 mg/m^3 and a_2 = 4.559.

In the following **Figure 5**, is reported the scatterplot of chlorophyll-a (Chl-a) *in situ* measurements (mg m^{-3}) with respect to Sentinel-2 MSI water-leaving blue-to-green maximum band ratio (MBR) in the Tyrrhenian and the Adriatic Sea. **Figure 5** highlights the non-linearity that characterises the relationship between Chl-a concentrations and MBR.

However, statistical regression algorithms show limitations in handling non-linearity and non-monotonicity, and to overcome these limitations, Marzano et al. [76] used also neural networks. This allowed the exploitation of data contained in several MSI spectral channels of Sentinel-2 products (from B1 to B8A) and spatio-temporal information. In the same way as the previously described methods, also in this case, two neural network-based algorithms were tested:

• empirically trained algorithms, for which the inputs were constituted by atmospherically corrected satellite data, extracted in the point closest to observation, for bands B1 to B8A, together with latitude-longitude information;

• model-trained algorithms, for which the input was made of results obtained from the model-based regressive method (radiative transfer).

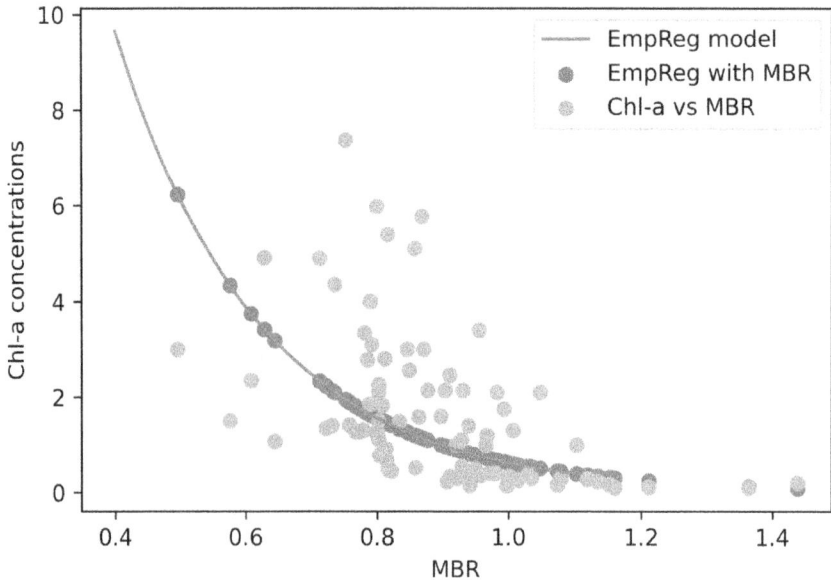

Figure 5.
Measured MBR with respect to in situ Chl-a concentration for the whole training dataset.

The experiments conducted in the work lead to observing better results for NN algorithms when trained with empirical data, rather than with synthetic ones. Although a test to be performed on a wider dataset would be needed. The results obtained through the empirical regressive algorithms with MBR, instead, did not always provide an accurate estimation of the Chl-a concentration, depending on the higher turbidity of Tyrrhenian coastal waters. This was probably related to turbidity conditions of water, which can impact the effectiveness of estimation.

4. Implications for farmed and wild bivalves

Coastal water quality is strictly correlated to the food safety of bivalves. In fact, mollusc bivalves are filter feeders, and they can accumulate microbiological and chemical contaminants from surrounding water. Bivalves are generally cultivated or collected along with the coastal areas where their nutrients are abundant and flow from inland water. But the rivers can also discharge faecal bacteria, mostly from untreated wastewater [80]. The malfunctioning of urban waste water treatment plants or their by-pass during the severe rain events can contribute to the release faecal bacteria into the river. This can pose a potential risk to the consumers of bivalves that can accumulate these bacteria.

To avoid any health human risks, according to the EU Regulation No 627/2019 [37] the competent authorities firstly classify the production areas (Class A, B, and C) through specific monitoring campaigns. Then, they continue to control the level of faecal contamination of the bivalves according to the specific surveillance monitoring plan. The bacterium *E. coli* is used as a faecal indicator organism. For the Class A assignment, for example, the samples (80%) shall not exceed 230 *E. coli* per 100 g of flesh and intravalvular liquid. From this Class, molluscs can be collected for direct human consumption. During the sanitary control, if the mollusc health standards are not met, the competent authorities shall close the production areas and/or reclassify them [37].

As a decision support system for the competent authority, several studies have investigated the correlation between the increase of bacterial concentration in molluscs and weather conditions [40–49, 81–83]. The prediction of local precipitation and river discharges have been used as early warning signals for mollusc bacterial contamination [84, 85]. The advantages are multiple—i) to avoid the collection of the potentially contaminated product; ii) to avoid any temporary closing of production areas; iii) to optimise the monitoring surveillance programme; iv) to ensure the health of consumers.

Generally, results demonstrated that the correlation is site-specific and it depends on numerous factors, such as the geographical location, land use, and catchment size.

From ancient times, the Adriatic basin is particularly devoted to the bivalve farming and fishing in the lagoons and along the coasts. Here, the influence of the weather condition and river run-off on the bivalve hygiene condition has been investigated [41, 86, 87].

In the central Adriatic coast of the Marche region, recently, Ciccarelli et al. [88] published the correlation between the concentration of *E. coli* in the natural banks of *Chamelea gallina* and the local precipitation from 2016 to 2020. The results showed that the rainfall events were significant for the increase of *E. coli* (> 230 MPN/100 g) in the molluscs collected from the south sampling points. In the same region, the increase of *Salmonella* spp. detected in bivalves was reconducted in 2015 and 2016 to the severe meteorological events [89].

In the Northern Adriatic Sea, the CADEAU project [90] developed specific indexes to evaluate the potential microbial pollution impact of urban waste water treatment plants on the farmed molluscs in the Municipality of Chioggia (Venice, Italy). It provides indexes of dilution for *E. coli* based on the bacterial decay due to salinity, temperature, and solar radiation [90].

In the following sections, we report some "site-specific" study cases carried out in the Abruzzo region, on the central Adriatic coast of Italy.

4.1 The study case of wild clams and farmed mussels in the Pescara province

In 2021, Colaiuda et al. [41] published the case study in the Pescara province (Abruzzo Region, Italy) that was already detailed in the paragraph 2.2. Here, two production areas of wild clams (*C. gallina*) and one farm of mussels (*Mytilus galloprovincialis*) facing the Pescara River were investigated (**Figure 1**). In **Figure 1**, Pescara 1 and 2 are the production areas of clams, the other is the farm of mussels. Thanks to the CapRadNet project, this study executed a correlation analysis between river discharge trough to the CHyM model, precipitation in the catchment area, and the concentrations of *E. coli* detected in the bivalves during the official monitoring programme. The referring period was from August 1, 2015 to July 31, 2016. The EU reference method to detect *E. coli* was ISO-16649-3 [91]. Results were expressed as the most probable number – MPN per 100 g of flesh and intravalvular liquid of mollusc. Microbiological data were downloaded from the database of the project CAPS2 developed also the informative tool "CAPS2 WEB GIS" useful for the management of the production areas. The classification of the three production areas (Class A) was viewable in the CAPS2 WEB GIS [92]. The competent authority was the unique authorised user to modify the classification and the boundaries of the production areas in the CAPS2 WEB GIS.

The results showed that the concentration of *E. coli* in molluscs increased within 6 days of a river discharge peak (**Figure 2**). Moreover, 87% of cases of high concentration of *E. coli* were consequent to the increased river flow, while 60% of cases to the precipitation. These results suggested that the Pescara River discharge was the potential hydrometeorological driver of *E. coli* in facing molluscs to be further evaluated with specific sampling before and after discharge peak at the river mouth.

4.2 The study case of mussel farm in the Teramo province

The research project FORESHELL was funded by the FLAG Costa Blu through the 2014-20 EMFF programme of the Abruzzo Region. It is aimed at developing sanitary/weather-environmental predictive technological tools to enhance the efficiency and sustainability of a mussel farm in the Teramo province (Giulianova city, Abruzzo region, Italy) [93]. This production area of *M. galloprovincialis* was classified as Class A, and it is facing the Salinello and Vibrata Rivers far away almost 3 miles from the coast (**Figure 6**).

The hydrological model (CHyM) has analysed the hydrographic basins of the rivers and it has been forecasting the discharge peaks. Before and after these events, a sample of freshwater at the river mouths, and of molluscs and sea water at the farm have been collected for the *E. coli* detection [91]. Preliminary results showed that until September 2021, there were four meteorological events (**Table 1**) that did not cause a peak discharge at the river mouth. Results did not register a significant increase of *E. coli* in the mussels (**Figure 7**). At the same time, the environmental parameters such as sea water temperature, salinity, Chl-a, sea currents, and wave motion are acquired by the satellites and *in situ* probes.

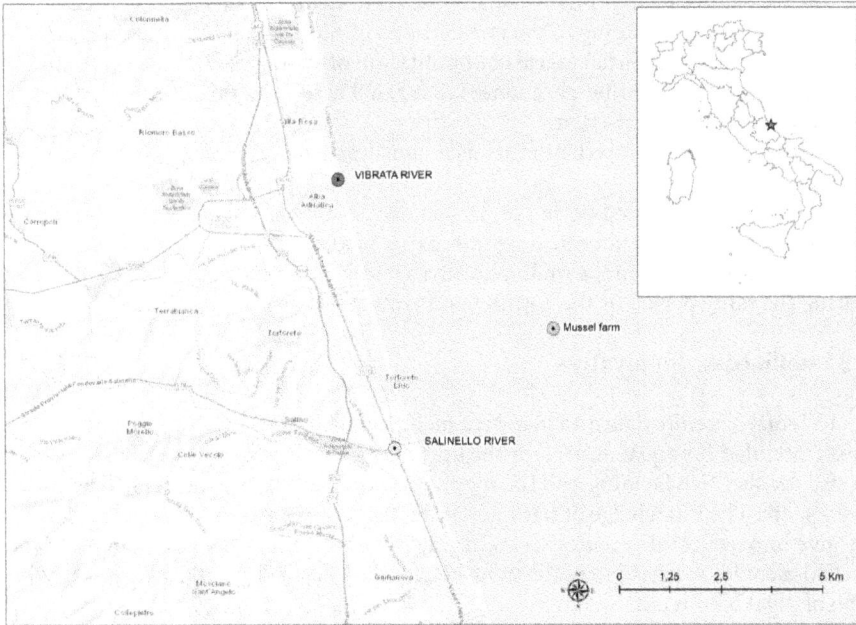

Figure 6.
Sampling points at river mouths and at the farm in the Abruzzo region.

Date of meteorological event	Description of the event
21/09/2020	Scattered rain in the internal area
10/10/2020	Severe event in the northern Adriatic Sea
17/07/2021	Rainfall in the coastal area
27/08/2021	Storm at the coastal area

Table 1.
FORESHELL project: Description of meteorological events.

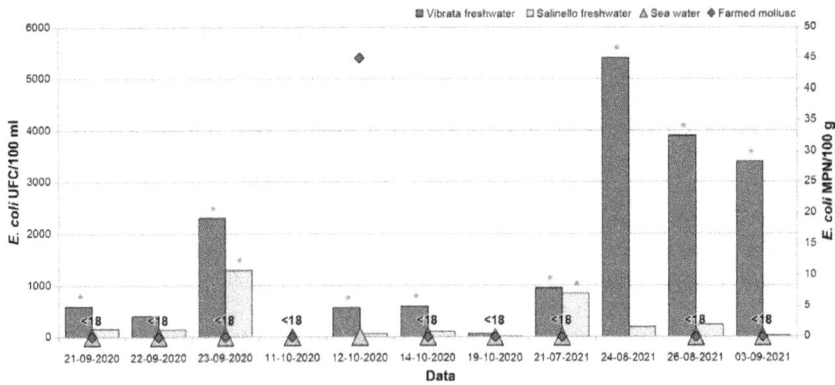

Figure 7.
FORESHELL project: E. coli concentrations before and after meteorological events.

The web application for data visualisation is under construction, as well as the early warning signalling to the farmer by mail/SMS/WhatsApp. The alerts are referred to the potential faecal contamination of molluscs predicted through hydrological data and other parameters that can damage the farm, such as high temperature and wave motion.

Furthermore, the growth of mussels is constantly monitored with biometric controls.

In conclusion, the first period of project execution was characterised by precipitation scarcity that did not cause any discharge peaks at the river mouths without any presence of *E. coli* in the molluscs. Further analyses are expected to be executed during the rainy period of the autumn and winter seasons.

4.3 Satellite data for bivalves

Recently, satellite data and maps are increasingly used for the identification of areas intended for aquaculture, for the knowledge of the environmental conditions useful for shellfish farming and fishing, for the prediction of potentially harmful events, etc. The knowledge of parameters such as temperature, salinity, and turbidity give important information to managing bivalve production. For example, data on Chl-a could be useful to understand the food disposal for molluscs or the prediction of algal bloom potential toxic.

The projects AQUACULTURE2000 and VALUE-SHELL analyse satellite data, such as pH, temperature, and CO_2, to assess the possible contribution of mussel farming in sequestering carbon from seawater through the biocalcification processes in the northern Adriatic Sea by mitigating the effects of climate change [94].

In 2021, the total suspended matter, temperature, and Chl-a estimated from satellite acquisitions have been used to predict the presence of radioactivity in molluscs [95].

Along the Adriatic coast facing the Abruzzo region where several farms are placed, two pilot studies were conducted. The goal was to calibrate algorithm coefficients, at a local scale, to set up a processing chain that derives accurate concentration maps of chlorophyll and suspended solids from the satellite, as said, taking advantage of the high frequency of revisit time and high spatial resolution of the satellite acquisitions.

The first study estimated Chl-a and sediment dispersions in the sea, derived from Sentinel-2 images, compared with *in situ* data acquired by means of a multiparametric probe in the monitoring stations that Agenzia Regionale per la Tutela dell'Ambiente (ARTA) Abruzzo monthly checks [96]. The Case-2 Regional Coast Color processor in ESA SNAP software was used, applying the C2RCC-Nets algorithm [97], whose parameters have been set using *in situ* measurements, specifically the salinity and temperature variables. This preliminary study provided encouraging results with only four sampling dates—for example, the concentration map of TSM of 09/03/2018 is reported (**Figure 8**).

In the second study [98], the authors developed the Water Color Data Analysis System (WC-DAS), a tool for the operational generation of maps and indicators useful in the monitoring of water quality. The tool, in its first release, allows the processing of satellite optical multispectral data acquired by Sentinel-3 OLCI, Sentinel-2 MSI, and Landsat-8 OLI sensors, using the algorithms Case 2 Regional CoastColour (C2RCC) [97] and Atmospheric Correction for OLI "lite" (ACOLITE) [99].

The tool was tested in the central Adriatic coastal zone, setting the local parameters according to the *ad hoc in situ* sampling campaign that ARTA Abruzzo carried out along the Abruzzo coast, simultaneously with the satellites overpasses (example in **Figure 9**).

The results show performances of calibrated algorithms and the data system's suitability to contribute to the production of monitoring maps and indicators, informing domain-specific decision-making and supporting services for integrated coastal zone management.

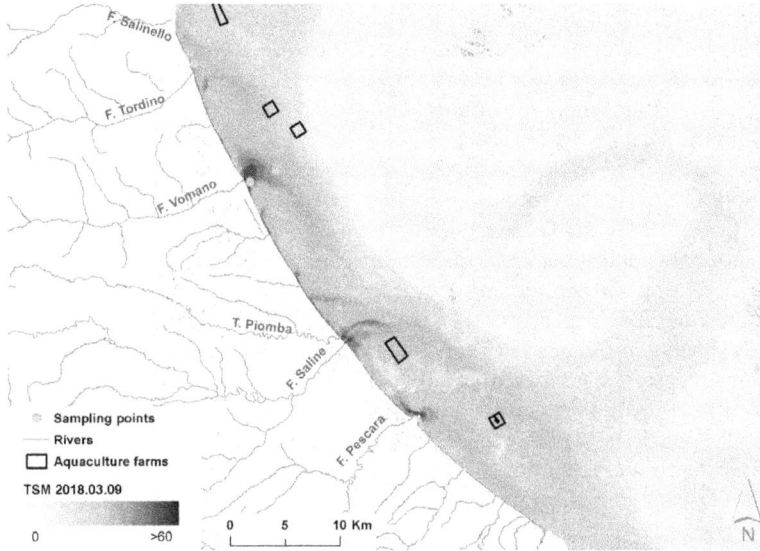

Figure 8.
Concentration maps of Total Suspended Matter along Abruzzo coast in the Adriatic Sea, as elaborated with the C2RCC processor from Sentinel-2 imagery of 09/03/2018.

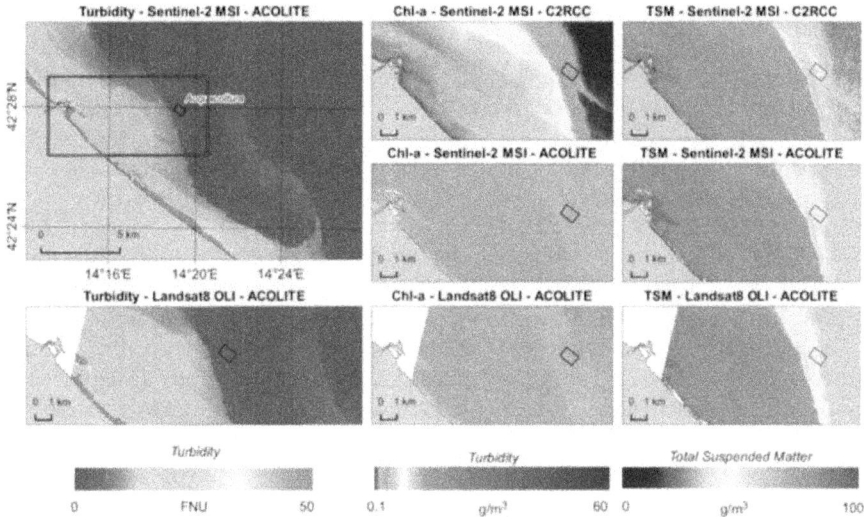

Figure 9.
Turbidity, Total Suspended Matter, Chlorophyll-a maps along Abruzzo coast, facing Pescara river mouth, as derived from Water Colour Data Analysis System Tool.

5. Conclusions

The human pressure on coastal areas is causing a significant impact on the resilience of the marine ecosystem in which natural and anthropogenic processes interact, modifying their geomorphological, physical, and biological characteristics. The combined effects from multiple pressures are not still considered in management or planning processes and this reduces the overall resilience of marine ecosystems. The assessment of the sea and coastal systems and their interaction, based on scientific knowledge, are the indispensable basis for the management of human activities, in view of promoting the sustainable use of the seas and coasts and conserving marine ecosystems and their sustainable development. A few research is available, where the impacts of human wastewater on coastal ecosystems and community health are assessed.

To overcome these limitations, an Early Warning System for health risk assessment should be optimised through a holistic approach based on the combined analysis of abiotic factors (physical and chemical), hydrometeorological components, and biotic factors also provide health information ecosystem.

The prediction and monitoring of water quality are among the main activities to be carried out for the protection of coastal areas. As for the prediction, water quality is strongly influenced by atmospheric events that could affect the pollution management systems, such as rainfall-dependent sewage drains and tributary river flow. For this reason, river mouths act as critical links between the hinterland and the sea. The prediction of river discharges and overflows using hydrometeorological models can be fundamental for indirect estimation of water quality, given that the drainage network runoff is closely related to the supply of marine nutrients and faecal bacteria. These can be accumulated by the farmed and wild bivalves in the coastal areas posing a risk for the human consumer. Therefore, the development of Early Warning Systems integrating predictive satellite data, could improve both the sanitary surveillance by competent authorities and the daily farming/fishing operations by workers. The economic loss could be reduced by improving the protection of consumer health.

The use of satellite data for water quality monitoring is gaining high importance, see the limitations that remote sensing techniques allow to overcome with respect to *in situ* observations (wide-area coverage, more frequent data availability). Several approaches have been followed for years to perform marine monitoring through satellite observations, depending on the specific parameter to be estimated. For what concerns the detection of quality of water, the parameters estimated are connected to its bio-optical properties. The Sentinel-2 data played a crucial role in the detection of water quality in the coastal areas, indeed, water quality is mainly conditioned by Chl-a and TSM.

A wide number of capabilities should be connected to work together to find ICT solutions: biologists, physicists, engineers, and economists, for an example. Different competencies may be required to achieve this purpose; on one hand, prediction of environmental processes requires deep knowledge of earth system modelling and data interpretation. On the other hand, environmental implications on food security monitoring are requested to adequately assess the design of useful tools or instruments able to really ameliorate the bivalves' productions.

The study cases detailed in this chapter have been conducted by a multidisciplinary team that has developed several tools, also predictive, useful for stakeholders, such as competent authorities, farmers, researchers, veterinary services, and fishermen, consumers. The overall aim was to acquire knowledge and develop innovative technological tools focused on the enhancement of regional services.

Conflict of interest

The authors declare no conflict of interest.

Author details

Annalina Lombardi[1*], Maria Paola Manzi[2], Federica Di Giacinto[3],
Valentina Colaiuda[4], Barbara Tomassetti[1], Mario Papa[2], Carla Ippoliti[3],
Carla Giansante[3], Nicola Ferri[3] and Frank Silvio Marzano[2]

1 CETEMPS, University of L'Aquila, L'Aquila, Italy

2 University of Rome, La Sapienza, Italy

3 Istituto Zooprofilattico Sperimentale dell'Abruzzo e del Molise "G. Caporale,
Teramo, Italy

4 Department of Physical and Chemical Sciences - CETEMPS, University of
L'Aquila, L'Aquila, Italy

*Address all correspondence to: annalina.lombardi@univaq.it

IntechOpen

References

[1] Halpern BS, Walbridge S, Selkoe KA, Kappel CV, Micheli F, D'Agrosa C, et al. A global map of human impact on marine ecosystems. Science. 2008;**319**:948-952

[2] Jackson JBC et al. Historical overfishing and the recent collapse of coastal ecosystems. Science. 2001;**293**(5530):629-637

[3] Micheli F et al. Cumulative human impacts on Mediterranean and Black Sea marine ecosystems: Assessing current pressures and opportunities' Meador, J. P. (ed.). PLoS ONE. 2013;**8**(12):e79889

[4] Vitousek PM, Aber JD, Howarth RW, Likens GE, Matson PA, Schindler DW, et al. Human alteration of the global nitrogen cycle: Sources and consequences. Ecological Applications. 1997;7:737-750

[5] Halpern BS, Ebert CM, Kappel CV, Madin EMP, Micheli F, Perry M, et al. Global priority areas for incorporating land—sea connections in marine conservation. Conservation Letters. 2009;**2**:189-196

[6] Malone TC, Newton A. The globalization of cultural eutrophication in the Coastal Ocean: Causes and consequences. Frontiers in Marine Science. 2020;**7**:670

[7] Diaz RJ, Rosenberg R. Spreading dead zones and consequences for marine ecosystems. Science. 2008;**321**:926-929

[8] Mekonnen MM, Hoekstra AY. Global gray water footprint and water pollution levels related to anthropogenic nitrogen loads to fresh water. Environmental Science & Technology. 2015;**49**: 12860-12868

[9] Rabalais NN, Turner RE, Díaz RJ. Global change and eutrophication of coastal waters. Marine Science. 2009

Available from: https://academic.oup.com/icesjms/article-abstract/66/7/1528/656749

[10] Howarth RW. Coastal nitrogen pollution: A review of sources and trends globally and regionally. Harmful Algae. 2008;**8**:14-20

[11] Breitburg DL, Hondorp DW, Davias LA, Diaz RJ. Hypoxia, nitrogen, and fisheries: Integrating effects across local and global landscapes. Annual Review of Marine Science. 2009;**1**:329-349

[12] Milieu Ltd et al. Study for the Strategy for a Non-Toxic Environment of the 7th Environment Action Programme. Brussels, Belgium: European Commission (DG ENV); 2017

[13] Wear SL, Acuña V, McDonald R, Font C. Sewage pollution, declining ecosystem health, and cross-sector collaboration. Biological Conservation. 2021;**255**:109010

[14] Wear SL, Thurber RV. Sewage pollution: Mitigation is key for coral reef stewardship. Annals of the New York Academy of Sciences. 2015;**1355**:15-30

[15] Wear SL. Battling a Common Enemy: Joining Forces in the Fight against Sewage Pollution. Bioscience. 2019;**69**:360-367

[16] Tuholske C, Halpern BS, Blasco G, Villasenor JC, Frazier M, Caylor K. Mapping global inputs and impacts from of human sewage in coastal ecosystems. PLoS One. 2021;**16**(11):e0258898. DOI: 10.1371/journal.pone.0258898

[17] ISPRA. Mare ed Ambiente costiero. 2011. Available from: https://www.isprambiente.gov.it/files/pubblicazioni/statoambiente/tematiche2011/05_%20Mare_e_ambiente_costiero_2011.pdf

[18] EC. Directive 2000/60/EC of the European Parliament and of the Council of 23 October 2000 establishing a framework for Community action in the field of water policy. 2000

[19] EEA. The European Environment – State and outlook 2010, Report 1/2010 14. EEA, 2018a. Contaminants in Europe's seas. EEA Report No 25/218. 2018

[20] EEA. European Waters. Assessment of Status and Pressures 2018. EEA Report No 7/2018. European Environment Agency; 2018

[21] EEA. Marine Messages II, EEA Report No 17/2019. European Environment Agency; 2020

[22] Büttner G, Feranec J, Jaffrain G, Mari L, Maucha G, Soukup T. The European CORINE land cover 2000 project. In: Paper presented at the XXth Congress of ISPRS; 12-13 July 2004; Istanbul, Turkey. 2004

[23] Morel A, Prieur L. Analysis of variations in ocean color. Limnology and Oceanography. 1977;**22**:709-722

[24] Manos A. An international programme for the protection of a semi-enclosed sea—The Mediterranean Action Plan. Marine Pollution Bulletin. 1991;**23**:489-496

[25] Nations, United and Unepmed WG.473Inf. United Nations Environment Programme Mediterranean Action Plan. 2015

[26] UNEP/MAP, Integrated Monitoring and Assessment Programme of the Mediterranean Sea and Coast and Related Assessment Criteria (IMAP), Official document series 1/2016; ISBN 978-92-807-3592-5. 2016. Available from: https://wedocs.unep.org/bitstream/handle/20.500.11822/10576/IMAP_Publication_2016.pdf?sequence=1&isAllowed=y

[27] Papathanasopoulou E, Simis SG, Alikas K, Ansper A, Anttila J, Barillé A, Barillé L, Brando V, Bresciani M, Bučas M, Gernez PM. Satellite-assisted monitoring of water quality to support the implementation of the Water Framework Directive. EOMORES White Paper. 2019

[28] Damania R, Desbureaux S, Rodella AS, Russ J, Zaveri E. Quality Unknown: The Invisible Water Crisis. World Bank. License: CC BY 3.0 IGO; 2019. Available from: https://openknowledge.worldbank.org/handle/10986/32245. DOI: 10.1596/978-1-4648-1459-4

[29] Mobley CD, Stramski D, Bissett WP, Boss E. Optical Modeling of Ocean Waters: Is the Case 1—Case 2 Classification Still Useful; Oceanography. 2004;**17**(2):60-67. DOI: 10.5670/oceanog.2004.48

[30] González et al; Review of the Commission Decision 2010/477/EU concerning MSFD criteria for assessing Good Environmental Status, Descriptor 7; EUR 27544 EN; doi:10.2788/435059

[31] EC. Directive 2014/89/EU of the European Parliament and of the Council of 23 July 2014 establishing a framework for maritime spatial planning. 2014

[32] EC. The EU Blue Economy Report. 2019. Luxembourg: European Commission, Publications Office of the European Union; 2019

[33] OSPAR Commission, MSFD Advice document on Good environmental status - Descriptor 7: Hydrographical conditions, A living document—Version 17 January 2012. ISBN 978-1-909159-16-7 Publication Number: 583/2012

[34] EC. Directive 2008/56/EC of the European Parliament and of the Council of 17 June 2008 establishing a framework for community action in the field of marine environmental policy

(Marine Strategy Framework Directive) (Text with EEA relevance). 2008

[35] EC. Commission Directive (EU) 2017/845 of 17 May 2017 amending Directive 2008/56/EC of the European Parliament and of the Council as regards the indicative lists of elements to be taken into account for the preparation of marine strategies (Text with EEA relevance). 2017

[36] UNEP MAP, 2017. Mediterranean Quality Status Report

[37] European Commission, Commission Implementing Regulation (EU) 2019/627 of 15 March 2019 laying down uniform practical arrangements for the performance of official controls on products of animal origin intended for human consumption in accordance with Regulation (EU) 2017/625 of the European Parliament and of the Council and amending Commission Regulation (EC) No 2074/2005 as regards official controls.

[38] Cabelli VJ, Heffernan WP. Accumulation of *Escherichia coli* by the northern quahaug. Applied Microbiology. 1970;**19**:239-244

[39] Pommepuy M, Hervio-Heath D, Caprais MP, Gourmelon M, Le Saux JC, Le Guyader F. Fecal contamination in coastal areas: An engineering approach. In: Belkin S, Colwell RR, editors. Oceans and Health: Pathogens in the Marine Environment. New York, USA: Springer. pp. 331-359

[40] Prieur D, Mével G, Nicolas JL, Plusquellec A, Vigneulle M. Interactions between bivalve molluscs and bacteria in the marine environment. Oceanography and Marine Biology: An Annual Review;**28**:277-352

[41] Colaiuda V, Di Giacinto F, Lombardi A, Ippoliti C, Giansante C, Latini M, et al. Evaluating the impact of hydrometeorological conditions on *E.*

coli concentration in farmed mussels and clams: Experience in Central Italy. Journal of Water Health. 2021;**19**(3): 512-533

[42] Ferguson CM, Coote BG, Ashbolt NJ, Stevenson IM. Relationships between indicators, pathogens and water quality in an estuarine system. Water Research;**30**(9):2045-2054

[43] Kelsey H, Porter DE, Scott G, Neet M, White D. Using geographic information systems and regression analysis to evaluate relationships between land use and fecal coliform bacterial Estuaries and Coasts pollution. Journal of Experimental Marine Biology and Ecology. 2004;**298**:197-209

[44] Krogh M, Robinson L. Environmental variables and their association with faecal coliform and faecal streptococci densities at thirteen Sydney beaches. Marine Pollution Bulletin. 1996;**33**(7-12):239-248

[45] Lipp EK, Kurz R, Vincent R, Rodriguez-Palacios C, Farrah SR, Rose JB. The effects of seasonal variability and weather on microbial fecal pollution and enteric pathogens in a subtropical estuary. Estuaries. 2001;**24**(2):266-276

[46] Campos CJA, Hargin K, Kershaw S, Lee RJ, Morgan OC. Rainfall and river flows are predictors for β-glucuronidase positive *Escherichia coli* accumulation in mussels and Pacific oysters from the Dart Estuary (England). Journal of Water and Health. 2011;**9**(2):368-381

[47] Coulliette AD, Money ES, Serre ML, Noble RT. Space/time analysis of fecal pollution and rainfall in an eastern North Carolina estuary. Environmental Science and Technology. 2009;**43**(10): 3728-3735

[48] Brock RL, Galbraith GR, Benseman BA. Relationships of rainfall, river flow, and salinity to faecal coliform levels in a mussel fishery. New Zealand

Journal of Marine and Freshwater Research. 1985;**19**:485-494

[49] Fiandrino A, Martin Y, Got P, Bonnefont JL, Troussellier M. Bacterial contamination of Mediterranean coastal seawater as affected by riverine inputs: Simulation approach applied to a shellfish breeding area (Thau lagoon, France). Water Research. 2003;**37**:1711-1722

[50] Crowther J, Kay D, Wyer MD. Faecal-indicator concentrations in waters draining lowland pastoral catchments in the UK: Relationships with land use and farming practices. Water Research. 2002;**36**:1725-1734

[51] Kay D, Edwards AC, Ferrier RC, Francis C, Kay C, Rushby L, et al. Catchment microbial dynamics: The emergence of a research agenda. Progress in Physical Geography. 2007;**31**(1):59-76

[52] Taraglio S, Chiesa S, La Porta L, Pollino M, Verdecchia M, Tomassetti B, et al. Decision Support System for smart urban management: Resilience against natural phenomena and aerial environmental assessment. International Journal of Sustainable Energy Planning and Management. 2019;**24**:135-146

[53] Tomassetti B, Coppola E, Verdecchia M, Visconti G. Coupling a distributed grid based hydrological model and MM5 meteorological model for flooding alert mapping. Advances in Geosciences. 2005;**2**:59-63

[54] Verdecchia M, Coppola E, Faccani C, Ferretti R, Memmo A, Montopoli M, et al. Flood forecast in complex orography coupling distributed hydro-meteorological models and in-situ and remote sensing data. Meteorology and Atmospheric Physics. 2008;**101**:267-285

[55] Verdecchia M, Coppola E, Tomassetti B, Visconti G. Cetemps Hydrological Model (CHyM), a distributed grid-based model assimilating different rainfall data sources. In:

Sorooshian S, Hsu KL, Coppola E, Tomassetti B, Verdecchia M, Visconti G, editors. Hydrological Modelling and the Water Cycle. Vol. 63. Berlin/Heidelberg, Germany: Springer; 2009. pp. 165-201

[56] Colaiuda V, Lombardi A, Verdecchia M, Mazzarella V, Antonio R, Ferretti R, et al. Flood prediction: Operational hydrological forecast with the Cetemps Hydrological Model (CHyM). International Journal of Environmental Sciences and Natural Resources;**24**(3):201-208

[57] Lombardi A, Colaiuda V, Verdecchia M, Tomassetti B. User-oriented hydrological indices for early warning systems with validation using post-event surveys: Flood case studies in the Central Apennine District. Hydrology and Earth System Sciences. 2021;**25**:1969-1992

[58] EC N. ESA. The ever growing use of Copernicus across Europe's Regions: A selection of 99 user stories by local and regional authorities. 2019

[59] GEOmedia R. The Growing Use of GMES across Europe's Regions. GEO [Internet]. 15 March 2013;**16**(5). Available from: https://mediageo.it/ojs/index.php/GEOmedia/article/view/218

[60] Donlon C, Berruti B, Buongiorno A, Ferreira MH, Féménias P, Frerick J, et al. The global monitoring for environment and security (GMES) sentinel-3 mission. Remote Sensing of Environment. 2012;**120**:37-57

[61] Neukermans G, Ruddick K, Loisel H, Roose P. Optimization and quality control of suspended particulate matter concentration measurement using turbidity measurements. Limnology and Oceanography: Methods. 2012;**12**:1011-1023

[62] Topp SN, Pavelsky TM, Jensen D, Simard M, Ross MR. Research trends in the use of remote sensing for inland

water quality science: Moving towards multidisciplinary applications. Watermark. 2020;**12**(1):169. DOI: 10.3390/w12010169

[63] Pollard JA, Spencer T, Brooks SM. The interactive relationship between coastal erosion and flood risk. Progress in Physical Geography: Earth and Environment. 2019;**43**(4):574-585. DOI: 10.1177/0309133318794498

[64] Brown CW, Connor LN, Lillibridge JL, Nalli NR, Legeckis RV. An introduction to satellite sensors, observations and techniques. In: Remote Sensing of Coastal Aquatic Environments. Dordrecht: Springer; 2007. pp. 21-50

[65] Stumpf RP, Tomlinson MC. Remote sensing of harmful algal blooms. In: Remote Sensing of Coastal Aquatic Environments. Dordrecht: Springer; 2007. pp. 277-296

[66] Werdell PJ, McKinna LI, Boss E, Ackleson SG, Craig SE, Gregg WW, et al. An overview of approaches and challenges for retrieving marine inherent optical properties from ocean color remote sensing. Progress in Oceanography. 2018;**160**:186-212

[67] Werdell PJ, Franz BA, Bailey SW, Feldman GC, Boss E, Brando VE, et al. Generalized ocean color inversion model for retrieving marine inherent optical properties. Applied Optics. 2013;**52**(10):2019-2037

[68] Werdell PJ, McKinna LI. Sensitivity of inherent optical properties from ocean reflectance inversion models to satellite instrument wavelength suites. Frontiers in Earth Science. 2019;**7**:54. DOI: 10.3389/feart.2019.00054

[69] Hu C. A novel ocean color index to detect floating algae in the global oceans. Remote Sensing of Environment. 2009;**113**(10):2118-2129

[70] Matthews MW, Bernard S, Robertson L. An algorithm for detecting trophic status (chlorophyll-a), cyanobacterial-dominance, surface scums and floating vegetation in inland and coastal waters. Remote Sensing of Environment. 2012;**124**:637-652. DOI: 10.1016/j.rse.2012.05.032

[71] Elnabwy MT, Elbeltagi E, El Banna MM, Elshikh MM, Motawa I, Kaloop MR. An approach based on Landsat images for shoreline monitoring to support integrated coastal management—a case study, Ezbet Elborg, Nile Delta, Egypt. ISPRS International Journal of Geo-Information. 2020;**9**(4):199

[72] Trinh RC, Fichot CG, Gierach MM, Holt B, Malakar NK, Hulley G, et al. Application of Landsat 8 for monitoring impacts of wastewater discharge on coastal water quality. Frontiers in Marine Science. 2017;**4**:329. DOI: 10.3389/fmars.2017.00329

[73] Lim J, Choi M. Assessment of water quality based on Landsat 8 operational land imager associated with human activities in Korea. Environmental Monitoring and Assessment. 2015;**187**(6):1-7

[74] Drusch M, Del Bello U, Carlier S, Colin O, Fernandez V, Gascon F, et al. Sentinel-2: ESA's optical high-resolution mission for GMES operational services. Remote Sensing of Environment. 2012;**120**:25-36. DOI: 10.1016/j.rse.2011.11.026

[75] Hedley JD, Roelfsema C, Brando V, Giardino C, Kutser T, Phinn S, et al. Coral reef applications of Sentinel-2: Coverage, characteristics, bathymetry and benthic mapping with comparison to Landsat 8. Remote Sensing of Environment. 2018;**216**:598-614

[76] Marzano FS, Iacobelli M, Orlandi M, Cimini D. Coastal water remote sensing from Sentinel-2 satellite data using physical, statistical, and neural network retrieval approach. IEEE

Transactions on Geoscience and Remote Sensing. 2020;**59**(2):915-928

[77] Morel A, Prieur L. Analysis of variations in ocean color 1. Limnology and Oceanography. 1977;**22**(4):709-722

[78] Sauer MJ, Roesler CS, Werdell PJ, Barnard A. Under the hood of satellite empirical chlorophyll a algorithms: Revealing the dependencies of maximum band ratio algorithms on inherent optical properties. Optics Express. 2012;**20**(19):20920-20933

[79] Blondeau-Patissier D, Gower JF, Dekker AG, Phinn SR, Brando VE. A review of ocean color remote sensing methods and statistical techniques for the detection, mapping and analysis of phytoplankton blooms in coastal and open oceans. Progress in Oceanography. 2014;**123**:123-144

[80] World Health Organization (WHO). In: Rees G, Pond K, Kay D, Bartram J, Domingo JS, editors. Safe Management of Shellfish and Harvest Waters. London, UK: IWA Publishing; 2010 ISBN: 9781843392255

[81] Tryland I, Robertson L, Blankenberg AB, Lindholm M, Rohrlack T, Liltved H. Impact of rainfall on microbial contamination of surface water. International Journal of Climate Change Strategies and Management. 2011;**3**(4):361-373

[82] Zimmer-Faust AG, Brown CA, Manderson A. Statistical models of fecal coliform levels in Pacific Northwest estuaries for improved shellfish harvest area closure decision making. Marine Pollution Bulletin. 2018;**137**:360-369

[83] Jung A-V, Le Cann P, Roig B, Thomas O, Baurès E, Thomas M-F. Microbial contamination detection in water resources: Interest of current optical methods, trends and needs in the context of climate change. International Journal of Environmental Research and Public Health. 2014;**11**(4):4292-4310

[84] Mälzer H-S, Aus der Beek T, Müller S, gebhardt J. Comparison of different model approaches for a hygiene early warning system at the lower Ruhr River, Germany. International Journal of Hygiene and Environmental Health. 2015;**219**(7):671-668

[85] de Souza RV, de Campos CJA, Garbossa LHP, Vianna LFDN, Seiffert WQ. Optimising statistical models to predict faecal pollution in coastal areas based on geographic and meteorological parameters. Marine Pollution Bulletin. 2018;**129**:284-292

[86] Ciccarelli C, Semeraro AM, Leinoudi M, Di Trani V, Murru S, Capocasa P, et al. Assessment of relationship between rainfall and *Escherichia coli* in clams (Chamelea gallina) using the Bayes Factor. Italian Journal of Food Safety. 2017;**6**(6826): 99-102

[87] Tabanelli G, Montanari C, Gardini A, Maffei M, Prioli C, Gardini F. Environmental factors affecting *Escherichia coli* concentrations in striped Venus Clam (*Chamelea gallina* L.) Harvested in the North Adriatic Sea. Journal of Food Protection. 2017;**80**(9): 1429-1435. DOI: 10.4315/0362-028X. JFP-17-058

[88] Ciccarelli C, Semeraro A, Di Trani V, Murru S, Carboni S, Ciccarelli E. Banchi naturali di vongole (*Chamelea gallina*) sulla costa del Piceno: contaminazione da *E. coli* e rapporti con le precipitazioni atmosferiche. In: Proceedings of the Conference of Società Italiana di Ricerca Applicata alla Molluschicoltura (SIRAM 2021). 2021. pp. 19-23

[89] Barchiesi F, Napoleoni M, Ciccarelli C, Ferraro R, Calandri E, Rocchegiani E, et al. Molluschi bivalvi bioindicatori di contaminazione da Salmonella spp.: uno strumento One Health, tra salute e ambiente. In: Proceedings of the Conference of Società Italiana di Ricerca Applicata alla

Molluschicoltura (SIRAM 2021). 2021. pp. 15-16

[90] Petochi T., Bruschi A., Cossarini G., Marino G., Querin S., Solidoro C. Stima del potenziale impatto degli impianti di trattamento delle acque reflue urbane sui livelli di contaminazione da *Escherichia coli* nelle aree di mitilicoltura di Chioggia (VE). In: Proceedings of the VIII National Conference of Società Italiana di Ricerca Applicata alla Molluschicoltura (SIRAM 2019); 2019. 52-53. La Spezia

[91] ISO-16649-3. 2015 Microbiology of the Food Chain–HorizontalMethod for the Enumeration of β-glucuronidase Positive *E. Coli*- Part 3: Detection and Most Probable Number (MPN) Technique Using 5-Bromo-4-chloro-3-indolyl-β-D-glucuronide.

[92] Tora S, Sacchini S, Listeš E, Bogdanović T, Di Lorenzo A, Smajlović M, et al. A geographical information system for the management of the aquaculture data in the Adriatic Sea – the Strengthening of Centres for Aquaculture Production and Safety surveillance in the Adriatic countries experience: Present capabilities, tools and functions. Geospatial Health. 2017;**12**(2):300-308

[93] Conti F, Mascilongo G, Colaiuda V, Tomassetti B, Lombardi A, Capoccioni F, et al. Sviluppo di strumenti tecnologici predittivi sanitario/meteo-ambientali per potenziare l'efficienza e la sostenibilità degli impianti di molluschicoltura: avvio del progetto FORESHELL. In: In: Proceedings of the Conference of Società Italiana di Ricerca Applicata alla Molluschicoltura SIRAM (SIRAM 2021). 2021. pp. 24-27

[94] Martini A, Pulcini D, Capoccioni F, Martinoli M, Buttazzoni L, Rossetti E, et al. Mitilicoltura e cambiamento climatico: ruolo del sequestro di carbonio nelle conchiglie nella valutazione di impatto ambientale. In:

Proceedings of the Conference of Società Italiana di Ricerca Applicata alla Molluschicoltura SIRAM (SIRAM 2021). 2021. pp. 32-34

[95] Hurtado-Bermúdez SJ, Expósito JC, Villa-Alfageme M. Correlation of phytoplankton satellite observations and radiological doses in molluscs. Marine Pollution Bulletin. 2021;**172**:112911

[96] Ippoliti C, Tora S, Giansante C, Salini R, Filipponi F. Sentinel-2 e campionamenti in situ per il monitoraggio delle acque marine dell'Abruzzo: primi risultati. In: Proceeding of Eighth International Symposium "Monitoring of Mediterranean Coastal Areas". Problems and Measurement Techniques. Livorno; 2020. pp. 557-568. DOI: 10.36253/978-88-5518-147-1.56

[97] Brockmann C, Doerffer R, Peters M, Kerstin S, Embacher S, Ruescas A. Evolution of the C2RCC neural network for Sentinel 2 and 3 for the retrieval of ocean colour products in normal and extreme optically complex waters. In: Proceeding of Living Planet Symposium. 2020

[98] Filipponi F, Ippoliti C, Tora S, Giansante C, Scamosci E, Petrini M, et al. Water color data analysis system for coastal zone monitoring. In: Proceeding of X International Conference AIT "Planet Care from Space", Trends in Earth Observation. 2021. DOI: 10.978.88944687/00

[99] Vanhellemont Q, Ruddick K. ACOLITE for Sentinel-2: Aquatic applications of MSI imagery. Proceeding of Living Planet Symposium, 2016. Volucella 740.

Chapter 15

Biological Treatment of Cannery Wastes

Yung-Tse Hung, Seyedkiarash Sharifiilierdy, Howard H. Paul,
Christopher R. Huhnke and Rehab O. Abdel Rahman

Abstract

This chapter reviews various methods of cannery wastewater biological treatment, namely up-flow anaerobic sludge blanket (UASB), sequencing batch reactor (SBR), three-stage aerobic rotating biological contactor (RBC), three sequentially arranged reactors (anaerobic-anoxic-aerobic reactors), lagooning, and anaerobic digestion. The general principles for dealing with the uncertainty of general wastewater treatment plants are applied to control the uncertainty in cannery wastewater treatment. An overview of on the application of Monte Carlo, support vector machine (SVM), artificial neural network (ANN), and genetic algorithm to manage the uncertainty in the biological treatment of wastewater is provided.

Keywords: cannery wastes, up-flow anaerobic sludge blanket, anaerobic digestion, sequencing batch reactor, three-stage aerobic rotating biological contactor, uncertainty in wastewater treatment

1. Introduction

Canned foods were well known and widely used for feeding the armies since the mid-eighteenth century. Nowadays, they play a crucial role in the everyday nutrition of everyone all over the globe, where they provide food with good quality that can last for a long time compared with fresh food. The canned food production industry, as any other industry, includes material processing, storing, and transportation. These activities lead to waste and emission generation and can affect the environment negatively if not well planned and applied. These negative effects might include air and water pollution and soil contamination. Of the major pollutants generated by this industry, the organic pollutants are very crucial [1].

In general, food processing from raw materials requires large volumes of high-grade water, which will become wastewater after usage. In particular, it requires a large volume of potable water for several usages, e.g. raw materials cleaning, fluming, blanching, pasteurizing, processing equipment cleaning, and cooling of the final products. These vast usages require the enforcement of quality criteria for the water used in each application; the best quality usage often requires independent treatment to assure complete freedom from odor and taste and to ensure uniform conditions [1]. The wastewater effluents from this industry are characterized by their large volumes. On average, some 10–20 m³ wastewaters are produced per tonne of products. The precise characteristics of these wastewaters are highly dependent on the performed processes during the canned food production, i.e. the process of vegetable washing leads to the generation of wastewaters with high loads of some dissolved organics, and particulate matter [2].

The organic content in the wastewater generated as a result of the operation of different processes in the food canning industries is characterized by high concentrations of biodegradable contaminants and variable pH levels. When an environmental reservoir, e.g. a stream or waterway, receives these wastewater effluents, the organic pollutants will consume some of the dissolved oxygen (DO) that exists in the reservoir during their stabilization. This will reduce considerably the DO to levels below that required for the sustainability of lives of the aquatic organisms. The extent of pollution caused by these effluents can be characterized based on the plant capacity, the utilized process, and the characteristics of the raw materials. In this respect, it is beneficial to categorize the plant capacity in terms of population, where seasonal plants are likely to generate waste loads equivalent to 15,000 to 25,000 people, and large plants generate loads up to 250,000 people. The processing of fruit and vegetables is one of the sources of wastewaters, which contain organic matters. Fruit and vegetable canning companies generate wastewaters with high levels of biochemical oxygen demand (BOD), total solid (TS), and suspended solids (SS) [1]. This chapter aims to introduce the available technologies for secondary wastewater treatment that are widely investigated to prevent and control pollution from the food industry. In this respect, the features of the aerobic and anaerobic biological treatment technologies are summarized. Then, an overview of the uncertainty management in biological treatment plants is provided.

2. Cannery wastewater treatment

Due to the nature of the food industry, the preparatory and operational processes of the raw animals, vegetables, and fruits into edible products do not include the application of chemicals. Subsequently, the organic maters in most of the cannery wastewater effluents are best treated using biological treatment, where these matters are rarely present toxicant or inhibitory compounds in their composition. Yet in some operations, e.g. sterilizing and cleaning the equipment, chemicals are used. In particular, disinfectants and caustic soda are used at the end of the processed batch. These effluents could be characterized as short-time concentrated discharges. They may cause shock loads in the wastewater treatment plants that are not designed to deal with these effluents. In this case, the use of equalization unit can achieve acceptable flow equalization and pH adjustment and dilute the high concentration to a nominal concentration that allows safe operation for the biological treatment unit [3].

3. Aerobic treatment of cannery wastewaters

Aerobic wastewater treatment processes could be applied using several technologies, i.e. pond and lagoon-based treatments; surface and spray aeration; oxidation ditches; trickling filters; septic or aerobic tanks; activated sludge; and aerobic digestion. In this section, sequencing batch reactor (SBR), activated sludge (AS), rotating biological contactor (RBC), and aerobic lagoons (AELs) for treating food processing wastewater are discussed.

3.1 Sequencing batch reactor (SBR)

In general, SBR is a fill-and-draw activated sludge system for wastewater treatment. In that system, wastewater effluent is added to a single batch reactor, where treatment is achieved by removing undesirable components, and then, the effluent is discharged. In the same single batch reactor, equalization, aeration, and clarification are conducted.

The formation of granules in aerobic conditions has been possible and appears as a promising technique for treating high-strength or highly toxic wastewaters. It appeared that aerobic granules were successfully cultivated only in SBR. The cyclic operation of SBR consisted of influent filling, aeration, settling, and effluent removal [4].

The development of aerobic granular sludge to achieve simultaneous removal of COD, phosphorous (P), and nitrogen (N) from saline fish-canning wastewater was investigated by Campo [5]. In that work, a 1.6-L SBR with a hydraulic retention time (HRT) of 0.25 d and a volumetric exchange ratio (VER) of 50% was used. The wastewater fed to the SBR was collected from a fish-canning factory located in the south of Galicia (Spain). The SBR was operated in 3-hour cycles comprising 60-min anaerobic feeding, 112-min aeration, 7–1-min settling, and 1–7-min effluent discharge. The salt concentration was approximately 10.4 ± 0.8 g NaCl/l, and the applied organic loading rate (OLR) equals 5.4 ± 1.9 kg COD/(m^{3}d). Under these conditions, aerobic granules were detected after operational time equals 34 days. Some filamentous bacteria were detected on the surface of the aggregates. The granular biomass has a volatile suspended solids (VSS) concentration of 1.34 gVSS/l, density near 11.5 gVSS/l granule, and mean diameter of 1.35 mm. After 41 days of operation, fluffy-flocculent suspension was formed in the presence of the granules. This behavior was attributed to the salinity and the respectively high fraction of slowly biodegradable COD in the influent (35% of total COD). The study reported good removal efficiencies of soluble COD nearly equal to 80%. The phosphorus and ammonium were mainly concluded to be removed to cover the minimum metabolic demand of heterotrophic bacteria. The study indicated that the enrichment of the biomass with slow growing autotrophic and phosphorus-accumulating bacteria in a saline environment requires a longer operational time [5].

3.2 Activated sludge (AS)

The most commonly used biological wastewater treatment technology is the activated sludge. In that technology, the activated sludge (bacterial biomass suspension) is used to degrade the organic pollutants. Over years, various activated sludge processes have been developed. Depending on the design of the AS unit, the wastewater treatment plant (WWTP) can degrade organic carbon substances and remove nutrients, i.e. N and P. In some biological wastewater treatment systems, activated sludge was attached to a surface to form a bio-film. Examples of these systems are integrated fixed-film activated sludge systems, rotating biological reactors, trickling filters, and moving bed bio-film reactors [6].

3.3 Rotating biological contactor (RBC)

RBC process entails the contact between the wastewater and the biological medium that is used to degrade the organic contaminants. A RBC is described as a device that "consists of a series of closely spaced, parallel discs mounted on a rotating shaft which is supported just above the surface of the wastewater" [4, 7]. RBCs are used to remove biodegradable organic matter and convert ammonia-N and organic-N to nitrate-N. Operational problems caused by high organic loading rates restrict the use of RBCs for partial removal of organic matter (i.e. for "roughing" treatment). However, they can be used quite effectively for the substantial removal of organic matter. Process effluent (i.e. clarified) five-day biochemical oxygen demand (BOD5) and total suspended solids (TSS) concentrations can easily be reduced to less than 30 mg/l each, and even lower can be obtained in some instances [8].

3.4 Aerobic lagoons (AELs)

Aerobic lagoons (AELs) are designed and operated to exclude algae. This is accomplished by two means. First, sufficient mixing is used to keep all biomass from the treatment system in suspension, thereby providing turbidity that restricts penetration of light into the water column. The mixing also has the effect of making the solid retention time (SRT) equal to the hydraulic retention time (HRT). Second, the HRT is controlled to values less than the minimum SRT for algal growth (about two days). Because algae are excluded, oxygen must be delivered by mechanical means [8].

4. Anaerobic treatment of cannery wastewaters

This class of biological wastewater treatment technology utilizes microorganisms to degrade the organic pollutants in the absence of oxygen. The sludge in the anaerobic biological reactor consists of anaerobic bacteria and other microorganisms. Food processing wastewaters are particularly suitable for anaerobic treatment processes, firstly because of their high organic load and secondly because they rarely contain toxicants or inhibitory compounds [2].

4.1 Up-flow anaerobic sludge blanket (UASB)

The UASB process makes use of suspended growth biomass, but the gas-liquid-solids separation system is integrated with the bioreactor. The operating conditions within the reactor could be adjusted to allow the formation of large, dense, and readily settleable particles that can lead to the accumulation of very high concentrations of SS, on the order of 20 to 30 g/l as VSS. These high suspended solids concentrations allow significant separation between the SRT and HRT, and operation at relatively short HRTs, often on the order of two days or less, even when the SRT is long. The three-phase UASB reactor allows the achievement of compact and cheaper units due to its ability to separate the gas, water, and sludge mixtures under high turbulence conditions. It has multiple gas hoods for the separation of biogas and can be operated in a one-metre-height reactor that prevents the formation of floating layers [9]. Due to the extremely large gas/water interfaces, the turbulence is greatly reduced, so it is possible to operate the treatment process with relatively high loading rates of 10–15 kg/m^{3}d [8].

UASB technology is known for its efficiency in treating wastewaters with high carbohydrate content. In this respect, the wastewater effluents from the canning industry are efficiently treated by microbes to produce a nutrient-rich starting material for anaerobic hydrogen production. This has led to the wide application of up-flow anaerobic sludge blanket (UASB) reactor for the treatment of the wastewater effluents from food processing plants [10]. These reactors are well known for their ability to withstand variations in wastewater quality and complete shutdown of the reactor in off season [11].

Anaerobic treatment of a highly alkaline fruit-cannery lye-peeling wastewater was investigated, using an up-flow anaerobic sludge blanket (UArSB) reactor. Only a short initializing period was required before COD reduction and OLR had stabilized at 85 to 90% and 2.40 kg COD/m^{3}d, respectively. With subsequent increases in OLR to 8.1kgCOD/m^{3}d, the COD reduction remained between 85 and 93% and biogas production peaked at 4.1 l/d (63% methane). After 111 days, the COD and reactor pH started to decrease and the gas production was reported to decrease after 102 days and continue to decrease to reach the lowest value of 0.93 l/d after 129 days. Subsequent reductions in the OLR, by reducing influent COD, had no effect on reactor stability. This reduction in the reactor performance was attributed to the inhibition of methanogenesis due to the sodium accumulation of sodium (potentially >20,000 mg/l) in the biomass [10, 12].

4.2 Anaerobic digestion

Anaerobic digestion (AD) is used for the stabilization of particulate organic matter. An anaerobic digester is well mixed with no liquid-solids separation. Consequently, the bioreactor can be treated as a continuous stirred tank reactor (CSTR) in which the HRT and SRT are identical. An SRT of 15 to 20 days is typically used, although SRTs as low as 10 days have been used successfully and longer SRTs are employed when greater waste stabilization is required [8].

For several cannery waste streams, the recovery of useful by-products could be achieved by anaerobic digestion. High COD content fruit and vegetable wastes (>50,000 mg/l) have been treated successfully by AD using a HRT of 10 days and a sludge age of 80 days. For elder sludges, the SS build-up within the reactor reached 30,000 mg/l, but at higher concentrations settling became a problem. Generally, successful treatment of food processing wastes could be achieved using AD with a retention time greater than 10 days and gas production of up to 0.75 m^3/kg volatile solids [9, 11].

4.3 SHARON-Anammox process

A Single reactor system for High activity Ammonium Removal Over Nitrite (SHARON) is a treatment process, which utilizes partial nitrification process for the degradation of ammonia and organic nitrogen components from wastewaters. The process results in stable nitrite formation, rather than complete oxidation to nitrate. The process relies on controlling the pH, temperature, and retention time to prevent the nitrate formation by nitrite-oxidizing bacteria, e.g. Nitrobacter. The wastewater denitrification that employs SHARON reactors can proceed with an anoxic reduction, such as Anammox. In the Anammox process (anaerobic ammonium oxidation), nitrite and ammonium are converted into nitrogen gas under anaerobic conditions without the need to add an external carbon source. In comparison with conventional N-removal processes, the SHARON process results in a reduction of required aeration energy and carbon source.

The application of the successive SHARON-Anammox processes was tested to treat the wastewater from a fish cannery plant. The effluents generated from the anaerobic digestion are characterized by their salinity up to 8000–10,000 g NaCl m^{-3}, organic carbon content (1000–1300 g TOCm^{-3}), and high ammonium content (700–1000 g NH$_{,4}^-$Nm^{-3}). In the SHARON reactor, nearly half the ammonia is oxidized to NO$^-$$_2$-N via partial nitrification. Then, SHARON effluent was directed to feed the Anammox reactor. The system was reported to attain average nitrogen removal of 68%. The bacterial population distribution in the Anammox reactor, followed by FISH analysis and batch activity assays, did not change significantly despite the continuous entrance to the system of aerobic ammonium oxidizers coming from the SHARON reactor. Most of the bacteria corresponded to the Anammox population and the rest with slight variable shares to the ammonia oxidizers. Despite the continuous variations in the amounts of ammonium and nitrite in the feed wastewater, the Anammox reactor showed an unexpected robustness. Only in the period when NO^-_2-N concentration was higher than the NH^+_4-N concentration did the process destabilized and it took 14 days until the nitrogen removal percentage decreased to 34% with concentrations in the effluent of 340 g NH^+_4-Nm^{-3} and 440 g NO^-_2-N m^{-3}, respectively. That study concluded that this successive application of SHARON-Anammox reactors is successful in treating high nitrogen and saline effluents with acceptable control on the ratio between the NO^-_2-N and NH^+_4-N [13].

4.4 Anaerobic membrane bioreactors (An-MBR)

An-MBR can be simply defined as a biological treatment process operated without oxygen and using a membrane to provide solid–liquid separation. The advantages offered by this process over conventional anaerobic systems and aerobic MBR are widely recognized [12].

Saline wastewaters are known for their negative impact on the performance of the biological treatment units. Sodium chloride is widely used, not only for cooking and to melt snow and ice, but also in a wide variety of food industries including food canning, seafood processing, milk processing, etc. In particular, the operation of the seafood processing industry leads to the generation of wastewaters with high soluble and colloidal pollutants and a high concentration of N and SS. For these effluents, the application of conventional biological treatment is not efficient. It was reported that the efficiency could be enhanced by reducing the sodium toxicity with compatible solutes that can increase the sludge activity. Moreover, the anaerobic membrane bioreactor reduces the COD concentration in the wastewater [14].

4.5 Anaerobic filter (AF)

AF consists of a fixed bed biological reactor with one or more series of filtration chambers. This technology relies on the entrapment of the particles in the wastewater on the filter media and the subsequent degradation of the organic matter by the active biomass attached to the surface of the filter media [15]. As the anaerobic biomass should grow on the filter media, 6–9-month start-up period is required for AF to attain the full treatment capacity. The filter can be inoculated with anaerobic bacteria to reduce the start-up time, and the flow should be gradually increased [13, 16].

4.6 Blanket/anaerobic filter

Hybrid UASB/AF systems combine aspects of the UASB process with aspects of the AF process. Influent wastewater and recirculated effluent are distributed across the bioreactor cross section and flow upward through granular and flocculent sludge blankets where anaerobic treatment occurs. The effluent from the sludge blanket zone enters a section of media similar to that used in AF systems where gas-liquid-solids separation occurs. Treated effluent then exits the media section and is collected for discharge from the bioreactor. Gas is collected under the bioreactor cover and is transported to storage and/or use. The hybrid UASB/AF process primarily uses suspended biomass, and process loadings are similar to those used with the UASB process. The solids removal system is similar to that used with the UASB process [8].

A research study on the treatment of wastewater generated from vegetable processing was conducted. In this project, an anaerobic filter, a fluidized bed reactor, and an up-flow anaerobic sludge blanket reactor associated with an anaerobic filter were designed, constructed, and tested [5]. For the anaerobic filter, the removal efficiency for COD was reported to exceed 80% for HRT of 16 h, at temperatures ranging from 20 to 31°C. The FBAR was operated at HRT of 2 h with mean COD removal efficiency of 63%. The UASB/AF achieved mean COD removal efficiency of 80% at HRT of 6 h [17].

5. Cannery wastewater treatment with anaerobic-anoxic-aerobic system

Anoxic process is widely used in wastewater treatment. Anoxic means depletion or deficiency of oxygen. Anoxic process is a biological treatment process by which NO_3-N is converted to molecular nitrogen gas in the absence of oxygen.

A system comprised of anaerobic-anoxic-aerobic reactors was tested to treat the wastewater from tuna cooker. This wastewater stream is characterized by high COD and N concentrations. The up-flow anaerobic sludge blanket reactor was used to achieve the anaerobic digestion in a two-step process. In the first, the COD concentration was varied and ORLs up to 4 g COD/(l.d) were achieved. In the second step, the 6 g COD/l and the HRT were varied between 0.5 and 0.8 day, and this step led to ORLs less than 15 g COD/(l.d). The denitrification process was carried out in an up-flow anoxic filter, and the result of the project indicated that the efficiency of this process is dependent on the supplied carbon content. For optimum carbon content, the ratio between the COD and N equalled 4 and the denitrification percentage equalled 80%. Finally, the nitrification was reported to be fixed at 100% ammonia removal regardless of the amount of carbon in the range of 0.2–0.8 g TOC/l. The variation of the recycling ratios between the denitrification and nitrification reactors in the range of 1–2.5 was found to affect the efficiency of the COD and N-removal percentage, where 90 and 60% removal for COD and N was reported at recycling ratio between 2 and 2.5 [15].

6. Uncertainty in cannery wastewater

There are several types of uncertainties that should be addressed during the design of a wastewater treatment plant, e.g. the variation in strength and quantity of wastewater entering into the plant, the diversity, and the dynamics of the microbial community. An uncertainty analysis for a pre-denitrification plant that uses an activated sludge unit was performed [18]. The unit consists of five compartments: the first two are anoxic and the last three are aerobic. Three scenarios were considered in that study that cover the uncertainty due to stoichiometric, bio-kinetic and influent parameters; uncertainty due to hydraulic behavior of the plant and mass transfer parameters; and uncertainty due to the combination of both scenarios. The study concluded that parameters related to the first and second scenarios introduce significant uncertainties in the plant performance measures. In addition, it was stated that the applied uncertainty farming technique largely affects the uncertainty estimates.

The Monte Carlo simulation was intensively used to simulate the design and upgrade of wastewater treatment plants under uncertainty in balancing effluent costs, violating effluent quality standards, predicting the disinfection performance, generating different influent compositions for posterior process performance evaluation or as a pragmatic procedure to automate the calibration of ASM models, and considering the impact of the input parameter uncertainty on the multi-criteria evaluation of control strategies at wastewater plant [19–22].

Due to the complexity and non-linearity of wastewater treatment plant operations, mathematical models are generally not sufficient to predict the performance of WWTPs. Therefore, AI models have been proposed as an alternative model to linear methods. The methods for minimizing the effect of uncertainty in wastewater characteristics and wastewater flow on wastewater treatment reported in the literature have included support vector machine (SVM) and artificial neural network (ANN) [23]. An optimization model to control uncertainty in operation of wastewater treatment from the shale gas production has been reported in the literature [24]. In addition, genetic algorithms have been developed to model and optimize a biological wastewater treatment plant [25].

Information regarding uncertainty in the wastewater treatment plants in treating cannery wastewater is lacking in the literature. However, the principles governing uncertainty in wastewater treatment plants can be applied to control uncertainty in cannery wastewater treatment.

7. Conclusions

The following conclusions can be reached based on the review of literature in cannery wastewater treatment:

- The wastewater that is generated in food canning industries contains high quantities of organic material, a high level of biodegradables and variable pH levels.

- The food processing industry requires a large amount of potable water for a variety of non-consumption usages, such as for initial cleaning of raw material, fluming, blanching, pasteurizing, cleaning of processing equipment, and cooling of finished product.

- The nature of the organic matter of cannery industry wastewater makes it suitable for biological treatment.

- Flow equalization and influent pH control normally have enough diluting and neutralizing effect to permit the use of biological processes for cannery wastewater treatment.

- Various treatment processes using aerobic and anaerobic treatment can be applied for the treatment of cannery wastewater, depending upon wastewater strength.

- Information regarding uncertainty in wastewater treatment plants in treating cannery wastewater is lacking in the literature. However, the principles governing uncertainty in wastewater treatment plants can be applied to control uncertainty in cannery wastewater treatment.

Author details

Yung-Tse Hung[1]*, Seyedkiarash Sharifiilierdy[1], Howard H. Paul[2],
Christopher R. Huhnke[1] and Rehab O. Abdel Rahman[3]

1 Department of Civil and Environmental Engineering, Cleveland State University,
Cleveland, Ohio, USA

2 Department of Information Systems Cleveland State University, Cleveland, Ohio,
USA

3 Hot Lab. Center, Egyptian Atomic Energy Authority, Cairo, Egypt

*Address all correspondence to: yungtsehung@gmail.com

IntechOpen

References

[1] Teo A, Teo S. Food Canning Waste in Industrial Processes. UCSI University, 2016. 0.13140/RG.2.1.2769.4487. Available online: https://www.researchgate.net/profile/Swee-Teo/publication/301693328_Food_Canning_Waste_in_Industrial_Processes/links/5722ebae08ae262228a5f60c/Food-Canning-Waste-in-Industrial-Processes.pdf?_sg%5B0%5D=vIAIBcs9LF3epjp0U62oZrNNFZNoGGcar8mhZKGwv8zjj-MI7jdOLnLFTs1La4Gc_uDKg25RATGpycSv-QjvDA.5J1Tw5SImTkmqddE_XdO7ZWmYp9QbjpH-BBCeU8La2QeMe-C1JF-ScnyvTWJ-h1Pat3HmGm1-hKwAfOrLuhyAQ&_sg%5B1%5D=gaDk8VGa_lHlJWi6lN_8Ff_iCFmRgKYCg8imCS4KWlU2Szsm4I4SAscTfySz0OIZYHLZ-lQb2Nt-HZA6X5hAyx0oPOuRKP6LYAoO2AOVp5yZ.5J1Tw5SImTkmqddE_XdO7ZWmYp9QbjpH-BBCeU8La2QeMe-C1JF-ScnyvTWJ-h1Pat3HmGm1-hKwAfOrLuhyAQ&_iepl=(last accessed 4 June 2022)

[2] Bolzonella D, Cecchi F. Treatment of food processing wastewater. In: Handbook of Waste Management and co-Product Recovery in Food Processing. Elsevier; Cambridge, UK. 2007. pp. 573-596

[3] Bell JW. Proceedings of the 45th Industrial Waste Conference May 1990. Purdue University: CRC Press; 2018

[4] Adav SS, Lee K-Y, Show J-HT. Aerobic granular sludge: recent advances. Biotechnology Advances. 2008;**26**(5):411-423

[5] Campo R, Carrera-Fernandez P, Di Bella G, Mosquera-Corral A, Val del Rio A. Fish-Canning Wastewater Treatment by Means of Aerobic Granular Sludge for C, N and P Removal, Lecture Notes in Civil Engineering, 2017;**4**;530-535. DOI: 10.1007/978-3-319-58421-8_83

[6] Gernaey KV, Sin G. Wastewater treatment models. In: Reference Module in Earth Systems and Environmental Sciences. 2013. DOI: 10.1016/B978-0-12-409548-9.00676-X

[7] Mba D. Mechanical evolution of the rotating biological contactor into the 21st century. Proceedings of the Institution of Mechanical Engineers, Part E: Journal of Process Mechanical Engineering. 2003;**217**(3):189-219

[8] Grady CL Jr, Daigger GT, Love NG, Filipe CD. Biological Wastewater Treatment. New York, USA: CRC press; 2011

[9] Van den Berg L, Lentz C. Food processing waste treatment by anaerobic digestion. In: Paper Presented at the Proceedings-Industrial Wastes Conference. USA: Purdue University; 1978

[10] Sigge G, Britz T. UASB treatment of a highly alkaline fruit-cannery lye-peeling wastewater. Water SA. 2007;**33**(2):275-278

[11] Daud M, Rizvi H, Akram Ali MF, Rizwan M, Nafees M, Jin ZS. Review of upflow anaerobic sludge blanket reactor technology: Effect of different parameters and developments for domestic wastewater treatment. Journal of Chemistry. 2018:1596319. DOI: 10.1155/2018/1596319

[12] Lin H, Peng W, Zhang M, Chen J, Hong H, Zhang Y. A review on anaerobic membrane bioreactors: Applications, membrane fouling and future perspectives. Desalination. 2013;**314**:169-188

[13] Dapena-Mora A, Campos A. Mosquera-Corral R. Méndez, Anammox process for nitrogen removal from anaerobically digested fish canning effluents. Water Science and Technology. 2006;**53**(12):265-274

[14] Yang J, Spanjers H, Jeison D, Van Lier JB. Impact of Na+ on biological wastewater treatment and the potential of anaerobic membrane bioreactors: A review. Critical Reviews in Environmental Science and Technology. 2013;**43**(24):2722-2746

[15] Mosquera-Corral A, Campos J, Sánchez M, Méndez R, Lema J. Combined system for biological removal of nitrogen and carbon from a fish cannery wastewater. Journal of Environmental Engineering. 2003;**129**(9):826-833

[16] Tilley E, Ulrich L, Lüthi C, Reymond P, Schertenleib R, Zurbrügg C. Compendium of sanitation systems and technologies: Eawag. 2014. Available online: https://www.eawag.ch/fileadmin/Domain1/Abteilungen/sandec/schwerpunkte/sesp/CLUES/Compendium_2nd_pdfs/Compendium_2nd_Ed_Lowres_1p.pdf (last accessed 5 June 2022)

[17] Fernandes JA. Applicability of a UASB/AF Reactor in the Treatment of Cannery Industry Wastewaters. (in Portuguese) MSc. Thesis. Brazil: São Carlos School of Engineering. The University of Sao Paulo; 1984

[18] Sin G, Gernaey KV, Neumann MB, van Loosdrecht MCM, Gujer W. Uncertainty analysis in WWTP model applications: A critical discussion using an example from design. Water Research. 2009;**43**(11):2894-2906

[19] Helton JC, Davis FJ. Latin hypercube sampling method and the propagation of uncertainty in analyses of complex systems. Reliability Engineering and System Safety. 2003;**81**(1):23-69

[20] Benedetti L, Bixio D, Vanrolleghem PA. Assessment of WWTP design and upgrade options: Balancing costs and risks of standards' exceedance. Water Science and Technology. 2006;**54**(6-7):371-378

[21] Neumann MB, von Gunten U, Gujer W. Uncertainty in prediction of disinfection performance. Water Research. 2007;**41**(11):2371-2378

[22] Alsina XF, Roda I, Sin G, Gernaey KV. Multi-criteria evaluation of wastewater treatment plant control strategies under uncertainty. Water Research. 2008;**42**:4485-4497

[23] Hejabi N, Saghebian SM, Aalami MT, Nourani V. Evaluation of the effluent quality parameters of wastewater treatment plant based on uncertainty analysis and post-processing approaches (case study). Water Science and Technology. 2021;**83**(7):1633-1648

[24] Tosarkani BM, Amin SH. A robust optimization model for designing a wastewater treatment network under uncertainty: Multi-objective approach. Computers & Industrial Engineering. 2020;**146**:106611

[25] Do HT, Bach NV, Nguyen LV, Tran HT, Nguyen MT. A design of higher-level control based genetic algorithms for wastewater treatment plants. Engineering Science and Technology, an International Journal. 2021;**24**(4):872-878